Statistics for the Quality Control Chemistry Laboratory

Statistics for the Quality Control Chemistry Laboratory

Eamonn Mullins
Trinity College, Dublin, Ireland

ISBN 0-85404-671-2

A catalogue record for this book is available from the British Library

Published by The Royal Society of Chemistry,
Thomas Graham House, Science Park, Milton Road, Cambridge CB4 0WF, UK
Registered Charity Number 207890

For further information see our web site at www.rsc.org

Typeset by Alden Typeset, Northampton, UK
Printed by TJ International Ltd, Padstow, Cornwall, UK

Preface

This book is intended to be a simple, though not simplistic, introduction to the statistical methods that are routinely applied in analytical chemistry Quality Control laboratories. While it is strongly influenced by my experiences of teaching short courses to laboratory scientists in pharmaceutical companies, the ideas are quite general and would find application in virtually every type of analytical chemistry laboratory. I have included data from different application areas, based on a wide range of analytical techniques. I hope, therefore, that the book will have a broad appeal and serve the needs of a large laboratory audience.

The book is oriented towards the needs of analysts working in QC laboratories rather than towards a wider research-oriented readership. Accordingly, it focuses on a small number of statistical ideas and methods and explores their uses in the analytical laboratory. The selected methods are important aids in method development, method validation and trouble-shooting. The book strongly emphasises simple graphical methods of data analysis, such as control charts, which are a key tool in Internal Laboratory Quality Control and which are a fundamental requirement in laboratory accreditation. A large part of the book is concerned with the design and analysis of laboratory experiments. The coverage ranges from the simplest studies, requiring a single system parameter change, to robustness/ruggedness studies, involving simultaneous changes to many system parameters as an integral part of the method validation process. The approach taken focuses on the statistical ideas rather than on the underlying mathematics. Practical case studies are used throughout to illustrate the ideas in action. A short introduction to the eight chapters is given below.

Chapter 1 establishes the statistical terminology which is basic to the description of the quality of analytical results. It introduces the idea of a statistical model (the Normal distribution) as a basis for thinking about analytical data and describes some of the more commonly used measures of analytical precision, which are related to this model.

Chapter 2 is mainly about control charts. The chapter focuses entirely on Shewhart charts, because these are the most widely used and most easily understood charts. Interested readers will be able to develop their knowledge of other charting techniques, having established a sound understanding of the basic material covered here. This approach characterizes the book as a whole. I have chosen to provide an extended discussion, including a detailed case study, of the use of one type of chart, rather than to provide succinct introductions to several different types of chart. The chapter also contains a short introduction to proficiency tests.

Chapter 3 is an introduction to some of the ideas of statistical inference. It covers three main areas: statistical tests, confidence intervals and the determination of sample size. These ideas are fundamental and are developed in various ways in the remaining chapters.

Chapter 4 is an introduction to the statistical aspects of experimental design. First, it builds on the methods introduced in Chapter 3 by showing how statistical tests and the associated confidence intervals may be used to analyze the data generated by designed experiments – the studies are of the simplest type, involving a change to a single aspect of the analytical system. Next it discusses how sample sizes appropriate for such studies can be determined in advance. It then discusses design aspects such as randomization, pairing and appropriate measures of precision. Throughout, there is a focus on validating the assumptions of the simple statistical models that underlie the tests and confidence intervals. Residual plots and tests for Normality are key tools in doing so.

Chapter 5 develops the arguments of Chapter 4 to discuss how complex systems may be investigated using two-level factorial designs. These designs are important tools for method development, method validation and trouble-shooting. Full factorial designs for investigating relatively small numbers of system parameters are discussed in detail. The fractional factorial designs that form the basis of the designs commonly used for robustness/ruggedness testing are then discussed.

Chapter 6 is concerned with the use of regression analysis for modelling relationships between variables. The technique is introduced in the context of stability testing of pharmaceutical products – a case study concerned with the establishment of the shelf life of a drug is described. The application of regression analysis to calibration is then discussed. The use of residual analysis for validating the statistical models is emphasized. Examples are shown of how the commonly encountered problems of non-linearity and changing variability may be detected. Methods for dealing with these problems are then introduced:

weighted least squares to allow for changing response variability and multiple regression to model non-linearity. A short introduction to the fitting of response surfaces is also given.

Chapter 7 extends the discussion of experimental designs. It introduces (fixed effects) Analysis of Variance (ANOVA), which is used to analyze multi-level factorial designs – these are a simple extension of the two-level designs of Chapter 5. In this context, the idea of blocking is discussed–this is an extension of the discussion of paired comparisons, introduced in Chapter 4. Nested or hierarchical designs are then introduced. These designs implicitly underlie the discussion of control charts in Chapter 2 and are also key to the estimation of the various measures of precision discussed in Chapter 1.

Chapter 8 discusses quantitative measures of the quality of measurements produced by a laboratory. The long-standing approach of estimating the repeatability and reproducibility standard deviations of a method by conducting a collaborative inter-laboratory trial is discussed first. This is followed by an introduction to the statistical ideas used in estimating 'measurement uncertainty'. The use of collaborative trial data in the estimation of measurement uncertainty is then discussed. The final section reviews many of the ideas discussed earlier in the book using the framework of a hierarchical statistical model of the measurement process. Sophisticated readers may find this section a useful introductory overview of some of the important issues discussed in the book. Novices would probably find it abstract and unhelpful. Careful study of the rest of the book will, I hope, change this.

I have assumed that virtually everyone working in a technical environment will have access to computers. All the calculations required for this book may be implemented in a spreadsheet, though I would recommend, in preference, use of a validated statistics package. Accordingly, while formulae suitable for hand calculations are presented, the book assumes that calculations will be carried out using a computer, and focuses on the interpretation of the results. The data analyses in the book were carried out using Minitab, but any statistical package might be used for the purpose – the book is not intended to teach Minitab, *per se*.

The book is not a comprehensive account of all the statistical methods that are likely to be of value in the analytical laboratory. If it is successful, it will give the reader a strong grasp of the concept of statistical variation in laboratory data and of the value of simple statistical ideas and methods in thinking about and manipulating such data. I have deliberately limited the range to include only those topics that I have encountered in use in the laboratories with which I have had dealings. Of course, this means that experienced readers will find that topics they consider important are absent, but a line had to be drawn and this seemed a reasonable decision criterion. For some, the omission of any discussion of multivariate statistical methods (chemometrics) will be notable. However, a good understanding of the simple ideas discussed in this book is, in my opinion, a minimum requirement before the more sophisticated methods are likely to be used confidently or correctly. Accordingly, I have preferred to focus the discussion on the simpler ideas.

Acknowledgements

My principal debt is obviously to those who developed the statistical ideas discussed in the book. Next, I am indebted to those who taught me these ideas – both in the classroom and through their writing. Of these, the first group is composed of colleagues and former colleagues in Trinity College, Dublin. In particular, I want to single out Professor Gordon Foster, who set up the Department of Statistics, and from whom I learned so much over many years of interaction. I would find it difficult to produce a list of the statistical writing that has most influenced me, though some books are given in the references. However, the writings of Professor George Box would undoubtedly be at the top of any such list. The references in the text list papers from which I learned much about the application of statistical methods to analytical chemistry data. I am conscious of how often the statistical sub-committee of the Analytical Methods Committee of the RSC is listed and, in particular, of the number of times I have made reference to the work of its Chairman, Professor Michael Thompson and his co-workers. Readers will find it profitable to follow-up these references.

This book arises directly out of my teaching in-house short courses in industry. These have covered a wide range of industries and statistical topics and have involved teaching both manufacturing engineers and chemists, and laboratory scientists. All the courses have helped to shape the current book, since the statistical ideas are essentially the same whether the objective is to evaluate, monitor, trouble-shoot or optimize either a manufacturing or an analytical system. Accordingly, I am indebted first to those who commissioned the courses and then to the attendees. I doubt if I have ever taught a short course without learning something from the participants. It would be impossible though to thank by name all those who have influenced me in this way. However, some people who have very directly influenced the book must be thanked.

I am indebted to the following friends, clients and colleagues who have either commissioned courses, provided me with data, or who have clarified analytical chemistry ideas for me. To some I am heavily indebted, but I am grateful to all of them for what they have given me: Jill Ahearne, Norman Allott, Norah Blount, John Bohan, John Buckley, Dorothy Claffey, Margaret Connolly, Marion Cullinane, Martin Danaher, Tom Dempsey, Robert Dunne, Mary English, Lynn Feery, Marion Finn, Jacintha Griffin, Rosemary Hayden, Denise Heneghan, Arlene Hynes, Fintan Keegan, Jim Kelly, Kevin Kinnane, Des McAteer, Ger McCann, Ken McCartney, Kevin McNamara, Michael Metzler, Liam Murphy, John O'Connor, Michael O'Dea, Tom O'Hara, Michael O'Keeffe, Marie O'Rourke, John Prendergast, Joe Rowley, Colette Ryan, Eva Ryan, Tom Ryan, Kevin Shelly and Sarah Tait. Andrew Mullins and Paul McNicholas provided research assistance for which I am grateful. I remember with pleasure many conversations on topics relevant to the book with my former MSc student Joe Vale. My colleague Myra O'Regan, by taking on the headship of the Department of Statistics in Trinity College for six months, allowed me to take sabbatical leave in order to concentrate on finishing the book – I am grateful for her generosity in doing so.

I would like to thank Janet Freshwater, Katrina Turner and the editorial and production staff of the Royal Society of Chemistry for turning what was undoubtedly a difficult typescript into the finished book you are now holding. The book is heavily dependent on data extracted from two analytical chemistry journals, The Analyst and The Journal of the Association of Official Analytical Chemists – I am grateful to the editors for permission to reproduce the data extracted from the articles cited in the references. I am grateful to Martin Danaher and Michael O'Keeffe of the Irish Food Centre for providing me with a copy of Figure 6.29 and to Dr A. Lamberty of the Institute for Reference Materials and Measurements, Belgium, for providing a copy of Figure 1.2. I am grateful to Pearson Education for permission to reproduce Figure 3.15 from the book Statistics for the Life Sciences by Professor Myra L. Samuels.

Two friends deserve special thanks. Jack Doyle of the Irish State Laboratory has been a constant source of information on matters related to analytical chemistry, over many years. He also provided valuable comments on the first draft of Chapter 8. Michael Stuart, as a teacher, colleague and friend has had more influence than anyone else on my thinking about statistics. He read earlier drafts of all the chapters and provided detailed criticisms and helpful suggestions. It is certainly a better book than it would otherwise have been without his many contributions.

Finally, I must mention my family. First, my sisters for their friendship and support, as Tom Lehrer might say, from childbirth! My brother-in-law Tom Mayhew was a more than generous host when, as a student, I needed to earn enough to survive the next academic year. For this and all his kindness to my family I will always be grateful. My immediate family, Andrew, Deirdre and Genevieve, have had to live with the writing and re-writing of this book for too long, but finally, here it is, and it is dedicated to them.

Eamonn Mullins

Contents

Chapter 1 Variability in Analytical Measurements **1**

1.1 Introduction 1
1.2 An Example of Measurement Variability 3
1.3 Describing Measurement Error 4
1.3.1 A Schematic Inter-laboratory Study 6
1.4 Sources of Analytical Variability 8
1.5 Measuring Precision 10
1.5.1 The Standard Deviation as a Measure of Precision 13
1.5.2 Variation of Precision with Concentration 15
1.5.3 Measures of Repeatability and Reproducibility 19
1.6 Case Study: Estimating Repeatability from Historical Data 22
1.7 Improving Precision by Replication 26
1.8 Conclusion 29
1.9 Review Exercises 29
1.10 References 33

Chapter 2 Control Charts in the Analytical Laboratory **35**

2.1 Introduction 35
2.2 Examples of Control Charts 36
2.3 The Theory Underlying the Control Limits 39
2.4 Setting Up Control Charts 42
2.4.1 Calculating the Limits 42
2.4.2 Data Scrutiny 45
2.4.3 Sample Size 47
2.5 Monitoring Precision 50
2.5.1 Range Charts 50
2.5.2 The Nature of Replicates 51
2.5.3 Standard Deviation Charts 52
2.6 Case Study 55
2.7 Control Chart Performance 61

2.7.1 Average Run Length Analysis 61
2.7.2 How Many Control Samples? 64
2.8 Learning from Control Charts 66
2.8.1 Improving Precision by Replication: Revisited 66
2.8.2 Obtaining Measures of Precision from Control Charts 72
2.8.3 Using Control Charts 74
2.8.4 Concluding Remarks 76
2.9 Proficiency Testing 76
2.9.1 Overview 76
2.9.2 Technical Issues 78
2.9.3 Concluding Remarks 79
2.10 Conclusion 80
2.11 Review Exercises 80
2.12 References 84

Chapter 3 Some Important Statistical Ideas 87
3.1 Introduction 87
3.2 Statistical Significance Tests 88
3.2.1 Example 1: A Method Validation Study 88
3.2.2 Example 2: Acceptance Sampling 93
3.2.3 Summary 95
3.3 Determining Sample Size 96
3.3.1 The Nature of the Problem 96
3.3.2 Using the Sample Size Table 98
3.3.3 Discussion 100
3.3.4 Some Useful Graphs: Power Curves 101
3.4 Confidence Intervals for Means 103
3.4.1 Example 1: Estimating the Average Potency of a Pharmaceutical Material 104
3.4.2 Example 2: The Method Validation Study Revisited–Estimating Bias 108
3.4.3 Example 3: Estimating the Potency of a Pharmaceutical Material: Revisited 110
3.4.4 Example 4: Error Bounds for Routine Test Results 112
3.5 Sampling 114
3.6 Confidence Intervals for Standard Deviations 116
3.7 Checking Normality 121
3.7.1 Normal Probability Plots 122
3.7.2 A Significance Test for Normality 125
3.7.3 Departures from Normality 126
3.7.4 Transformations 128

3.8 Concluding Remarks 130
3.9 Review Exercises 131
3.10 References 132

Chapter 4 Simple Comparative Studies 135
4.1 Introduction 135
4.2 A Typical Comparative Study 135
4.2.1 A Statistical Significance Test for Comparing Method Means 137
4.2.2 Estimating the Difference in Recovery Rates 140
4.2.3 Comparing Standard Deviations 143
4.2.4 Comparing Means when Standard Deviations are Unequal 145
4.2.5 Validating the Assumption of Normality 147
4.3 Paired Comparisons 148
4.3.1 A Trouble-shooting Exercise 149
4.3.2 Case Study 154
4.4 Sample Size for Comparative Studies 159
4.4.1 Comparing Means of Two Independent Groups 159
4.4.2 Paired Studies of Relative Bias 164
4.4.3 Sample sizes for Comparing Standard Deviations 166
4.5 Some Comments on Study Design 168
4.5.1 Experimental Run Order 169
4.5.2 Appropriate Measures of Precision 171
4.5.3 Representativeness 179
4.6 Concluding Remarks 179
4.7 Review Exercises 180
4.8 References 183

Chapter 5 Studying Complex Systems 185
5.1 Introduction 185
5.2 Statistical or Traditional Designs? 186
5.3 The 2^2 Design 189
5.3.1 An Example 190
5.3.2 Model Validation 197
5.3.3 Organizing the Calculations 198
5.4 The 2^3 Design 201
5.4.1 An Example 201
5.4.2 Data Analysis 207
5.4.3 Model Validation 210
5.5 Sample Size for Factorial Designs 212

5.6 Experiments with many Factors 215
5.7 Fractional Factorial Designs 219
5.7.1 A Simple Example 219
5.7.2 The 2^{5-1} Design 224
5.7.3 Blocking 228
5.8 Ruggedness Testing 230
5.8.1 Designing Ruggedness Tests 231
5.8.2 Example 1 233
5.8.3 Example 2 235
5.9 Concluding Remarks 241
5.10 Review Exercises 242
5.11 References 246

Chapter 6 Fitting Equations to Data 247
6.1 Introduction 247
6.2 Regression Analysis 248
6.2.1 Introductory Example 248
6.2.2 Using the Regression Line 253
6.2.3 Analysis of Variance 260
6.3 Calibration 264
6.3.1 Example 265
6.3.2 Error Bounds for the Estimated Concentration 267
6.3.3 Zero-intercept Calibration Lines 270
6.4 Detection Limit 275
6.5 Residual Analysis 278
6.6 Weighted Regression 286
6.6.1 Fitting a Calibration Line by WLS 287
6.6.2 Is Weighting Worthwhile? 290
6.7 Non-linear Relationships 295
6.7.1 A Single Predictor Variable 296
6.7.2 A 'Lack-of-fit' Test 299
6.7.3 Response Surface Modelling 301
6.8 Concluding Remarks 304
6.9 Review Exercises 305
6.10 References 307

Chapter 7 The Design and Analysis of Laboratory Studies Re-visited 309
7.1 Introduction 309
7.2 Comparing Several Means 310
7.2.1 Example 1: A Laboratory Comparison Study 310

7.2.2 Multiple Comparisons 320
7.2.3 Example 2: A Method Development Study 321
7.3 Multi-factor Studies 324
7.3.1 Example 1: The GC Study Re-visited 324
7.3.2 Example 2: A 3×3 Study 330
7.3.3 Example 3: A 2^3 Study 334
7.4 Blocking in Experimental Design 338
7.4.1 Example 1: The GC Development Study Re-visited, Again! 338
7.4.2 Example 2: Paired *t*-Tests Revisited 342
7.5 Estimating Components of Test Result Variability 343
7.5.1 Example 1: Control Charts 344
7.5.2 Example 2: Three Variance Components 351
7.6 Conclusion 360
7.7 Review Exercises 361
7.8 References 364

Chapter 8 Assessing Measurement Quality 365
8.1 Introduction 365
8.2 Inter-laboratory Collaborative Trials 366
8.2.1 Estimating the Reproducibility Standard Deviation 366
8.2.2 Data Scrutiny 369
8.2.3 Measuring the Trueness of a Method 375
8.3 Measurement Uncertainty 377
8.3.1 Example: Preparing a Stock Solution 379
8.3.2 Discussion 384
8.4 An Integrated Approach to Measurement Uncertainty 386
8.5 Concluding Remarks 388
8.6 Review Exercises 393
8.7 References 396

Solutions to Exercises 397

Appendix: Statistical Tables 431

Subject Index 447

CHAPTER 1

Variability in Analytical Measurements

1.1 INTRODUCTION

All measurements are subject to measurement error. By error is meant the difference between the observed value and the true value of the quantity being measured. Since true values are invariably unknown, the exact magnitude of the error involved in an analytical result is also invariably unknown. It is possible, however, to estimate the likely magnitude of such errors by careful study of the properties of the analytical system. The term 'analytical system' refers to everything that impinges on a measurement: the method, the equipment, the reagents, the analyst, the laboratory environments, *etc.* It is fair comment that a measurement is of no value unless there is attached to it, either explicitly or implicitly, some estimate of the probable error involved. The Analytical Methods Committee of the Royal Society of Chemistry[1] has taken a very clear position on this question in saying that "analytical results must be accompanied by an explicit quantitative statement of uncertainty, if any definite meaning is to be attached to them or an informed interpretation made. If this requirement cannot be fulfilled, there are strong grounds for questioning whether analysis should be undertaken at all." It gives a simple example which illustrates the necessity for measures of uncertainty: "Suppose there is a requirement that a material must not contain more that 10 $\mu g\ g^{-1}$ of a particular constituent. A manufacturer analyses a batch and obtains a result of 9 $\mu g\ g^{-1}$. If the uncertainty... in the results is 0.1 $\mu g\ g^{-1}$ (*i.e.*, the true result falls within the range 8.9–9.1 $\mu g\ g^{-1}$ with a high probability) then it can be accepted that the limit is not exceeded. If, in contrast, the uncertainty is 2 $\mu g\ g^{-1}$ there can be no such assurance. The 'meaning' or information content of the measurement thus depends on the uncertainty associated with it."

In recent years the question of uncertainty in measurements has been of major concern to the measurement community in general, and has stimulated much activity among analytical chemists. This has been driven, in part, by the requirement by laboratory accreditation bodies for explicit statements of the quality of the analytical measurements being produced by accredited laboratories. The term ‘uncertainty’ has been given a specific technical meaning, *viz*., “a parameter associated with the result of a measurement, that characterizes the dispersion of the values that could reasonably be attributed to the measurand”[2] and guidelines have been developed to help chemists estimate uncertainties.[3] Uncertainty will be discussed as a technical concept in Chapter 8, but the remainder of the book may be read as a discussion of how to describe and analyze uncertainty in its usual sense of indicating a lack of certainty regarding the trueness of measured values.

Consultancy experience suggests that the importance of measurement error is often not fully appreciated. In pharmaceutical manufacturing, for instance, crucially important management decisions can be made on the basis of a single measurement or on the average of a small number of measurements, often poorly determined. Examples of such measurements would be: the purity of a raw material on the basis of which the conversion rate of a process is to be determined; the strength of an intermediate product, which will determine the dilution of the next stage of production; the potency of a final product, which will determine the amount of product to be filled into vials intended to contain a given strength of product. In the latter case lack of confidence in the quality of the measurement may lead to overfill of the vials in order to ensure conformance to statutory or customer requirements. The problem in all of these cases is not one of product variability, though this may be present also, but rather the uncertainty in a decision criterion (*i.e.*, a test result) due to measurement variability. Since the costs of improving such measurements will generally be small compared to the costs of poor production decisions, it is clear that the economic consequences of measurement error are, as yet, not fully appreciated.

In industrial experimentation measurement error can be vitally important also. Studies are often carried out to investigate proposed process changes which it is hoped will result in small but economically important yield improvements. In such situations the ability of the experimenter to detect small yield improvements will be influenced in an important way by the number of measurements made. If the measurement error is relatively large and the number of measurements small, the effect of the process change may be missed. The determination of appropriate sample sizes for experiments is an important topic which will be discussed in Chapter 4.

This chapter begins with an example illustrating the effects of random measurement error on the results produced by a stable analytical system.

The terms bias and precision are introduced to describe the nature of measurement errors; this is done in the context of an inter-laboratory study. The Normal curve is discussed as the most important statistical model for describing and quantifying analytical error. Various measures of the magnitude of analytical error are then considered: these include the standard deviation, the relative standard deviation or coefficient of variation, repeatability and reproducibility. Finally, the effect on the precision of an analytical result of averaging several measurements is discussed.

1.2 AN EXAMPLE OF MEASUREMENT VARIABILITY

Figure 1.1 is a histogram based on 119 determinations of the potency of a quality control material used in monitoring the stability of an analytical system; this system is used for measuring the potency of batches of a pharmaceutical product. The measurements were made by high performance liquid chromatography (HPLC). Here the heights of the bars represent the frequencies of values in each of the potency intervals on the horizontal axis. Very often the areas of the bars (height and area are equivalent when the bases are the same) are drawn to represent the relative frequencies, *i.e.*, the fractions of the total number of data points that fall into each interval. In such a case the total area is one – this will be discussed again later.

The results are distributed in a roughly symmetrical bell-shaped form, with most values clustered around the centre and few values very far from the centre in either direction. This shape is often seen in laboratory

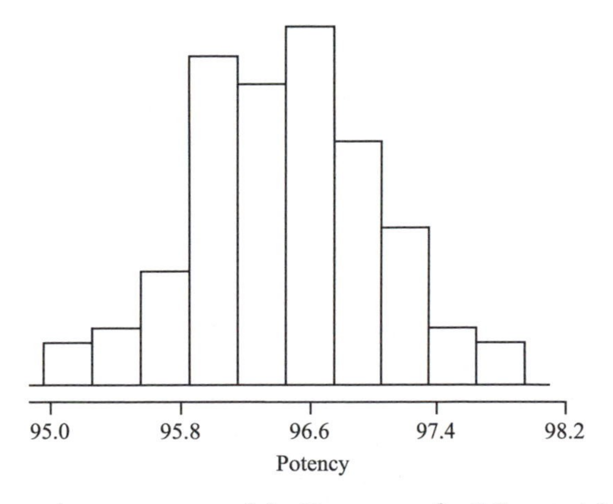

Figure 1.1 *Repeated measurements of the % potency of a QC material*

data and is typical of the type of variation that results from the combined effects of a myriad of small influences. It can be shown mathematically that the cumulative effect on a system of very many small chance perturbations will be to produce a Normal distribution for the measured responses. A Normal curve is symmetrical and bell-shaped and its exact shape can be varied to fit data from many different contexts; consequently, it is the statistical model most often used to describe laboratory data. The Normal model will be discussed later.

It is clear that this HPLC measurement process is subject to some variation. While the average result obtained is 96.5%, individual measurements range from 95.1 to 97.9%. Errors of this magnitude would be significant in characterizing the potency of valuable pharmaceutical product and, undoubtedly, decisions should not be based on single measurements of batch potency in this case. We will return to this point later.

1.3 DESCRIBING MEASUREMENT ERROR

When it comes to assessing the likely magnitude of the analytical error in any test result, it is useful to distinguish between the types of error that may occur. These are often described as *systematic* and *random* errors, though the distinction between them is not always as clearcut as the words might imply. The two types of variability are often clearly seen in inter-laboratory studies such as, for example, that shown in Figure 1.2. These results come from an international study, *viz.*, IMEP – the International Measurement Evaluation Programme[4] carried out by the European Union's Institute for Reference Materials and Measurements (IRMM). Laboratories from fifteen countries each measured the amounts of ten different elements in a synthetic water whose certified value was determined jointly by IRMM and NIST (the United States' National Institute for Standards and Technology). The shaded area in Figure 1.2 gives the error bounds on the certified value for lead, while the error bars attached to each result represent each laboratory's assessment of the uncertainty of its result. The laboratories employed a wide range of analytical techniques, ranging from atomic absorption spectrometry to inductively coupled plasma spectrometry, with a mass spectrometer detector; different techniques gave results of similar quality.

The extent of the inter-laboratory variation is shocking, especially when it is realized that the data were produced by highly reputable laboratories. Less than half of the error bars intersect the shaded area in the graph, indicating that very many laboratories are over-optimistic

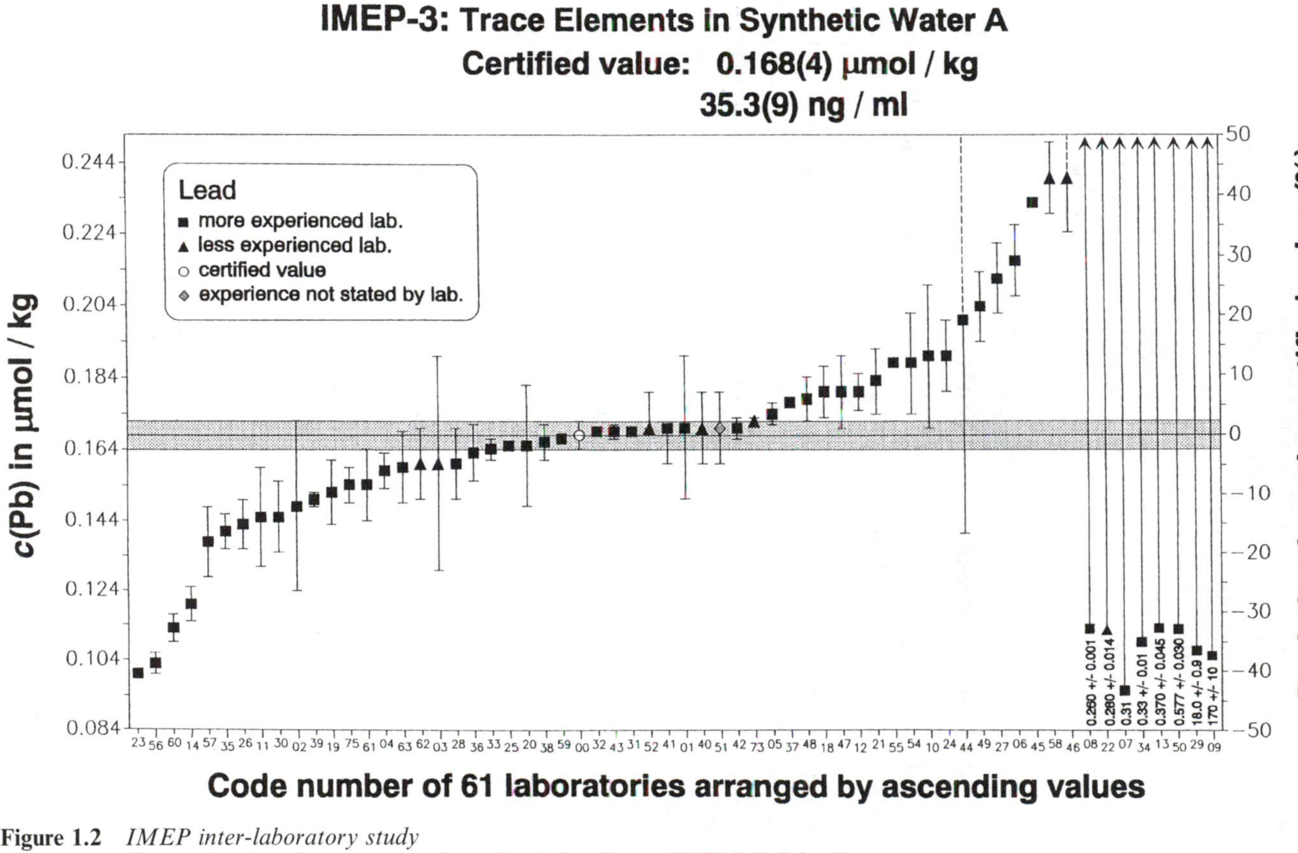

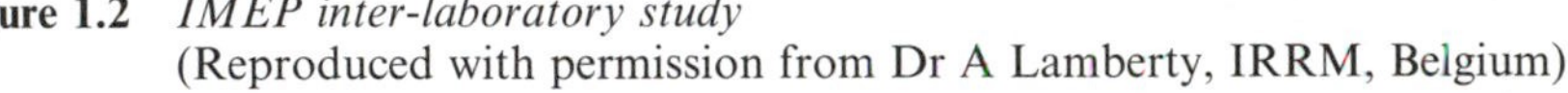

Figure 1.2 *IMEP inter-laboratory study*
(Reproduced with permission from Dr A Lamberty, IRRM, Belgium)

about the levels of uncertainty in the results they produce. This picture is a convincing argument for laboratory accreditation, proficiency testing schemes and international cooperation, such as IMEP, to help laboratories produce consistent results.

Figure 1.2 illustrates the need to distinguish between different sources of variation. The error bars that bracket each laboratory's result clearly reflect mainly within-laboratory variation, *i.e.*, that which was evident in Figure 1.1. The differences between laboratories are much greater than would be expected if only the random within-laboratory variation were present. The terminology used to distinguish between the two sources of error will be introduced in the context of a schematic inter-laboratory study.

1.3.1 A Schematic Inter-laboratory Study

Suppose a company with a number of manufacturing plants wants to introduce a new (fast) method of measuring a particular quality characteristic of a common product. An inter-laboratory trial is run in which test material from one batch of product is sent to five laboratories where the test portions are each measured five times. The standard (slow and expensive) method measures the quality characteristic of the material as 100 units; this will be taken as the true value. The results of the trial are shown schematically below (Figure 1.3).

It is clear that laboratories differ in terms of both their averages (they are 'biased' relative to each other and some are biased relative to the

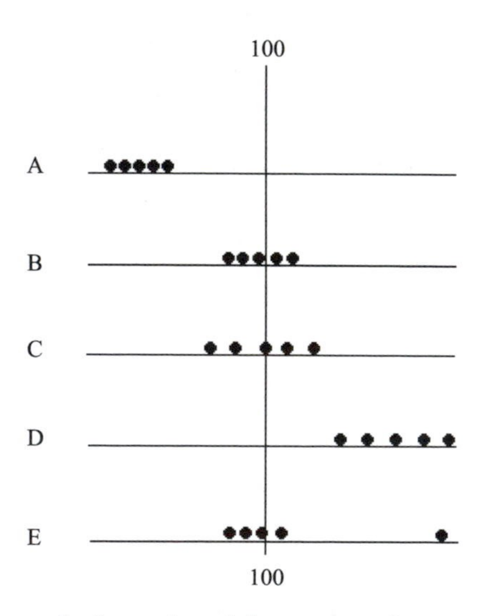

Figure 1.3 *Schematic results for an inter-laboratory study*

true value) and the spread of their measurements (they have different 'precision').

Bias. Even though individual measurements are subject to chance variation it is clearly desirable that an analytical system (instrument, analyst, procedure, reagents, *etc.*) should, *on average*, give the true value. If this does not happen then the analytical system is said to be *biased*. The bias is the difference between the true value and the long-run average value obtained by the analytical system. Since true values are never available, strictly it is only possible to talk about the relative bias between laboratories, analysts or analytical systems. While certified reference materials provide 'true values', even they are subject to some uncertainty. The most that can be said about them is that they are determined to a much higher precision than is required by the current analysis and, where their characteristics are determined by different methods, they are likely to be more free of any method bias that may be present in any one analytical method. As such, they are often treated as if they provide true values.

In our example, laboratories B and C give average results close to 100 so they show no evidence of bias. On the other hand, laboratory A is clearly biased downwards while D is biased upwards. Laboratory E is not so easy to describe. Four of the measurements are clustered around 100 but there is one extremely high value (an 'outlier'). This looks like it was subject to a gross error, *i.e.*, one which was not part of the stable system of chance measurement errors that would lead to a bell-shaped distribution. It could also be simply a recording error. If the reason for the outlier can be identified then laboratory E can probably be safely assumed to be unbiased, otherwise not.

Biases between two laboratories, two analytical methods or two analysts arise because of persistent differences in, for example, procedures, materials or equipment. Thus, two analysts might obtain consistently different results while carrying out the same assay because one allows 5 minutes for a solution to come to room temperature while the other always allows half an hour. The problem here is that the method is inadequately specified: a phrase such as 'allow the solution to come to room temperature' is too vague. Since bias is persistent there is no advantage to be gained from repeating the analysis under the same conditions: bias does not average out. Irrespective of the number of replicate analyses carried out in laboratory A, in our schematic example above, the average result will still be too low.

Precision. Even if an analytical system is unbiased it may still produce poor results because the system of chance causes affecting it may lead to a very wide distribution of measurement errors. The width of this distribution is what determines the *precision* of the analytical system.

Although laboratory A is biased downwards it has, nevertheless, good precision. Laboratory C, on the other hand, has poor precision, even though it is unbiased. D is poor on both characteristics.

The precision of an analytical system is determined by the myriad of small influences on the final result. These include impurities in the various reagents involved in the analysis, slight variations in environmental conditions, fluctuations in electrical quantities affecting instrumentation, minor differences between weights or volumes in making up solutions, minor differences between concentrations of solvents, flow rates, flow pressure, constitution and age of chromatographic columns, stirring rates and times, *etc*. If the influences are more or less random then they are equally likely to push the result upwards or downwards and the bell shaped distribution of Figure 1.1 will be seen if sufficient data are observed. Because the influences are random they can be averaged out: if a large number of measurements is made positive random errors will tend to cancel negative ones and the average result will be more likely to be closer to the true value (assuming no bias) than individual results would be.

1.4 SOURCES OF ANALYTICAL VARIABILITY

Figure 1.4 represents the results of a brainstorming exercise on the sources of variability affecting the results of a HPLC assay of an in-process pharmaceutical material. The exercise was conducted during an in-house short course and took about fifteen minutes to complete – it is not, by any means, an exhaustive analysis. The diagram is known as a 'Cause and Effect', 'Fishbone' or 'Ishikawa' diagram (after the Japanese quality guru who invented it). The sixth 'bone', would be labelled 'measurements' when tackling production problems; since the subject of analysis here is a measurement system, this is omitted. Use of the diagram in brainstorming focuses attention, in turn, on each of the major sources of variability.

The content of the diagram should be self-explanatory and will not be elaborated upon further. While some of the identified causes will be specific to particular types of assay, the vast majority will be relevant to virtually all chromatographic analyses. The diagram is presented as it emerged from the brainstorming session: no attempt has been made to clean it up. Arguments could be made to move individual items from one bone to another, but as it stands the diagram illustrates very clearly, the large number of possible sources of variability in a routine assay.

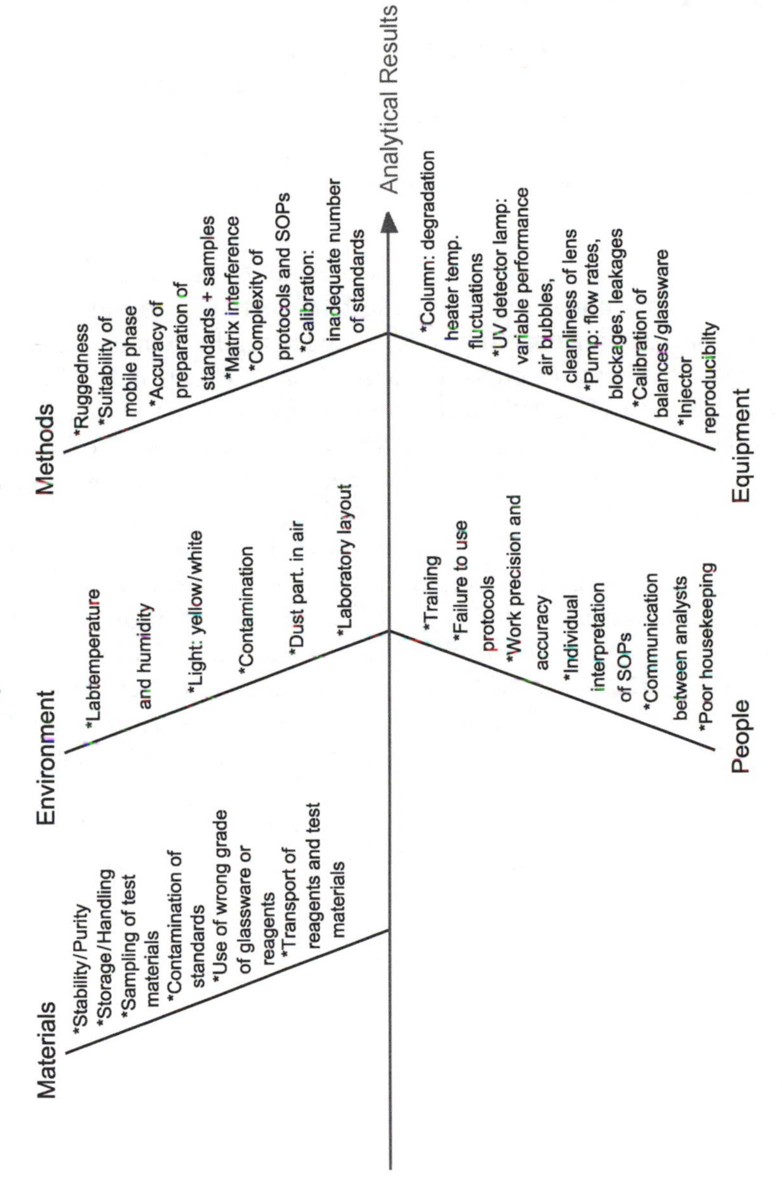

Figure 1.4 *Results of a brainstorming exercise*

Diagrams such as Figure 1.4 are a useful starting point when troubleshooting problems that arise in an analytical system.

1.5 MEASURING PRECISION

Before discussing how to quantify the precision of an analytical system the properties of the Normal curve, which is the standard model for measurement error, will be considered. This model implicitly underlies many of the commonly used measures of precision and so an understanding of its properties is essential to understanding measures of precision.

The Normal Curve. If Figure 1.1 were based on several thousand observations, instead of only 119, the bars of the histogram could be made very narrow indeed and its overall shape might then be expected to approximate the smooth curve shown below as Figure 1.5. This curve may be considered a 'model' for the random variation in the measurement process. The area under the curve corresponding to any two points on the horizontal axis represents the relative frequency of test results between these two values – thus, the total area under the curve is one.

The curve is symmetrical about the centre, the mean (often denoted μ, the Greek letter 'm'), and its width is measured by the standard deviation (denoted σ, the Greek letter 's'). The standard deviation is formally defined through a mathematical formula, but can usefully be thought of as one sixth of the width of the curve. In principle the curve stretches from minus infinity to plus infinity but in practice (as can be seen from Figure 1.5) the area in the tails becomes infinitesimally small very quickly.

Figure 1.6 shows three Normal curves with mean μ and standard deviation σ. Approximately sixty eight percent of the area lies within

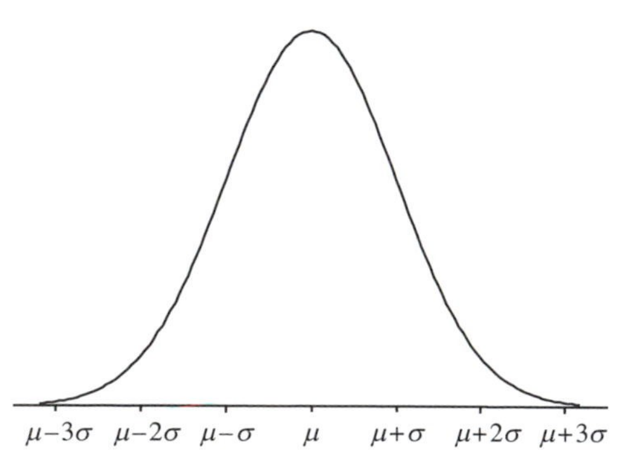

Figure 1.5 *The Normal curve*

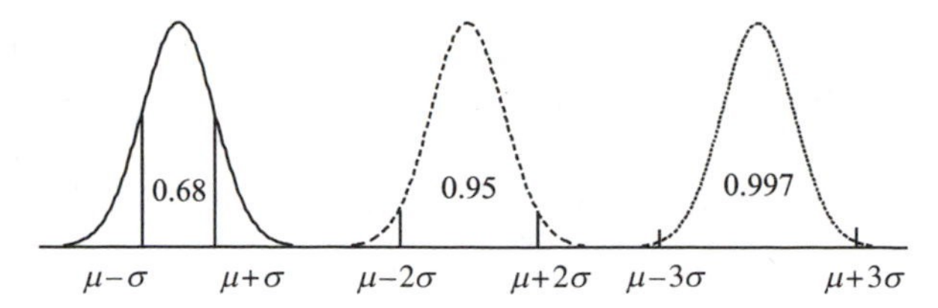

Figure 1.6 *Areas within one, two and three standard deviations of the mean*

one standard deviation of the mean; the corresponding values for two and three standard deviations are 95 and 99.7%, respectively. These areas will be the same for all Normal curves irrespective of the values of the means and standard deviations.

The implications of this figure are that if a measurement system can be described by a Normal measurement error model, then we know that 68% of all measurements lie within one standard deviation of the mean, 95% within two, and 99.7% within three standard deviations of the mean. If the analytical system is unbiased, the mean of the distribution will be the 'true value' of the measurement. Visualizing the output of an analytical system in this way allows us to assess its fitness for purpose both quickly and simply.

Standard Normal Curve. Calculations for Normal distributions are usually carried out using a Standard Normal curve, *i.e.* one for which the mean is zero and the standard deviation is one. Table A1 (Appendix) gives the areas under the Standard Normal curve up to any point on the axis (labelled z): for example, inspection of the table shows that the area up to $z = 1$ is 0.84. Since the area up to zero (the mean) is 0.5 this implies that the area between the centre and 1 standard deviation is 0.34. By symmetry, the area enclosed between the mean and the point on the axis which is 1 standard deviation below the mean is also 0.34. Since the total area under the curve is one, the area that lies within one standard deviation of the mean is 68% of the total, as shown in Figures 1.6 and 1.7.

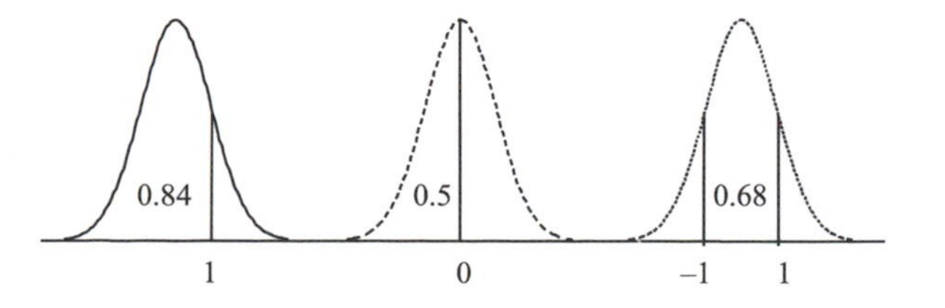

Figure 1.7 *Finding areas under the standard Normal curve*

If we want to know the area between any two values, x_1 and x_2, on a given Normal curve, we simply calculate the corresponding Standard Normal values, z_1 and z_2, and find the area between them – this will be the required area. The correspondence between the x and z values is given by:

$$z = \frac{x - \mu}{\sigma}$$

so that z may be interpreted as the number of standard deviations by which x deviates from the mean of the distribution, μ.

Example 1. Suppose an unbiased analytical system is used to measure an analyte whose true value is 70 units, and that replicate measurements have a standard deviation of 2 units. What fraction of test results will be between 71 and 73 units?

Figure 1.8 shows that $x_1 = 71$ and $x_2 = 73$ correspond to Standard Normal values of:

$$z_1 = \frac{71 - 70}{2} = 0.5$$

$$z_2 = \frac{73 - 70}{2} = 1.5$$

Table A1 (Appendix) shows that the area to the left of $z_2 = 1.5$ is 0.9332, while that to the left of $z_1 = 0.5$ is 0.6915. Accordingly the area between these two values is 0.2417, which means that approximately 24% of test results will fall between 71 and 73 units.

Example 2. If we ask the question: how many standard deviations do we need to go out from the mean in each direction in order to enclose 95% of the area?, then Table A1 can give us the answer. If 95% of the

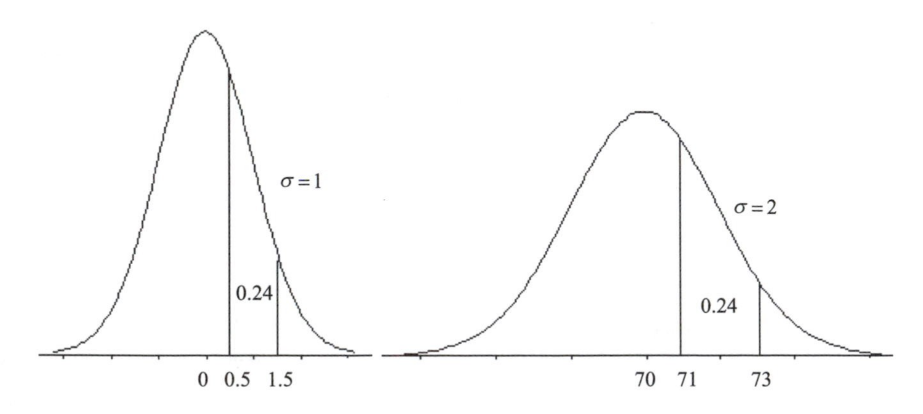

Figure 1.8 *Using the Standard Normal curve*

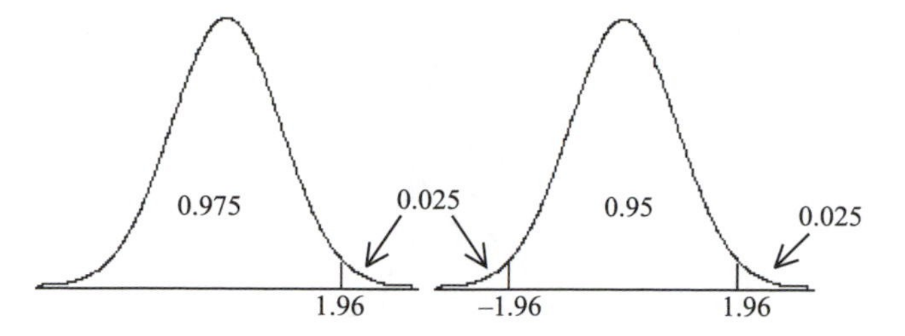

Figure 1.9 *Limits within which a given fraction of the area lies*

area is enclosed in the centre, it will leave 2.5% in each tail, *i.e.*, a fraction 0.025 in each tail, as shown in Figure 1.9. This means that the upper bound on the axis will have a fraction 0.975 to its left. Inspection of the body of Table A1 shows that a fraction 0.975 corresponds to $z = 1.96$. Hence 95% of the area lies between the two points that are ± 1.96 standard deviations from the centre, as shown in Figure 1.9. This is often rounded to ± 2, as was done in drawing Figure 1.6.

Exercises

1.1 What values symmetrically enclose 90 and 99% of the area under a Normal curve?

1.2 An unbiased analytical system is used to measure an analyte whose true value for the parameter measured is 100 units. If the standard deviation of repeated measurements is 0.80 units, what fraction of measurements will be further than one unit from the true value? Note that this fraction will also be the probability that a single measurement will produce a test result which is either less than 99 or greater than 101, when the material is measured. What fraction of measurements will be further than 1.5 units from the true value? Give the answers to two decimal places.

1.5.1 The Standard Deviation as a Measure of Precision

Variation may be quantified in various ways. One of the simplest is to use the standard deviation as a measure of precision: the larger the standard deviation, the worse the precision. Figure 1.10 shows two Normal curves (representing two analytical systems) with the same mean but quite different standard deviations.

While both of these analytical systems give the same results *on average*, *i.e.* they are not biased relative to one another, A has much better precision ($\sigma_A < \sigma_B$). Figure 1.10 shows that the standard deviation σ is a simple direct measure of the precision of a measurement system. Other commonly used measures of precision are based on the standard

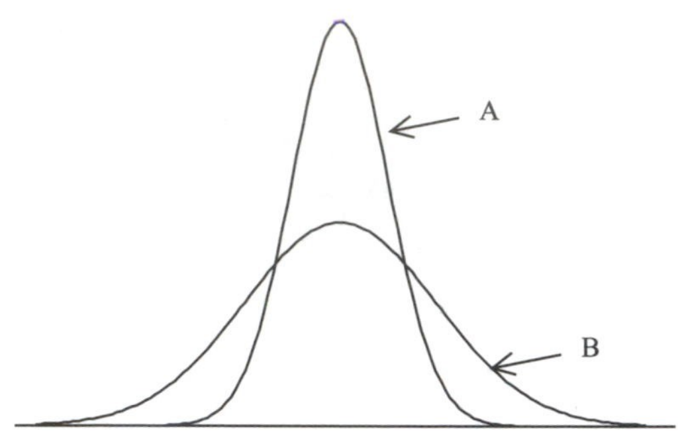

Figure 1.10 *Effect of size of standard deviation on the shape of the Normal curve*

deviation, as we will see below. Figure 1.6 tells us how to interpret σ as a measure of precision: approximately 95% of all measurements will lie within 2σ of the mean result, which is the true value for an unbiased system.

Calculating the Standard Deviation. Suppose a test material has been measured n times and the results are $x_1, x_2, \ldots, x_n$. The average of these, $\bar{x}$, gives an estimate of the true value, μ, which is the objective in carrying out the measurement exercise. The standard deviation of the set of measurements gives an estimate of the precision of the analytical system. This is calculated as:

$$s = \sqrt{\frac{\sum_{i=1}^{n} (x_i - \bar{x})^2}{n-1}} \tag{1.1}$$

i.e. the mean is subtracted from each value, the deviation is squared and these squared deviations are summed. To get the average squared deviation, the sum is divided by $n-1$, which is called the 'degrees of freedom'.* The square root of the average squared deviation is the standard deviation. Note that it has the same units as the original measurements.

It is important always to distinguish between system parameters, such as the standard deviation, and estimates of these parameters, calculated from the sample data. These latter quantities are called 'sample statistics' (any number calculated from the data is a 'statistic') and are themselves subject to chance variation. System parameters, such as μ and σ, are considered fixed, though unknown, quantities. In practice, if the sample size n is very large, the result calculated using equation (1.1) will often be labelled σ, since a very large set of measurements will give the 'true'

*See Exercise 1.3 for a discussion of this terminology.

standard deviation. If the sample size is small, the result is labelled s (or sometimes $\hat{\sigma}$) as above; this makes clear to the user that the calculated value is itself subject to measurement error, *i.e.* if the measurements were repeated a different value would be found for the sample standard deviation.

Many calculators incorporate a second formula for standard deviation which divides by n instead of $n-1$. If the sample size n is large the calculated values will be, effectively, the same. For small sample sizes $n-1$ is, for technical reasons, the more appropriate divisor.

Exercises

1.3 Calculate by hand the standard deviation of the three numbers 5, 6, 7. If you do not get the answer 1 – do it again! Now check that your calculator, spreadsheet and statistical package give that answer.

Note that the three deviations from $\bar{x}$ sum to zero, *i.e.*, $\sum_{i=1}^{n}(x_i - \bar{x}) = 0$. This is always the case for deviations about the sample mean. Therefore, once $n-1$ deviations are determined, the value for the last one must be such that the sum equals zero. Thus, only $n-1$ are free to vary at random: hence the terminology 'degrees of freedom'.

1.4 Munns *et al.*[5] describe methods for determining several quinolone residues in catfish muscle by liquid chromatography, with fluorescence and UV detection. Quinolones are antibacterial drugs widely used to combat various important diseases of farmed fish. Catfish muscle was fortified with each quinolone at ppb levels. The recovery results for two of the additives (at approximately 20 ng g^{-1}) are shown in Table 1.1. For each residue calculate the standard deviation for % recovery.

Table 1.1 *Residue recovery rates (%)*
(Reprinted from the Journal of AOAC INTERNATIONAL, 1995, Vol. 78, pp 343. Copyright, 1995, by AOAC INTERNATIONAL.)

Oxolinic acid		*Flumequine*	
95.4	103.0	81.2	86.3
97.0	94.9	81.3	78.7
96.5	95.9	78.7	75.7

1.5.2 Variation of Precision with Concentration

The standard deviation of measurements in many cases increases with the magnitude of the measured quantity: this situation is depicted schematically in Figure 1.11 and will hold for most analytical systems if a

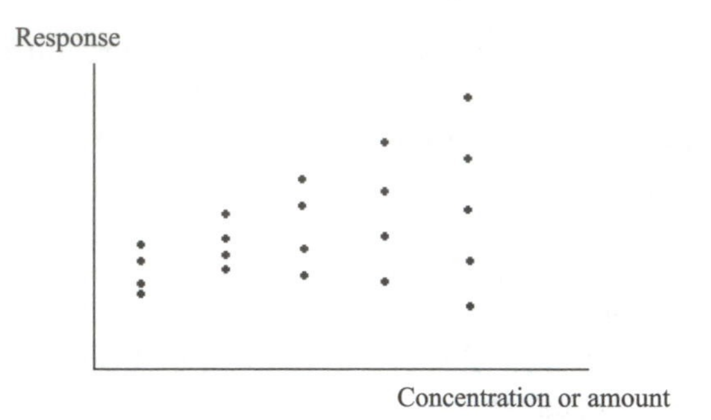

Figure 1.11 *Variability of a response increasing with concentration*

sufficiently wide concentration range is measured. If this is so, then the concentration (or amount) being measured must also be quoted if the standard deviation is to be interpreted meaningfully.

Often the ratio of the standard deviation to the concentration will remain roughly constant, for measurement levels that are reasonably far from zero concentration. In such cases the relative standard deviation (RSD, also called the coefficient of variation) is an appropriate measure of the precision of the system.

$$\text{Relative Standard Deviation} = \left(100\frac{\sigma}{\mu}\right)\% \tag{1.2}$$

This measure expresses the standard deviation as a percentage of the mean (usually regarded as the true value of the measurement). It is a useful summary in that it allows the reader to judge the importance of the likely analytical error in relation to the magnitude of the quantity being measured. In practice, the system parameters μ, σ, are replaced by sample values $\bar{x}$ and s. For an extended discussion of the variation of measurement precision with magnitude of measurement levels see Thompson.[6]

Figure 1.12 derives from a proficiency testing scheme which includes approximately two hundred laboratories involved in measuring blood alcohol levels (BAL).[7] Over a period of six months, test samples at two different concentration levels were measured each month by each of the participating laboratories; the measurements varied over a wide range of concentration levels (the units are mg per 100 mL). The standard deviation and mean result were calculated for each set of measurements and plotted against each other. Thus, each plotted point represents the standard deviation and mean of the results returned by between 186 and 204 laboratories, for each test sample.

It is clear that the precision of the measurements depends strongly on the level of ethanol in the blood. In this case the relationship is slightly

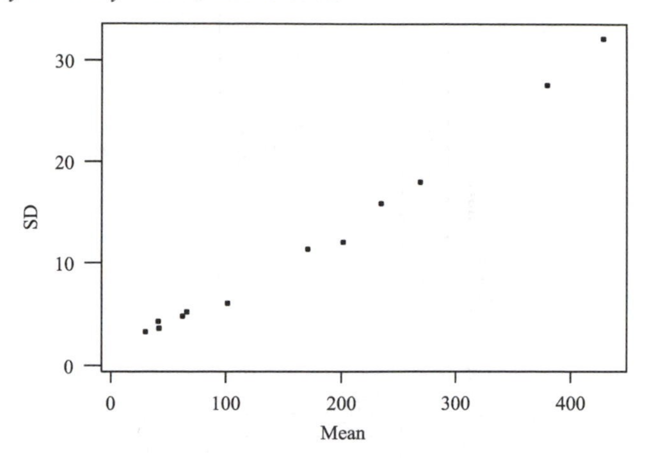

Figure 1.12 *Standard deviations versus means for BAL measurements (mg per 100 mL)*

curved rather than linear, though a linear approximation would probably be adequate for most purposes.

Note that the standard deviations in this case contain both within-laboratory and between-laboratory variation. However a similar graph (but with smaller standard deviation values) would be expected for the relationship between within-laboratory standard deviation and blood alcohol concentration levels. The effect of precision changing with concentration on the fitting of calibration lines is discussed in Chapter 6, where further examples showing the same features as Figure 1.11 are given.

Where the analytical range is narrow the precision would be expected to be more or less constant, and a single standard deviation will characterize the performance of the system. An example of such a situation is given in Section 1.6 where a case study from a pharmaceutical QC laboratory is described.

Exercises

1.5 Calculate the relative standard deviations for the data of Exercise 1.4, p 15.

1.6 Patel and Patel[8] report the data shown in Table 1.2 for Flow Injection Analysis of sodium lauryl sulfate (SLS), an anionic surfactant, in various surface, ground and municipal water samples from central India. The main sources of SLS are assumed to be household commodities and perfumed care products, *viz*., detergents, soaps, *etc*.

By multiplying the RSD values by the SLS results and dividing by 100, create a column of standard deviation values. Plot the SDs against the SLS results and show that the variability increases strongly.

Table 1.2 *SLS concentrations (ppm) in various water sample from India*
(Reprinted from the Analyst, 1998, Vol. 123, pp 1691. Copyright, 1998, by Royal Society of Chemistry.)

Ponds	*SLS*	*%RSD*	*Ground waters*	*SLS*	*%RSD*	*Rivers*	*SLS*	*%RSD*	*Municipal waste waters*	*SLS*	*%RSD*
Budha	14.2	1.3	Budharpara	2.0	2.0	Kharoon	1.7	1.6	Raipur	36.0	1.4
Kankali	13.5	1.4	Kankalipara	2.4	1.9	Hasdo	3.6	1.7	Bhilai	43.8	1.2
Raja	23.2	1.2	Raja talab	4.5	1.8	Shivnath	2.8	1.6	Durg	33.2	1.2
Katora	18.9	1.2	Katora talab	3.8	1.6	Arpa	3.4	1.8	Rajnandgaon	27.0	1.3
Handi	12.5	1.3	Handipara	5.3	1.5				Raigarh	25.8	1.3
									Bilaspur	32.0	1.4

1.5.3 Measures of Repeatability and Reproducibility

A common requirement in analytical method validation is that the repeatability and either the reproducibility or intermediate precision of the method be estimated; for example, see International Conference on Harmonization of Technical Requirements for Registration of Pharmaceuticals for Human Use (ICH) guidelines.[9] Repeatability and reproducibility are descriptions of the precision of test results under different specifications of the conditions under which measurements are made. Repeatability allows for as little variation as possible in analytical conditions, whereas reproducibility allows for maximum variation in conditions, by comparing results obtained in different laboratories, while still measuring the same analyte. Intermediate precision lies somewhere between these extremes, depending on the number of factors allowed to vary.

The design and analysis of studies suitable for estimating these different measures of precision are discussed in Chapter 8. Alternatively, repeatability and intermediate precision may be estimated from control chart data; this will be discussed in the next chapter. Repeatability may also be estimated from replicate test sample results where these come from the same analytical run; an example is given in Section 1.6 below.

Repeatability. The International Organization for Standardization[10] defines repeatability conditions as "conditions where independent test results are obtained with the same method on identical test items in the same laboratory by the same operator using the same equipment within short intervals of time". To provide a quantitative measure of repeatability the 'Repeatability Limit' is defined as "the value less than or equal to which the absolute difference between two test results obtained under repeatability conditions may be expected to be with a probability of 95%". On the assumption that under repeatability conditions individual test results follow a Normal distribution with standard deviation σ_r, the differences between pairs of such results will follow a Normal distribution with mean zero and standard deviation $\sqrt{2\sigma_r^2}$, as illustrated in Figure 1.13.

For all Normal distributions, 95% of values lie within 1.96 standard deviations of the mean. Figure 1.13 illustrates this fact for the distribution of the differences between pairs of test results. It shows that the magnitudes of differences (*i.e.*, ignoring the sign of the difference) between pairs of test results will be less than $1.96\sqrt{2\sigma_r^2}$ 95% of the time. So, if only a single pair of test results is obtained, the probability will be 0.95 that the magnitude of the difference between the two results will be less than $1.96\sqrt{2\sigma_r^2}$. Hence, the repeatability limit is given by:

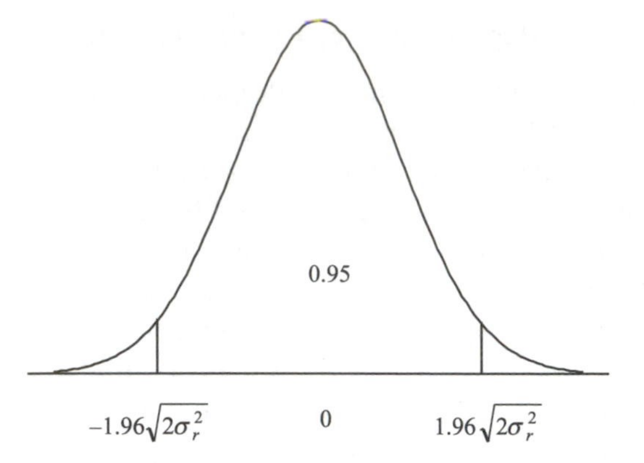

Figure 1.13 *The distribution of differences between two test results*

$$\text{Repeatability Limit} = 1.96\sqrt{2\sigma_r^2} \tag{1.3}$$

where σ_r is the repeatability standard deviation, *i.e.*, the standard deviation of individual test results obtained under repeatability conditions.

It is typical of work organization in an analytical laboratory that a single analyst will analyze a number of test materials together (perhaps together with blanks and control materials); such a batch of analyses is described as a 'run'. It is usually assumed that conditions remain constant during an analytical run, so that the set of materials may be considered to be measured under repeatability conditions. While the definition of repeatability conditions refers to a single analyst, in any practical context the definition will be understood to embody the assumption that all the analysts who carry out the analysis achieve a uniform level of precision, as a result of adequate training. Without such an assumption, the method being validated would require to be validated separately for every analyst. With the assumption, a combined standard deviation based on the data from a range chart (see Chapter 2) or from replicate test results (see Section 1.6 below) provides a realistic estimate of the level of repeatability precision that is being achieved routinely in the laboratory.

Exercise

1.7 Assume the oxolinic acid and flumequine % recoveries from catfish muscle (as described in Exercise 1.4, p 15) were all obtained within single analytical runs and estimate the repeatability limits for both residues.

Reproducibility. Repeatability is a very narrowly conceived measure of analytical precision, describing the performance capability of the analytical system under idealized conditions. Reproducibility, on the other hand, lies at the core of the question of measurement quality in situations where disputes are likely to arise about a test result, *e.g.*, when vendor and customer, or manufacturer and regulatory authority, disagree on the value of some measured quantity. The ISO[10] defines reproducibility conditions as "conditions where test results are obtained with the same method on identical test items in different laboratories with different operators using different equipment." The 'Reproducibility Limit' is defined as "the value less than or equal to which the absolute difference between two test results obtained under reproducibility conditions is expected to be with a probability of 95%". Again on the assumption of Normally distributed measurement errors, both within and between laboratories, the differences between two single test results, one from each of two laboratories, will follow a Normal distribution with mean zero and standard deviation $\sqrt{2(\sigma_r^2 + \sigma_L^2)}$, where σ_r is the repeatability standard deviation and σ_L is the standard deviation that measures variation between laboratories.

Figure 1.14 illustrates the distribution of such differences and shows that the magnitude of differences between single test results, one from each of two laboratories, will be less than $1.96\sqrt{2(\sigma_r^2 + \sigma_L^2)}$ with probability 0.95. Hence, the reproducibility limit is given by:

$$\text{Reproducibility Limit} = 1.96\sqrt{2(\sigma_r^2 + \sigma_L^2)} \tag{1.4}$$

where σ_r is the repeatability standard deviation and σ_L is a measure of the laboratory-to-laboratory variation. An inter-laboratory study, involving measurement of the same analyte in each of a large number

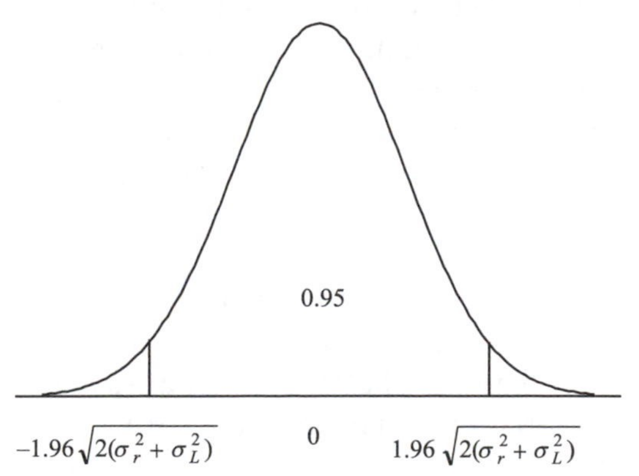

Figure 1.14 *The distribution of differences between two test results, one from each of two laboratories*

of laboratories, is required to estimate the reproducibility of a method (See Chapter 8).

Intermediate Measures of Precision. Where the assay is product dependent, as will be the case for many pharmaceutical products, the use of an inter-laboratory study will not, for commercial reasons, usually be an option, and, hence, estimates of reproducibility will not be available. As an alternative, measures of precision which are intermediate between repeatability and reproducibility are recommended.[9] Measures of intermediate precision are similar to reproducibility in that the various factors which are maintained constant for the definition of repeatability may be allowed to vary; the difference is that all measurements are made in only one laboratory. The principal factors which may be varied are time, operator, instrument and calibration.[10] Replacing σ_L^2 in the reproducibility standard deviation formula by a variance* associated with varying conditions within one laboratory, gives an intermediate measure of precision. Different measures of intermediate precision may be defined, depending on which combination of factors is allowed to vary over the course of the validation study; see ISO 5725-3[10] for a detailed discussion of such measures. These measures are usually obtained from special validation studies, but they may also be estimated from control chart data, as will be seen in the next chapter.

1.6 CASE STUDY: ESTIMATING REPEATABILITY FROM HISTORICAL DATA

Table 1.3 shows data from an in-process potency assay of a pharmaceutical product. Three independent replicate measurements were made on each of 24 consecutive batches of product. For each batch the three replicates were made by a single analyst within a single analytical run. Hence, each of the standard deviation values is an estimate of precision under repeatability conditions. Accordingly, they may be combined to produce a single overall estimate of the repeatability limit.

Before combining the standard deviations, it is worthwhile checking that the data are well-behaved and that the assumption of a common within-run standard deviation is appropriate. Since the potency of the material varies from batch to batch there is always the possibility that the measurement variability may increase with the magnitude of the measurements (as, for example, in Figures 1.11 and 1.12). Figure 1.15,

*The square of the standard deviation is called the 'variance'.

Table 1.3 *Twenty-four sets of three replicate in-process measurements*

Batch	*Potency*	*Mean*	*SD*	*Batch*	*Potency*	*Mean*	*SD*
1	244.21			13	234.70		
	241.32				236.93		
	243.79	243.11	1.56		239.17	236.93	2.24
2	241.63			14	243.70		
	241.88				243.49		
	243.67	242.39	1.11		244.29	243.83	0.41
3	244.33			15	243.15		
	245.43				242.03		
	247.20	245.65	1.45		242.39	242.52	0.57
4	241.23			16	248.40		
	244.74				245.67		
	242.73	242.90	1.76		244.47	246.18	2.01
5	244.92			17	235.01		
	245.40				236.85		
	244.80	245.04	0.32		234.66	235.51	1.18
6	245.09			18	232.11		
	244.89				249.51		
	244.98	244.99	0.10		244.02	241.88	8.90
7	238.92			19	238.81		
	239.35				236.40		
	240.64	239.64	0.90		240.35	238.52	1.99
8	237.86			20	242.66		
	236.72				243.66		
	238.66	237.75	0.97		243.93	243.42	0.67
9	240.30			21	241.10		
	237.00				241.20		
	238.15	238.48	1.68		239.20	240.50	1.13
10	235.73			22	239.38		
	235.97				241.92		
	236.63	236.11	0.47		238.81	240.04	1.66
11	237.72			23	243.86		
	237.67				240.98		
	240.84	238.74	1.82		242.35	242.40	1.44
12	239.36			24	234.99		
	239.61				234.58		
	240.57	239.85	0.64		236.04	235.20	0.75

in which the within-run standard deviations are plotted against the corresponding potency means, shows no tendency for this to occur. This is not surprising as the range of the measurements is fairly narrow. The plot includes a single large outlier, corresponding to batch 18. Inspection of Table 1.3 indicates that one of the three batch 18 values is exceptionally low. Ideally, we should return to the laboratory records to try to explain the unusual value; here, we will simply exclude all three measurements for batch 18, since it is obvious that the outlier results

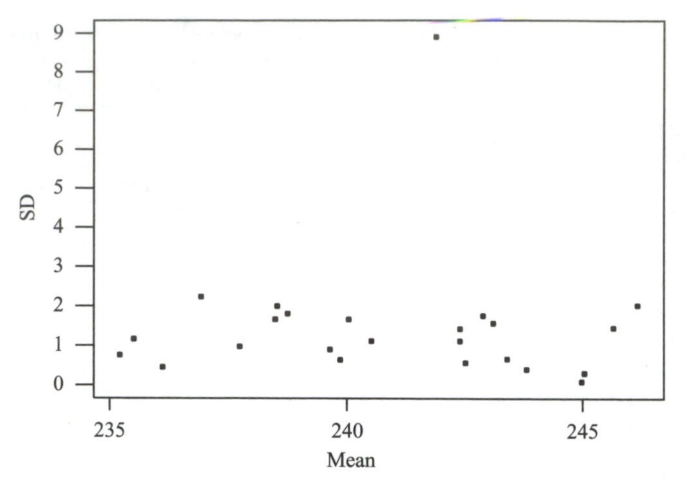

Figure 1.15 *Scatterplot of within-batch standard deviations versus batch means*

from a gross error. When this is excluded, the remaining points suggest constant precision and it appears justified to combine the standard deviations into a single estimate of precision for the purpose of estimating the repeatability limit.

The standard approach to combining standard deviations is first to square the individual standard deviations (the squared standard deviations are called 'variances'), take the average of the variances and then return to the original units by taking the square root of the average variance.* Thus, in this case we obtain:

$$s_r = \sqrt{(s_1^2 + s_2^2 + s_3^2.....s_{23}^2)/23} = \sqrt{1.72} = 1.31$$

From this we can calculate the repeatability limit:

$$\text{Repeatability Limit} = 1.96\sqrt{2s_r^2} = 3.64$$

The interpretation of this value is that duplicate measurements on the material, under repeatability conditions, would be expected to produce results which are within 3.64 units of each other. Differences much greater than this would be suspicious and should be investigated.

Production Versus Analytical Variation. The analysis described above of the within-batch standard deviations yielded information on the analytical variability, only. The batch means shown in Table 1.3, on the

*The underlying mathematics is similar to Pythagoras' theorem according to which, when calculating the length of the side of a triangle opposite to a right angle, the sum of the squares of the other two sides is found and then the square root is taken.

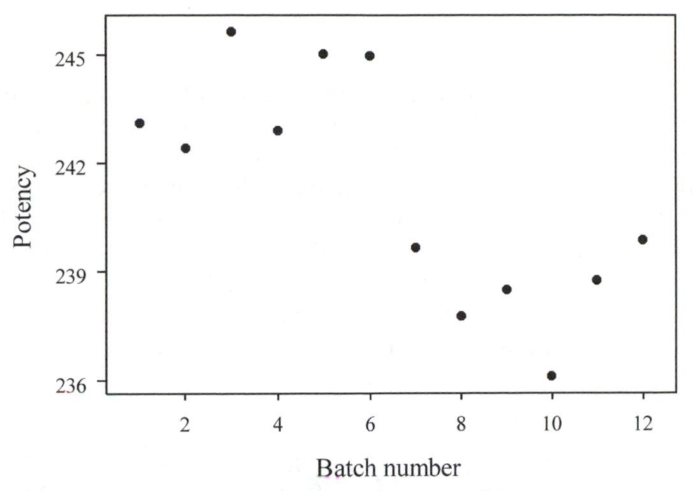

Figure 1.16 *First 12 batch means plotted in order of production*

other hand, contain information on product batch-to-batch variability, though, inevitably, they also include variation due to measurement error. In practice, the data were collected in order to monitor product quality – the production engineers and chemists routinely compare the batch mean results to product specifications, to ensure compliance with national and international regulatory standards. Plotting the means on a control chart (see Chapter 2) allows assessment of the stability of the production process. Figure 1.16 is a simple run chart (a plot of the data in time order) of the first twelve batch means. There is an obvious downwards shift in the test results after batch 6.

It is not clear, however, what caused this shift. It could reflect a change in production conditions, or, alternatively, it may be a result of a downwards bias in the analytical system. With only the current information, we cannot tell which of the two possibilities is responsible for the shift which is apparent in Figure 1.16. If, in addition to monitoring production results, the laboratory has simultaneously measured a control material, then it should be possible to disentangle the two possible causes of the shift. If the control measurements are stable, *i.e.*, exhibit only chance variation around the expected result, then the cause of the shift is likely to be a change in production conditions. If the control measurements show a downwards shift after batch 6, then the analytical system is at fault. This shows the benefits of monitoring the stability of the analytical system: it means that when problems arise, much time (and argument!) can be saved in searching for the likely cause. Chapter 2 is devoted to the use of control charts for monitoring analytical stability.

Exercise

1.8 Lynch *et al.*[11] carried out a collaborative study of the determination of the casein content of milk by Kjeldahl analysis, in which the direct and in-direct casein methods were compared. The study involved ten laboratories each measuring in duplicate nine portions of milk; the duplicates were coded such that the laboratory personnel were unaware which samples were duplicates (these are known as 'blind duplicates') and the order of analysis was randomized. Duplicate results for the direct method in one laboratory are given in Table 1.4. Calculate the standard deviations for the nine sets of duplicates and then combine them into a single standard deviation. From this, estimate the repeatability limit for this analytical system.

Table 1.4 *Casein content (%) of milk test portions*
(Reprinted from the Journal of AOAC INTERNATIONAL, 1998, Vol. 81, pp 763. Copyright, 1998, by AOAC INTERNATIONAL.)

Milk	*Casein content (%)*	
1	2.5797	2.5723
2	3.0398	3.0504
3	2.5771	2.5368
4	2.6462	2.6225
5	2.5026	2.4859
6	2.538	2.5358
7	2.5497	2.5651
8	2.4211	2.4305
9	2.5039	2.4852

1.7 IMPROVING PRECISION BY REPLICATION

Given time and resources analysts would always prefer to make a number of measurements and report the average, rather than make a single determination. It is intuitively obvious that if a measurement that is subject to chance variation is replicated a number of times, then positive errors will tend to cancel negative errors and the average value will be subject to less chance variation than a single result. The extent to which the average provides a more precise estimate of the true value than does a single test result is discussed in general terms below. We will return to this topic in Chapter 2, where a case study from a pharmaceutical QC laboratory will form the basis of a more detailed discussion.

The Effects of Averaging. Figure 1.17 illustrates the averaging effect. Figure 1.17(a) is an idealized histogram of individual measurements; as

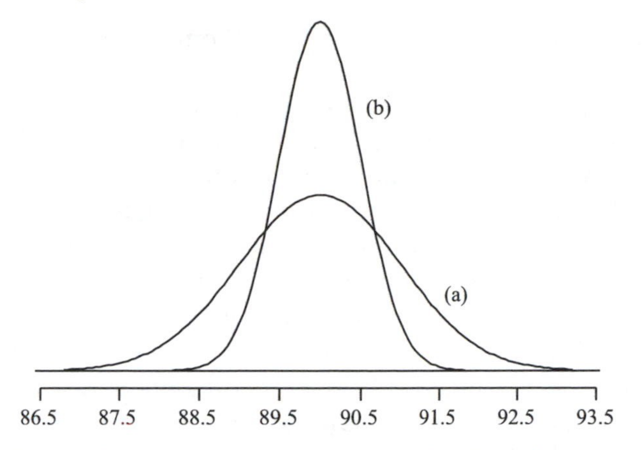

Figure 1.17 *Idealized histograms of the distributions of individual measurements (a) and of the averages of 4 measurements (b)*

such it represents what we might expect to get by drawing a histogram for many thousands of replicate measurements on a single test material using an entirely stable analytical system. Because of the large number of observations the bars of the histogram could be made very narrow and a smooth curve would be expected to fit the histogram very closely. In practice, of course, we would never have such a large quantity of data available to us, but that does not prevent us from visualizing what would happen under idealized conditions. The resulting curve may be thought of as a 'model' of the variability of the measurement process.

Figure 1.17(b) is also an idealized histogram but the data on which it is based are no longer single test results. Suppose we randomly group the many thousands of replicate measurements into sets of four, calculate the means for each set, and draw a histogram of the means. Four values selected at random from the distribution described by (a) will, when averaged, tend to give a mean result close to the centre of the distribution. In order for the mean to be very large (say greater than 91) all four of the randomly selected test results would have to be very large, an unlikely occurrence. Figure 1.17(b) represents an idealized version of this second histogram. Of course we would never undertake such an exercise in the laboratory, though we might in the classroom. The implications of the exercise are, however, very important. Figure 1.17 shows that means are very different from single measurements in one key aspect: they are much less variable. This difference is important, since virtually all data-based decision making is based on means rather than on single measurements. The narrower curve is called 'the sampling distribution of the mean', as it describes the variation of means, based on repeated sampling, and its standard deviation is called 'the standard error of the mean' or 'standard error' for short.

Mathematical theory shows that, if the standard deviation of the distribution of individual measurements is σ, then the standard error of the distribution of sample means is $\sigma/\sqrt{n}$, where n is the number of independent measurements on which each mean is based. Suppose the individual measurements represent purity values, in percentage point units, and that they vary about a long-run average value of 90 with a standard deviation of 1.0. It follows that averages of 4 measurements will also vary about a long-run value of 90, but their standard error will be 0.5. Recall that approximately 95% of the area under any Normal curve lies within two standard deviations (or standard errors) of the mean. This implies that while 95% of individual purity measurements would be between 88 and 92, 95% of means of 4 measurements would be between 89 and 91. Thus, the effect of the averaging process is that the sample mean result is likely to be closer to the long-run average value, which is the true value if the analytical system is unbiased.

Repeatability Critical Difference. In very many if not most cases, the reported result will be an average of several measurements; thus, in the case study discussed in Section 1.6 above, the reported result was the average of three within-run replicates. Where a measure of the repeatability of the reported value (the final value) is required, the ISO Standards define the 'Repeatability Critical Difference' to be "the value less than or equal to which the absolute difference between two final values each of them representing a series of test results obtained under repeatability conditions is expected to be with a specified probability." Again under the assumption of a Normal distribution and a specified probability of 0.95, this definition results in a slightly modified version of expression (1.3):

$$\textit{Repeatability Critical Difference} = 1.96\sqrt{2\sigma_r^2/n} \qquad (1.5)$$

where $n = 3$ in the case study. The squared repeatability standard deviation is divided by n to allow for the fact that the reported values are means, rather than individual test results.

Components of Analytical Variability. The simple description of the effect of averaging given above works for the definition of Repeatability Critical Difference. However, because of the manner in which analytical workloads are typically organized, the full benefits of the averaging process will not usually be realized. The data that form the basis of the case study discussed in Section 1.6 are typical QC laboratory data: the test portion sampled from each batch has been measured independently several times. However, the three replicates for each batch were all measured within one analytical run, *i.e.*, they were all measured by the same analyst, using the same set of equipment, close together in time – under repeatability conditions. Averaging them will reduce the random error present in any one analytical run, exactly as described by Figure 1.17. However, as the discussion of intermediate precision should, I hope, have made clear, the test results are also subject to other, between-run, sources of random variability. Since all three replicates are from one run, the averaging has no effect on the between-run random error. The importance of between-run error will be discussed and quantified in Chapter 2. In summary then, averaging always results in improved precision, but the extent to which it does depends critically on the conditions under which the replicates are measured.

1.8 CONCLUSION

In this chapter the variability that is an inescapable feature of analytical measurement was illustrated by several data sets generated by a variety of analytical methods. Terminology was introduced to distinguish between systematic errors (bias) and random errors (precision), and the distinction was illustrated by an international inter-laboratory study. Several measures of precision were introduced, including the standard deviation, the relative standard deviation or coefficient of variation, repeatability and reproducibility. All of these measures are related to the Normal curve, which is the basic statistical model for analytical error. Finally, the effects on measurement precision of using the average of several measurements was illustrated.

In conclusion, it is worth noting that the inherently statistical nature of the fundamental language of analytical measurement quality underlines the importance of statistical ideas for analytical chemists.

1.9 REVIEW EXERCISES

1.9 Kalra[12] reports the results of a collaborative study of the determination of pH of soils by different methods. The data in Table 1.5 are extracted from the paper; they represent blind duplicate potentiometrical analyses by one laboratory on each of ten soils.

Calculate the standard deviations and means of the duplicates and plot them against each other. Verify that there is no obvious relationship between the two summary statistics. Note that the single large standard deviation value is well within the limits that might be expected from sample standard deviation values calculated from duplicates and so it should be included in the calculations.

Calculate the repeatability limit and repeatability critical difference (average of two replicates) from pooled estimates of the repeatability standard deviation using, first, all the data and then the remaining nine values after the largest standard deviation is excluded. Note the impact of discarding the corresponding data from the estimation of the pooled standard deviation.

Table 1.5 *pH measurements on ten soils*
(Reprinted from the Journal of AOAC INTERNATIONAL, 1995, Vol. 78, pp 310. Copyright, 1995, by AOAC INTERNATIONAL.)

Soil	*pH*	
1	5.45	5.52
2	6.04	6.04
3	4.84	4.87
4	9.97	9.99
5	7.93	7.87
6	4.38	4.33
7	5.90	5.85
8	5.91	5.85
9	4.26	4.30
10	7.91	8.04

1.10 Routine analysis of in-coming raw material at a pharmaceutical plant involves making a single measurement using an unbiased analytical system which produces test results with a standard deviation of 0.5 (units are % points of potency). If a material with a potency of 85% is measured, what is the probability that a result either less than 84% or greater than 86% is obtained?

How many measurements should be made and averaged so that this probability is reduced to less than 0.001? Recall that the standard error of the mean of n values is $\sigma/\sqrt{n}$ where σ is the standard deviation for individual results.

1.11 Moffat *et al.*[13] report data on the percentage of paracetamol in batches of Sterwin 500 mg tablets. The tablets were assayed by both the BP1993 UV method and by an alternative NIR reflectance spectrophotometric method. Duplicate measurements on 25 production batches are reproduced in Table 1.6.

Carry out the following analyses for both the UV and NIR datasets.

- Calculate the means and standard deviations of the duplicate measurements for each of the 25 batches.

- Plot the standard deviations *versus* the means to verify that the measurement variability does not vary systematically with the magnitude of the measurements. Why would you be surprised if you found a relationship? Note that a statistical test does not suggest that the standard deviation of the duplicates for batch 1 of the NIR data is statistically significantly larger than those for the other batches.

Table 1.6 *Paracetamol content (%) in batches of tablets by two methods* (Reprinted from the *Analyst*, 2000, Vol. 125, pp 1341. Copyright, 2000, by Royal Society of Chemistry.)

Batch	*UV*		*NIR*	
1	84.06	84.08	84.63	83.67
2	84.10	84.58	84.54	84.64
3	85.13	85.87	84.77	84.88
4	84.96	85.10	84.24	84.36
5	84.96	85.18	83.78	84.17
6	84.87	84.95	84.90	84.91
7	84.75	85.47	84.48	84.76
8	83.93	83.95	84.79	84.86
9	83.92	84.19	85.31	85.36
10	84.39	84.42	84.37	84.49
11	84.62	84.96	84.03	84.54
12	83.32	83.69	84.18	84.19
13	83.97	84.44	85.08	85.23
14	85.65	85.87	84.79	84.92
15	84.38	84.66	84.64	84.82
16	84.71	85.46	84.73	84.88
17	85.34	85.94	85.33	85.57
18	84.18	84.28	84.07	84.34
19	84.18	84.73	84.36	84.51
20	84.05	84.73	84.99	85.36
21	83.97	84.32	84.34	84.44
22	83.92	84.72	85.11	85.25
23	85.65	85.69	85.02	85.10
24	83.86	84.01	84.35	84.38
25	84.57	84.81	84.01	84.32

- Combine the standard deviations into a single value using the method described in Section 1.6. Since the duplicates are within-run measurements of the same material, the standard deviations will be estimates of the repeatability standard deviations for the two analytical systems.

- Calculate repeatability limits in both cases.

- Assume the routine measurements will be based on the mean of two replicates and calculate the repeatability critical difference for each system.

1.12 Analysis of glucose is performed in a Public Analyst's laboratory in order to ensure compliance with legislation and for authentication of orange juice. The samples are analyzed by HPLC.

Two control charts are routinely monitored. An individuals chart to monitor bias is based on a control sample, which is either a spiked soft drink or clarified orange juice. A range chart to monitor precision is based on duplicate measurements of every tenth test sample. These charts will be discussed in Chapter 2.

Table 1.7 shows 44 pairs of results, representing duplicate measurements of the glucose content of routine test samples. The data are presented in the order of measurement; the units are g per 100 mL. Calculate the standard deviations and means of the duplicates. Plot the standard deviations in serial order. One relatively large value will stand out in the plot. No explanation for it being large was noted on the chart. Plot the standard deviations *versus* the sample means, to check that the standard deviations do not increase with the means.

Combine the individual standard deviations into a single combined value, both including and excluding the large value. Compare the results. Estimate the repeatability limit in both cases and compare the results.

Table 1.7 *Duplicate measurements of glucose content (g per 100 mL) of orange juices*

Run	*Glucose*		*Run*	*Glucose*	
1	2.27	2.30	23	2.56	2.56
2	2.30	2.28	24	2.55	2.51
3	2.23	2.23	25	2.50	2.50
4	2.27	2.32	26	2.60	2.55
5	2.27	2.25	27	2.52	2.55
6	2.30	2.27	28	2.62	2.57
7	2.30	2.26	29	2.64	2.58
8	2.26	2.25	30	2.57	2.57
9	2.34	2.28	31	2.56	2.60
10	2.34	2.36	32	3.03	2.93
11	2.50	2.54	33	2.20	2.24
12	2.46	2.41	34	2.94	2.93
13	2.12	2.17	35	2.99	2.94
14	2.09	2.15	36	1.95	2.01
15	2.16	2.18	37	1.86	1.87
16	2.13	2.17	38	3.07	3.12
17	2.18	2.22	39	2.15	2.21
18	2.28	2.24	40	2.48	2.47
19	2.28	2.24	41	2.21	2.27
20	2.42	2.56	42	2.26	2.22
21	2.43	2.39	43	1.47	1.49
22	2.53	2.50	44	1.52	1.43

1.13 Residues in animal liver of the anticoccidial drug lasalocid are subject to a regulatory upper limit of 100 ng g^{-1}. The analytical method used to test for the drug is considered unbiased and has a standard deviation for individual observations of 9.4 ng g^{-1}. Replicate observations are known to follow a Normal distribution.

- If a single measurement is made on a sample containing 110 ng g^{-1}, what is the probability the result will be less than the regulatory limit?

- If four measurements are made and averaged, how does this probability change?

- How many replicate measurements need to be made to ensure that the probability of the average result being less than the regulatory limit is less than 0.001?

1.14 A Public Analyst's laboratory routinely measures potable water samples by flame atomic absorption spectrometry to ensure compliance with the EU drinking water directive. One of the parameters that is measured is the chloride concentration. Every time a batch of test samples is measured the batch includes two replicates of a QC control sample spiked with 40 mg L^{-1} of chloride. Sample data are given in Table 1.8 (units are mg L^{-1}) – estimate the repeatability limit of the analytical system. Draw appropriate graphs to assess the validity of combining the information from all 42 runs.

Table 1.8 *Chlorine content of QC controls (mg L^{-1})*

Run	*Chloride*		*Run*	*Chloride*	
1	39.8	40.2	22	39.6	39.7
2	40.7	39.6	23	40.2	39.5
3	40.8	41.2	24	40.2	40.6
4	40.3	39.9	25	40.3	40.6
5	40.8	40.9	26	40.0	39.3
6	40.9	40.6	27	40.3	39.9
7	40.7	40.6	28	39.8	38.8
8	41.6	42.3	29	39.9	39.4
9	41.6	41.4	30	39.8	39.3
10	40.1	40.2	31	39.7	40.1
11	40.5	40.0	32	39.5	39.5
12	40.6	41.3	33	40.2	40.0
13	40.5	40.8	34	40.3	40.4
14	39.8	40.4	35	40.6	41.3
15	39.5	39.4	36	40.5	40.6
16	39.1	39.6	37	40.2	40.3
17	38.8	38.9	38	40.7	40.1
18	41.8	42.5	39	40.4	40.3
19	39.9	39.2	40	40.6	40.8
20	40.2	39.4	41	39.9	39.6
21	38.5	38.1	42	40.6	40.7

1.10 REFERENCES

1. Analytical Methods Committee, *Analyst*, 1995, **120**, 29.
2. *Guide to the Expression of Uncertainty in Measurement*, International Organization for Standardization, Geneva, Switzerland, 1993.

3. *Quantifying Uncertainty in Analytical Measurement*, Eurachem, London, 2000.
4. A. Lamberty, G. Lapitajs, L. Van Nevel, A. Gotz, J.R. Moody, D.E. Erdmann and P. De Bievre, *IMEP-3: Trace Elements in Synthetic and Natural Water*, Institute for Reference Materials and Measurements, Geel, Belgium, 1993.
5. R.K. Munns, S.B. Turnipseed, A.P. Pfenning, J.E. Roybal, D.C. Holland, A.R. Long and S.M. Plakas, 1995, *J. AOAC Int.*, **78**, 343.
6. M. Thompson, *Analyst*, 1988, **113**, 1579.
7. J. Doyle, Personal communication.
8. R. Patel and K. Patel, *Analyst*, 1998, **123**, 1691.
9. International Conference on Harmonization of Technical Requirements for Registration of Pharmaceuticals for Human Use (ICH), *Guidelines for Industry: Validation of Analytical Procedures*, ICH Q2A, ICH, Geneva, 1994.
10. *Accuracy (Trueness and Precision) of Measurement Methods and Results*, Parts 1 – 6, ISO 5725, International Organization for Standardization, Geneva, 1994.
11. J. Lynch, D. Barbano and R. Fleming, 1998, *J. AOAC Int.*, **81**, 763.
12. Y. Kalra, *J. AOAC Int.*, 1995, **78**, 310.
13. A.C. Moffat, A.D. Trafford, R.D. Jee and P. Graham, *Analyst*, 2000, **125**, 1341.

CHAPTER 2

Control Charts in the Analytical Laboratory

2.1 INTRODUCTION

This chapter is concerned with two aspects of laboratory quality control, *viz*., control charting, which is a form of *internal* quality control, and proficiency testing, which is a form of *external* quality control. The main focus is on control charts, as they are the simplest and cheapest method for obtaining assurance that the laboratory's analytical systems are well-behaved and, hence, that the data routinely produced by the laboratory are fit for purpose.

Statistical quality control charts are simple but powerful tools for monitoring the stability of analytical systems. A control material is measured regularly and the analytical responses are plotted in time order on a chart; if the chart displays other than random variation around the expected result it suggests that something has gone wrong with the measurement process. To help decide if this has happened control limits are plotted on the chart: the responses are expected to remain inside these limits. Rules are decided upon which will define non-random behaviour.

Control charts were developed to control engineering production processes rather than measurement systems. However, any apparent differences between the two situations disappear when we think of a measurement system as a production process whose output is measurements. The most commonly used charts were developed by Shewhart[1] in Bell Laboratories in the 1920s. In describing factors that affect any system he distinguished between 'chance causes' and 'assignable causes'. In an analytical context, chance causes are the myriad of small influences that lead to the Normal curve, which was discussed in Chapter 1 as a model for measurement error. Assignable causes are larger and lead to changes in measurement level (biases) or variability (precision). If a measurement system is stable such that it exhibits only chance variation around a given reference value and the size of that variation (as measured by the standard deviation) remains constant, in Shewhart's terminology it is said to be 'in statistical control' or simply 'in control'. The objective in using control charts is to achieve and maintain this state of statistical control.

The chapter begins by describing a control chart which is used to monitor the stability of an analytical system with respect to changes which would cause bias in test results. Two examples are presented and discussed, followed by a discussion of the assumptions and theory underlying the chart and of rules that are used to signal measurement problems. Following this, the calculations involved in setting up the control chart are described and explained. A second chart, used for monitoring analytical precision, is presented next. A case study from a pharmaceutical quality control laboratory is discussed. This is then used to illustrate the ability of charts to detect biases when they arise. It also forms the basis of a discussion of control charts as a source of information on assay precision and, in particular, of estimates of validation precision parameters, *viz.*, repeatability and intermediate precision limits.

The chapter concludes with a short introduction to proficiency testing. This involves many laboratories measuring the same material, thus providing a set of peer laboratory results with which each laboratory can compare its own performance. In this way an objective external measure of laboratory performance is available to provide assurance to customers and, increasingly, to laboratory accreditation bodies that the laboratory produces analytical results that meet customer requirements.

2.2 EXAMPLES OF CONTROL CHARTS

Example 1. Figure 2.1 shows a control chart for a HPLC potency assay of a pharmaceutical product.[2] The data displayed in the chart were collected over a period of several months. At each time point two replicate measurements were made on a control material. These results were averaged and it is the average that is plotted in the chart. For now, the centre line (CL) and control limits (upper and lower control limits, UCL and LCL) will be taken as given; later, the rationale for the limits and how they are set up in practice will be discussed.

There is no universal agreement on how control charts are to be used. In the literature, many different sets of rules are recommended by different authors, institutions and commercial companies. For example,

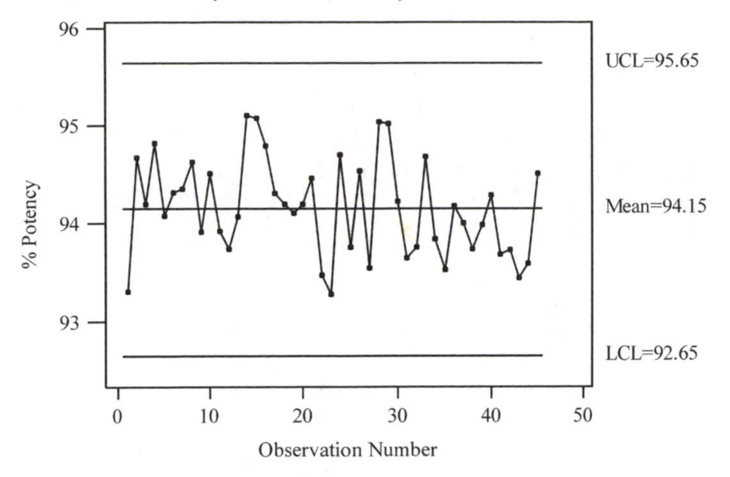

Figure 2.1 *A control chart for the HPLC potency assay*
(Reprinted from the *Analyst*, 1994, Vol. 119, pp 369. Copyright, 1994, Royal Society of Chemistry.)

the Ford Motor Company[3] suggested that action is required if any of the following occur:

- any point outside of the control limits;
- a run of seven points all above or all below the central line;
- a run of seven intervals up or down (*i.e.* 8 consecutive points running upwards or downwards);
- any other obviously non-random pattern.

The same basic principle underlies all rules: a system that is in statistical control should exhibit purely random behaviour – the rules correspond to improbable events on such an assumption. Accordingly, violation of one of the rules suggests that a problem has developed in the measurement system and that action is required. The rationale for the rules is discussed in the theory section, below.

The average level of the measurement system to which Figure 2.1 refers appears, by these rules, to be in statistical control. No points are outside the control limits and there are no long runs of points upwards or downwards or at either side of the central line.

Example 2. Figure 2.2, represents single measurements on water QC control samples made up to contain 50 ppb of Fe.[2] The samples were

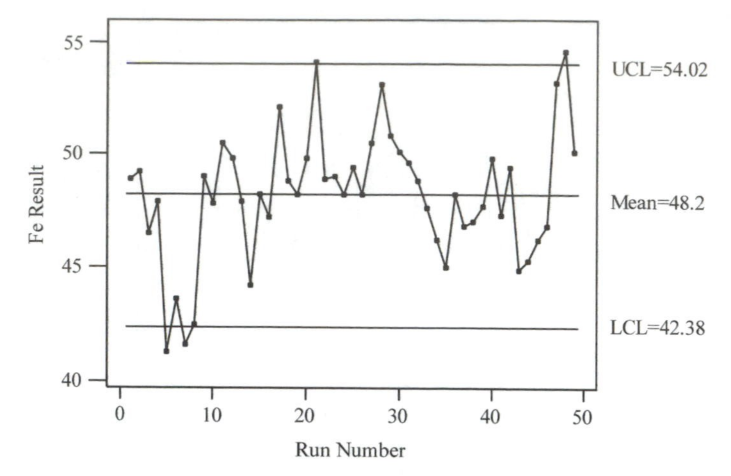

Figure 2.2 *A control chart for the Fe content of a water sample (ppb)* (Reprinted from the *Analyst*, 1994, Vol. 119, pp 369. Copyright, 1994, Royal Society of Chemistry.)

analyzed by an Inductively Coupled Plasma (ICP) spectrometer over a period of months.

The chart shows an analytical system which is clearly out-of-control; in fact, three gross outliers were removed before this chart was drawn. The data were collected retrospectively from the laboratory records as a first step towards implementing a measurement control system. They are presented here to illustrate the classic features of an out-of-control system: several points outside the control limits, a run of points above the centre line and a run of points downwards. There is an obvious need to stabilize this analytical system.

There are two ways in which control charts are used, *viz*., for assessing the performance of a measurement system (as has been done using Figures 2.1 and 2.2), and for monitoring the stability of a system, which is their routine use once stability has been established. Where a chart is being used for control purposes we would not expect to see a pattern such as is exhibited between observations 17 and 32 of Figure 2.2, where 16 points are all either on or above the centre line. Use of a control chart should lead to the positive bias suggested here being corrected, before such a long sequence of out-of-control points could develop.

Note that if the chart were being used for on-going control of the measurement process, the centre line could be set at the reference value of 50 ppb. Here the interest was in assessing the historical performance of the system, and so the data were allowed to determine the centre line of the chart.

2.3 THE THEORY UNDERLYING THE CONTROL LIMITS

The centre line (CL) of a control chart should be the mean value around which the measurements vary at random. Ideally, this would be the 'true value' of the control material being measured. If the control material is a certified reference material or an in-house reference material made up to have an assigned value for the determinand, then the assigned value may be used as the centre line, assuming the analytical system is unbiased. Where there is no assigned value, the mean of the most recent observations considered to be in-control should be used as the centre line.

The control limits are usually placed three standard deviations above and below the centre line (three standard errors if we are dealing with averages, see below). This choice is often justified on the assumption that the frequency distribution of measurement errors will follow a Normal curve, though it may be regarded simply as a sensible rule of thumb without such an assumption. As we have seen, a distribution curve can be thought of as an idealized histogram: the area under the curve between any two values on the horizontal axis gives the relative frequency with which observations occur between these two values. Thus, as shown in Figure 2.3, 99.74% of the area under any Normal curve lies within three standard deviations (3σ) of the long-run mean (μ) and so, while the system remains in control, 99.7% of all plotted points would be expected to fall within the control limits.

For individual measurements the (theoretical) control limits are:

$$\begin{aligned} \text{Upper Control Limit (UCL)} &= \text{CL} + 3\sigma \\ \text{Lower Control Limit (LCL)} &= \text{CL} - 3\sigma \end{aligned} \tag{2.1}$$

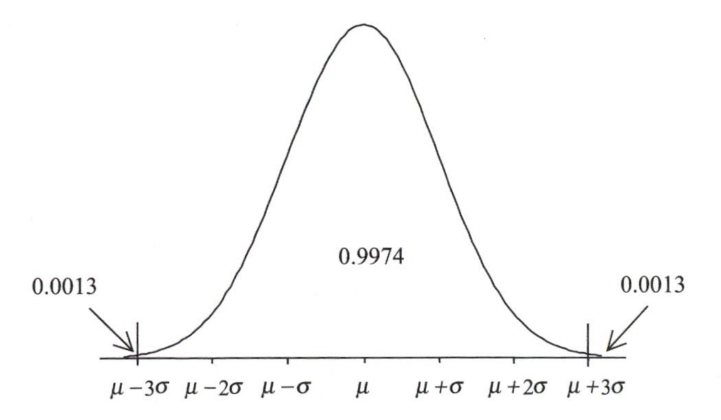

Figure 2.3 *Three sigma limits on the Normal curve*

where σ is the standard deviation of individual measurements. In practice, of course, σ will be unknown and will have to be estimated from historical data. If the estimated standard deviation is s, the control limits are given by:

$$\mathrm{CL} \pm 3s \tag{2.2}$$

If the plotted points are means of several results within each analytical run, then an adjustment has to be made to take account of the fact that means, rather than individual measurements, are being plotted. The simplest approach is to average the within-run results, treat these averages as single results and calculate the standard deviation (S_{X-bar}) of this set of values. Note, however, that this calculated value is known as the 'standard error' of the sample means. The control limits may then be calculated using:

$$\mathrm{CL} \pm 3S_{X-bar} \tag{2.3}$$

The distribution of the sample means will be Normal (approximately so, if the underlying raw measurements do not conform to a Normal curve), with smaller random variation than the raw data; recall the discussion in Chapter 1 of improving precision by averaging. Hence, the control limits will be tighter when means are plotted. The chart based on single measurements is sometimes called an 'Individuals or X-chart' while the chart based on means is called an 'X-bar chart'; the two charts are, however, essentially the same.

Justification for Out-of-control Rules. The control limits described above are known as 'action limits'. They were first proposed by Shewhart in the 1920s and are now almost universally used. Their principal justification is that if the measurement system is in-control then the probability of a point falling outside the control limits is very small: about three in a thousand, if the Normality assumption is correct. Accordingly, they give very few false alarm signals. If a point does fall outside the control limits it seems much more likely that a problem has arisen with the analytical system (*e.g.*, it has become biased upwards or downwards) than that the system is stable and, just by chance, the measurement errors have combined to give a highly improbable result. This is the rationale for the first rule, quoted above.

The basis for the second rule (a run of seven points all above or all below the central line) is that if the system is in-control the probability that any one measurement is above or below the central line is 1/2, irrespective of the values of previous measurements. Accordingly, the probability that seven in a row will be at one side of the central line is $(1/2)^7 = 1/128$. Again, such an occurrence would suggest that the analytical system had become biased upwards or downwards. The third rule (8 consecutive points running upwards or downwards) has a similar rationale: if successive measurements are varying at random about the centre line, long runs in any one direction would not be expected.

The last catch-all rule (any other obviously non-random pattern) is one to be careful of: the human eye is adept at finding patterns, even in random data. The advantage of having clear-cut rules that do not allow for subjective judgement is that the same decisions will be made irrespective of who is using the chart. Having said this, if there really is 'obviously non-random' behaviour in the chart (*e.g.*, cycling of results between day and night shift, perhaps indicating a need for better temperature control of a chromatographic column) it would be foolish to ignore it.

A second set of control limits called 'warning limits' at a distance of two standard errors from the centre line are often suggested.[4,5] Two consecutive points outside the warning limits are taken as an out-of-control signal. The rationale for this is that, since the probability of one point falling outside the warning limit on one side of the centre line is 0.023, when the system is in control, the probability of two consecutive (independent) points falling outside the warning limit (on the same side of the centre line) is the square of this, *i.e.*, 0.0005, indicating a very unlikely event.

The International Union of Pure and Applied Chemistry (IUPAC) has published 'Harmonized Guidelines for Internal Quality Control in Analytical Chemistry Laboratories'.[6] These guidelines recommend three rules: the action limits rule, the warning limits rule and a rule that regards nine points on the same side of the centre line as an out-of-control signal. The guidelines recommend that "the analytical chemist should respond to an out-of-control condition by cessation of analysis pending diagnostic tests and remedial action followed by rejection of the results of the run and reanalysis of the test materials".

The assumption that the data follow a Normal frequency distribution is more critical for charts based on individual measurements than for those based on averages. Averages tend to follow the Normal distribution unless the distribution of the measurements on which they are based is quite skewed. In principle, this tendency holds when the averages are based on very large numbers of observations, but, in practice, means of even four or five values will often be well behaved in this regard. If there is any doubt concerning the distribution of the measurements, a Normal probability plot may be used to check the assumption; this plot is discussed in Chapter 3.

2.4 SETTING UP CONTROL CHARTS

It is clear from the foregoing that two parameters are required for setting up a control chart. The first is a centre line, which is defined either by an assigned or certified value or by the mean of a 'training' set of measurements. The second is a standard deviation/standard error, which measures the variability of the measurements, and hence defines the control limits. Since the routine use of a control chart is to maintain the analytical system in statistical control, these parameters should be estimated from data that reflect the system in control.

The Control Material. As we have seen, the control material being analyzed may be a certified reference material, a control sample made up to have certain characteristics ('house reference material'), or a single batch of product whose 'true value' is considered already known from extensive testing, or can be so determined. Both the Analytical Methods Committee of the Royal Society of Chemistry[5] and the IUPAC guidelines[6] contain valuable discussions and examples of the important characteristics of control materials. Advice on the preparation of house reference materials is also given. Irrespective of the nature of the control material, it is important that it be stable. Appropriate measures should be taken to ensure that it is, since any instability in the control material may otherwise be falsely attributed to the analytical system. Because of the possibility of matrix interference effects, the matrix of the control material used for charting should be as close as possible to that of the test material. It is important, also, that the control material be treated in exactly the same manner as routine samples. If special care is taken in the analysis of control materials they will no longer be representative of routine analyses and their value will be diminished, if not entirely lost.

2.4.1 Calculating the Limits

The Centre Line. The method for choosing a centre line for an X-bar chart will depend on the status of the control material. If either an in-house or certified reference material is in use then the assigned value of the reference material may be used as the centre line of the chart. Using the assigned value as the centre line implies an assumption that the analytical system is unbiased, *i.e.*, that it will produce results for the control material that fluctuate randomly about the assigned value. If the control material does not have an assigned value (say, for example, it is a quantity of material from a single batch of pharmaceutical product, or a volume of a single blend of whiskey, set aside for use as a control material), then a value must be determined by making a series of measurements on the material and calculating the mean result. If there is only one control measurement in each of n analytical runs then, in (Equation 2.4) below, x_i represents the value of the result for run i. If several control measurements are made in each analytical run, then x_i represents the mean of the measurements in run i. Data should be obtained from at least 30 runs, though many more are desirable, especially if the data points are single observations. The centre line is given by:

$$\text{CL} = \bar{x} = \frac{\sum x_i}{n} \tag{2.4}$$

Where the control material has an assigned value, a difference between the assigned value and the centre line, as calculated by Equation (2.4), may indicate bias. A statistical significance test may be used to decide whether an observed difference is a reflection of bias or whether it could be attributed to chance variability; such tests are discussed in Chapter 3. If the assigned value is considered trustworthy, then the cause of any such bias should be sought in the analytical system. Provided the magnitude of the bias does not render the test results unfit for purpose, the control chart may be used to monitor the continued stability of the system while the source of bias is investigated. The calculated mean (Equation (2.4)), rather than the assigned value, should be used as the CL while the investigation is in progress.

The Standard Error/Deviation. In most cases the plotted points are averages of several replicates from each analytical run, and the control limits are placed three standard errors on either side of the centre line. Thus, the control limits are calculated as:

$$\text{CL} \pm 3S_{X\text{-}bar}$$

where the centre line, CL, is the overall mean of the data or the assigned value for a reference material. The standard error, $S_{X\text{-}bar}$, can be calculated directly from the replicate means using the usual formula for standard deviation.* Alternatively, $S_{X\text{-}bar}$ can be calculated as:

$$S_{X\text{-}bar} = \frac{\bar{R}_1}{1.128} \tag{2.5}$$

where the constant in the denominator is found in Table A2 (Appendix) and $\bar{R}_1$ is the average of the magnitudes of the differences between successive plotted points, *i.e.* $|x_2 - x_1|$, $|x_3 - x_2|$, *etc.* These 'moving ranges' are a reflection of the chance analytical variability when the system is

$^*S_{X\text{-}bar} = \sqrt{\frac{\sum_{i=1}^{n}(x_i - \bar{x})^2}{n-1}}$, where each x value is the mean of the test results from run i and there are n runs.

stable. This method of calculation survives in the QC literature from its early days (before hand calculators) when even simple statistical calculations were considered too complicated for shop-floor workers. It has its merits, beyond ease of calculation, as we will see below. The two methods of estimation should give similar results for in-control data, but the moving range method is preferable in establishing preliminary control limits, as it is less sensitive to large outliers. If the plotted points are individual values, rather than averages, the same formulas apply, but the resulting value would be labelled S (for standard deviation) rather than S_{X-bar}(for standard error).

Table 2.1 shows that the standard error for the HPLC data of Example 1 is $S_{X-bar} = 0.50$ from a direct calculation; this was used to draw Figure 2.1. The moving range method gives $S_{X-bar} = 0.46$; the two results are very similar because the data are in-control. The ICP data of Example 2 are not in control and give $S = 2.84$ by direct calculation, whereas $S = 1.94$ is obtained from the moving range calculation (Equation (2.5)); this was used to draw Figure 2.2. The control limits based on the larger standard deviation estimate are so wide that the large observations (numbers 21 and 48) are within the upper control limit. The instability of the ICP measurement system has affected the moving range calculation much less than the direct estimate of standard deviation. A reasonable strategy would be to calculate S or S_{X-bar} both ways and check that the results are approximately equal. If they differ substantially, it is most likely due to out-of-control points, which should be investigated and eliminated from the calculations.

Note that, irrespective of the method of calculation, S or S_{X-bar} should contain both within-run and between-run random variability. The IUPAC guidelines state that these two sources of variability are typically comparable in magnitude. It would be inadvisable, therefore, to set up a control chart using an estimate of the magnitude of the random variation based on n replicate measurements of a control material *all from the same analytical run*. Such an estimate would contain no information on between-run variability and would, most likely, give control limits which would be too narrow for controlling run-to-run variation. Consequently,

Table 2.1 *Calculating the standard error*

Dataset	*Method*	
	Direct	*Moving range*
Example 1 – HPLC	0.50	0.46
Example 2 – ICP	2.84	1.94

they would result in false alarm signals where the system was in-control: this would lead to much analytical effort being wasted trouble-shooting problems which simply do not exist. The effects of using only within-run variability as a basis for calculating control limits for the X-bar chart are illustrated in the case study presented in Section 2.6. The need for genuine replicates (*i.e.*, replicates that are truly independent of each other) is also discussed later.

2.4.2 Data Scrutiny

The requirement that the observations used for setting up a chart reflect an analytical system which is in-control is an important one: it would be unreasonable to expect a control chart based on unstable data to maintain stability in future data! To check that the system is in-control what is usually done is to calculate control limits using all the available data, draw the control chart and apply a set of rules (such as that discussed earlier) to identify out-of-control points. When these are identified, reasons for their being out of control should be sought. If explanations are forthcoming the points should be dropped and the centre line and control limits recalculated based on the remaining points. This process may have to be repeated as it can happen that points which previously appeared in-control will fall outside the revised control limits when outliers are discarded. This happens because very large or very small values can inflate the standard deviation estimate and thus give wider control limits than are justified by the in-control data, and also because the centre line may be biased upwards or downwards by outliers. If explanations are not found for the unusual values then hard decisions must be made. Including bad values will distort the charts, making them less sensitive and, therefore, less useful in detecting problems. Excluding large or small values, inappropriately, will narrow the control limits and lead to false alarm signals. The problem of how to deal with outliers is not confined to control charting. It arises in all areas of science and technology. For a general discussion of the problem, see the ASTM (American Society for Testing and Materials) guidelines.[7]

Figure 2.4 illustrates the effect outliers can have on the calculation of control limits. The data are the same as those of Example 2: measurements were made by ICP–MS over a period of months on QC samples made up to contain 50 ppb of Fe. The first chart shows two large outliers, both far above the upper control limit of 69 ppb. The rest of the data are well inside the control limits and are, for the most part, tightly clustered close to the centre line. Taken at face value, this chart would seem to suggest that, apart from the two outliers, the rest of the data are

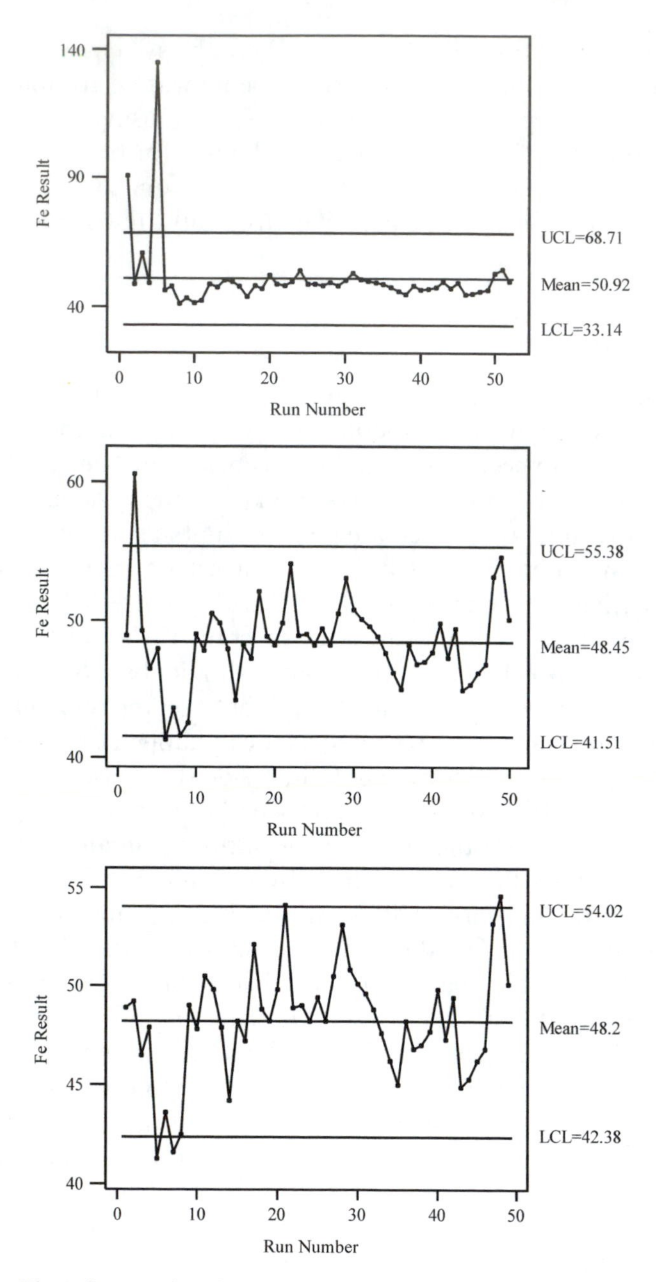

Figure 2.4 *The influence of outliers on the calculated control limits (Fe (ppb))*

well behaved, indicating a highly stable system after the sixth observation. However, when these two outliers are deleted the upper control limit decreases dramatically to 55 ppb and another point is indicated as being a long way out of control. The narrowing of the control limits shows the huge impact of the two outliers in the first chart on the estimate of the standard deviation (calculated in all cases using moving ranges, *i.e.* using Equation (2.5)). The same thing happens, though less dramatically, when the spike in the second chart is removed. The final chart is the same as Figure 2.2. We have already noted that it indicates a system not at all under statistical control. The apparently stable system, as indicated by the first chart, turns out to be very far from stable when (something approaching) correct control limits are placed on the chart. Note, also, that the large values distort the scaling of the first chart such that the variation in most of the data is hidden by the scale compression required to accommodate the very large values in a conveniently sized chart. A computer package is useful for the exploratory data analysis described above; Minitab[8] was used here.

2.4.3 Sample Size

The question as to how many observations are required when first setting up control charts is often asked. Putting a precise number on this is not easy, but the foregoing discussion of data scrutiny suggests a general qualitative rule: enough data should be collected to assure the analyst that the system is stable. The first phase in setting up control charts is exploratory – obtaining data on the system performance, discovering any problems with the system and eliminating them. This process continues until the system is seen to be stable. Unless a state of statistical control is achieved, then control charts cannot be implemented in a routine way and the data generated by the analytical system are unlikely to be fit for purpose.

Once statistical control is achieved and control charts are in routine use, then incremental improvements to the analytical system may mean that the control limits may need to be revised (made narrower) from time to time. Important system changes, such as a change of control material, may also require changing the centre line of the chart. Again, enough data should be allowed to accumulate in order to assure the analyst that chart revision is required.

Exercises

2.1 Table 2.2 contains three datasets from GC assays of distilled spirits (whiskies). The data, which are presented in time order, represent single measurements of the 'total fusel alcohol' (TFA) content of vials of a whiskey used as a QC control material in the laboratory. One vial was inserted into each analytical run. The three datasets were collected at different time periods,

Table 2.2 *Total fusel alcohol (ppm) content of a control material*

Run Number	*Set A*	*Set B*	*Set C*
1	1277	1335	1242
2	1322	1338	1248
3	1289	1336	1236
4	1264	1342	1266
5	1262	1335	1249
6	1279	1333	1268
7	1295	1414	1266
8	1280	1365	1297
9	1311	1352	1275
10	1265	1370	1237
11	1268	1359	1243
12	1350	1371	1311
13	1280	1334	1311
14	1277	1333	1253
15	1273	1337	1272
16	1270	1326	1245
17	1310	1358	1252
18	1306	1347	1250
19	1355	1349	
20	1269	1351	
21	1256	1287	
22	1277	1312	
23	1301	1311	
24	1266	1365	
25	1350	1330	
26	1287	1356	
27	1252	1352	
28	1288	1307	
29	1307	1304	
30	1287	1308	
31	1285	1311	

over several years. Here TFA comprises propanol, isobutanol, isoamylacetate, 2-methyl-1-butanol and 3-methyl-1-butanol. This is one of a number of summary measures routinely monitored by the laboratory.

For each set of data you should set up a control chart (individuals chart) and use the rules discussed earlier in the chapter to decide if the analytical system was in control during the three time periods. Thus, for each set you should:

- Plot the data in time order (a runs chart);
- Calculate the overall average TFA and use it to draw a centre line;
- Calculate the moving ranges (*e.g.*, Set A: $|1322-1277| = 45$, $|1289-1322| = 33$ *etc.*);

- Calculate the average range and estimate the standard deviation by dividing by 1.128 (see Table A2);
- Calculate and draw the control limits, use both action and warning limits;
- Examine the chart for evidence of lack of statistical control.

2.2 Analysis of glucose is performed in a Public Analyst's laboratory in order to ensure compliance with legislation and for authentication of orange juice. The samples are analyzed by HPLC.

Two control charts are routinely monitored. An individuals chart to monitor bias is based on a control sample, which is either a spiked soft drink or clarified orange juice. A range chart to monitor precision is based on duplicate measurements of every tenth test sample. The range chart will be discussed in the next section.

Table 2.3 shows 50 recovery results for glucose. Draw an individuals chart (use the guidelines in the previous exercise). Calculate the standard deviation of the 50 results directly and use it to calculate control limits for the chart. Do the two methods give substantially the same results? Does the system appear to be stable with respect to % recovery?

Table 2.3 *Glucose recovery (%) results*

Run	*Recovery (%)*	*Run*	*Recovery (%)*
1	99.00	26	96.50
2	94.00	27	101.68
3	101.00	28	98.35
4	97.00	29	99.90
5	98.00	30	100.53
6	96.50	31	95.83
7	95.50	32	99.03
8	96.10	33	101.35
9	96.50	34	99.90
10	101.55	35	97.80
11	97.45	36	99.45
12	96.20	37	97.85
13	100.05	38	98.40
14	104.10	39	98.90
15	97.40	40	98.05
16	101.60	41	101.10
17	93.50	42	95.80
18	92.20	43	96.65
19	98.10	44	101.55
20	99.78	45	95.70
21	98.18	46	101.20
22	100.85	47	99.15
23	102.68	48	97.60
24	98.10	49	100.35
25	96.05	50	98.15

2.5 MONITORING PRECISION

2.5.1 Range Charts

The charts discussed until now are useful primarily for detecting bias in the analytical system; this could, for example, be a short term shift (spike), a persistent shift, or a drift in the response of the system. A different chart is required for monitoring analytical precision. Analytical precision is concerned with the variability between repeated measurements of the same analyte, irrespective of the presence or absence of bias. When several replicate measurements are made within each analytical run, the range (*i.e.* the difference between the largest and smallest values) can be used to monitor the stability of analytical precision.

The range chart has the same general features and is used in the same way as X-bar charts: it has a centre line and upper and lower control limits and similar rules for detecting out of control conditions may be applied to the plotted points. The basis for calculating the control limits is different, since sample ranges do not follow a Normal distribution. This technical difference presents no difficulties in practice, as tables of constants have been prepared which simplify the calculation of the control limits. The centre line of the chart is the average, $\bar{R}_2$, of the within-run ranges; the control limits are obtained by multiplying the average range by multipliers shown in Table A2 (Appendix). These multipliers correspond to 3-standard error action limits (D_3 and D_4).

Table 2.4 gives 15 sets of three independent replicate measurements of the potency of a pharmaceutical product, a quantity of which was set

Table 2.4 *Fifteen sets of three replicate potency measurements*

Set	*Potency*	*Set*	*Potency*	*Set*	*Potency*
1	60.55	6	61.55	11	60.09
	60.97		61.20		59.94
	60.54		60.66		59.48
2	59.99	7	60.48	12	59.76
	59.54		60.29		58.48
	59.60		59.75		59.31
3	59.56	8	59.57	13	59.07
	59.33		59.55		58.45
	59.13		58.81		59.49
4	60.28	9	59.51	14	60.60
	59.14		60.58		59.55
	59.12		59.92		59.24
5	60.02	10	60.61	15	59.27
	59.10		59.90		59.58
	59.26		60.41		59.31

aside as control material for a particular assay. The data refer to 15 consecutive analytical runs. The average within-run range is $\bar{R}_2 = 0.81$. The lower control limit is calculated as:

$$D_3 * \bar{R}_2 = 0(0.81) = 0 \tag{2.6}$$

while the upper control limit is given by:

$$D_4 * \bar{R}_2 = 2.57(0.81) = 2.1 \tag{2.7}$$

Figure 2.5 shows the range chart for the potency data; it has three standard error action limits. The chart shows no signs of out-of-control behaviour. However, since it is based on only a fairly small number of data points the stability of the analytical system cannot be considered to be satisfactorily proven, as yet.

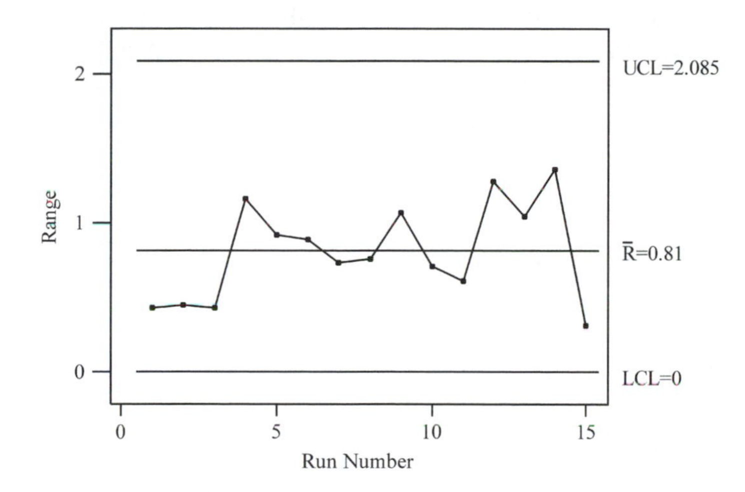

Figure 2.5 *A Range chart for the potency data*

2.5.2 The Nature of Replicates

It is important that the replicates used in drawing a range chart are full replicates, *i.e.* each is subject to all the possible sources of chance variation that may affect the assay. This means that all the steps in the preparation of a sample for analysis must be carried out separately for each replicate. If some are carried out jointly, differences between their measured values will not fully represent the possible random differences that can arise between replicate analyses of the routinely analyzed material that is of primary concern. An extreme example will, perhaps, make this point clearer.

Suppose the preparation and analysis of a control sample involves drying, weighing, dilution and subsequent injection onto a HPLC column and that what are considered to be two replicates simply involve two injections of a single sample prepared in this way onto the column. Clearly, the only factors that can produce differences between the measured results are those that operate from the injection stage onwards, *viz*. injection, separation and detection. Two full replicates, on the other hand, involve differences due to all these factors and, in addition, chance differences that arise in drying, weighing and dilution. Accordingly, the partial replicates may seriously underestimate the size of the random error present in the measurement and should not be used, for example, to estimate the repeatability of the system. Similarly, they do not provide a proper basis for judging whether or not the precision of the assay is in-control, since they do not replicate important aspects of the assay. Consequently, partial replicates should not be obtained for the purposes of plotting range charts. If, for some reason, partial rather than full replicates are obtained, they should be averaged and treated as individual observations for the purpose of setting up and plotting an individuals chart to detect possible bias in the analytical system.

When obtained for drawing a range chart, replicate measurements are intended to capture the random variability present in an analytical run. They must, therefore, be obtained in such a way as to allow such variability to manifest itself. If the replicates are placed beside each other in the analytical run only the very short-term random variability will be captured. Ideally, replicates should be placed in a randomized order throughout the run. This holds for replicates of test materials as well as those of the control material.

2.5.3 Standard Deviation Charts

Range charts are traditionally used to monitor process variability in preference to an alternative chart based on the standard deviation. This tradition survives from a time, before calculators were widely available, when the calculation of standard deviations was considered difficult and, therefore, prone to error. The range, on the other hand, is easy to calculate. If, however, the sample sizes are large (more than ten is often cited) the range uses the information in the data less efficiently than does the standard deviation. In such cases the standard deviation chart should be used in preference (see Howarth[9] or Montgomery[10] for further details on these charts).

In an analytical context there is a second reason why standard deviation charts may be of interest. Very often some, if not all, test materials will be measured at least in duplicate within the run.[6] In such circumstances it will often be possible to monitor the variability of the analytical process using the replicate results. The test materials will vary somewhat in their average results, due to batch-to-batch production variability. However, the variation is unlikely to be so large as to affect the analytical precision. Accordingly, the standard deviations of the within-run replicate results can form the basis for a standard deviation chart. The standard deviations of replicate measurements on different test materials can be combined, using the method described in Section 1.6, thus producing a single measure of precision for each run. Where there are varying numbers of test materials in different analytical runs, then a fixed number (say 5–10) might be decided upon for use in drawing the standard deviation chart. This chart will, in fact, be more powerful for detecting shifts in analytical precision than one based on only a small number of control material replicates. It will, however, require more computational effort, though this is of little concern where the charts are to be computer generated.

Exercises

2.3 The data shown in Table 2.5 came from an in-process assay of a pharmaceutical product; they are the same data as reported in Table 1.3 of Chapter 1.

Table 2.5 *Potency assay results*

Run	*Potency*	*Run*	*Potency*	*Run*	*Potency*
1	244.21	9	240.30	17	235.01
	241.32		237.00		236.85
	243.79		238.15		234.66
2	241.63	10	235.73	18	232.11
	241.88		235.97		249.51
	243.67		236.63		244.02
3	244.33	11	237.72	19	238.81
	245.43		237.67		236.40
	247.20		240.84		240.35
4	241.23	12	239.36	20	242.66
	244.74		239.61		243.66
	242.73		240.57		243.93
5	244.92	13	234.70	21	241.10
	245.40		236.93		241.20
	244.80		239.17		239.20
6	245.09	14	243.70	22	239.38
	244.89		243.49		241.92
	244.98		244.29		238.81
7	238.92	15	243.15	23	243.86
	239.35		242.03		240.98
	240.64		242.39		242.35
8	237.86	16	248.40	24	234.99
	236.72		245.67		234.58
	238.66		244.47		236.04

In each case three independent replicate measurements were made on the test materials. Calculate the within-sample ranges and plot them on a run chart. Calculate the average range $\bar{R}$. Using the constants D_3 and D_4 from Table A2, calculate action limits. Draw these and the centre line on the chart. Examine the chart for evidence of instability. Exclude any outliers from the chart and draw the revised lines on the chart. Does the second chart suggest stability?

2.4 Analysis of sucrose is performed in a Public Analyst's laboratory in order to ensure compliance with legislation and for authentication of orange juice. The samples are analyzed by HPLC. A range chart to monitor precision is based on duplicate measurements of every tenth test sample.

Table 2.6 shows 44 pairs of results, representing duplicate measurements of the sucrose content of routine test samples. The data are presented in the order of measurement; the units are g per 100 mL. Draw a range chart and assess the precision stability of the analytical system.

Note that two values exceed the UCL. When the measurements were made the range chart control limits in use were based on earlier validation data, rather than on the current data. Because the earlier data showed greater variation the control limits were further apart and the two points did not exceed the UCL. Thus, they were not regarded as outliers, and no investigation was carried out. Exclude the two values and reassess the stability of the remaining data.

Table 2.6 *Duplicate sucrose measurements (g per 100 mL)*

Run	*Sucrose*		*Run*	*Sucrose*	
1	3.98	3.99	23	0.18	0.20
2	4.05	3.99	24	0.20	0.17
3	4.02	4.05	25	0.23	0.21
4	4.04	4.06	26	0.22	0.19
5	4.07	3.91	27	0.29	0.25
6	4.02	4.00	28	0.25	0.22
7	3.95	3.99	29	0.35	0.25
8	3.94	3.91	30	0.23	0.25
9	3.94	3.85	31	0.26	0.30
10	3.96	3.89	32	5.11	5.05
11	4.01	3.94	33	3.79	3.95
12	4.03	3.96	34	4.94	4.93
13	1.01	1.02	35	5.13	5.05
14	0.89	0.92	36	3.70	3.67
15	0.90	0.86	37	3.58	3.59
16	0.76	0.75	38	4.87	4.83
17	0.62	0.68	39	3.45	3.41
18	0.58	0.62	40	4.21	4.16
19	0.53	0.58	41	4.11	4.17
20	0.50	0.51	42	4.00	4.06
21	0.56	0.52	43	4.46	4.54
22	0.27	0.26	44	4.64	4.64

2.5 The Public Analyst's Laboratory monitors several sugars simultaneously. The glucose measurements corresponding to these same samples were the basis for Exercise 1.12 (p 32). Carry out the same analyses for the glucose data as you did for the sucrose data in the previous exercise. The note above regarding the control limits in use when the data were generated is true for the glucose data also.

2.6 CASE STUDY

The data in Table 2.7 come from a potency assay of a control material which is to be used in monitoring the routine analysis of a pharmaceutical product to ensure stability of the analytical process.[11] The assay protocol is as follows. Ten tablets are weighed and transferred to an extraction vessel and 500 mL of methanol are added. The vessel is vibrated for twelve minutes. The material is then filtered. Dilution takes place in two stages, 5 + 45 and then 5 + 95. An internal standard is added at the second dilution and a vial is filled for HPLC analysis. Three independent

Table 2.7 *Twenty nine sets of three replicate potency measurements (mg per tablet)*
(Reprinted from the *Analyst*, 1999, Vol. 124, pp 433. Copyright, 1999, Royal Society of Chemistry.)

Run	*Potency*	*Run*	*Potency*	*Run*	*Potency*	*Run*	*Potency*
1	499.17	9	487.21	17	484.17	25	491.27
	492.52		485.35		490.72		488.90
	503.44		479.31		493.45		500.77
2	484.03	10	493.48	18	493.61	26	489.85
	494.50		496.37		488.20		488.42
	486.88		498.30		503.90		487.00
3	495.85	11	553.72	19	482.25	27	492.45
	493.48		554.68		475.75		484.96
	487.33		500.71		488.74		490.58
4	502.01	12	495.99	20	459.61	28	198.92
	496.80		499.36		465.03		479.95
	499.64		482.03		465.57		492.15
5	463.99	13	511.13	21	509.11	29	488.68
	457.61		504.37		510.18		476.01
	469.45		501.00		506.46		484.92
6	482.78	14	510.16	22	489.67		
	484.65		498.59		487.77		
	524.30		501.48		497.26		
7	492.11	15	479.57	23	487.82		
	485.58		462.64		489.23		
	490.24		479.57		493.45		
8	500.04	16	494.54	24	489.23		
	499.11		493.99		491.11		
	493.98		495.08		484.07		

replications of this procedure are carried out. Potency is determined in mg per tablet. The data were collected over several weeks, with varying numbers of measurements per day, as a preliminary data collection exercise with a view to setting up control charts for the assay. There are 29 sets of three replicates; each set was measured in a different analytical run. All analyses for any one run are the work of a single analyst. A total of eight analysts were involved in generating the data.

Figure 2.6(a) shows an X-bar chart of the full data set and Figure 2.6(b) the corresponding range chart. The control limits for Figure 2.6(a) are based on the moving ranges of the twenty-nine set means treated as individual data points (the moving ranges are the magnitudes of

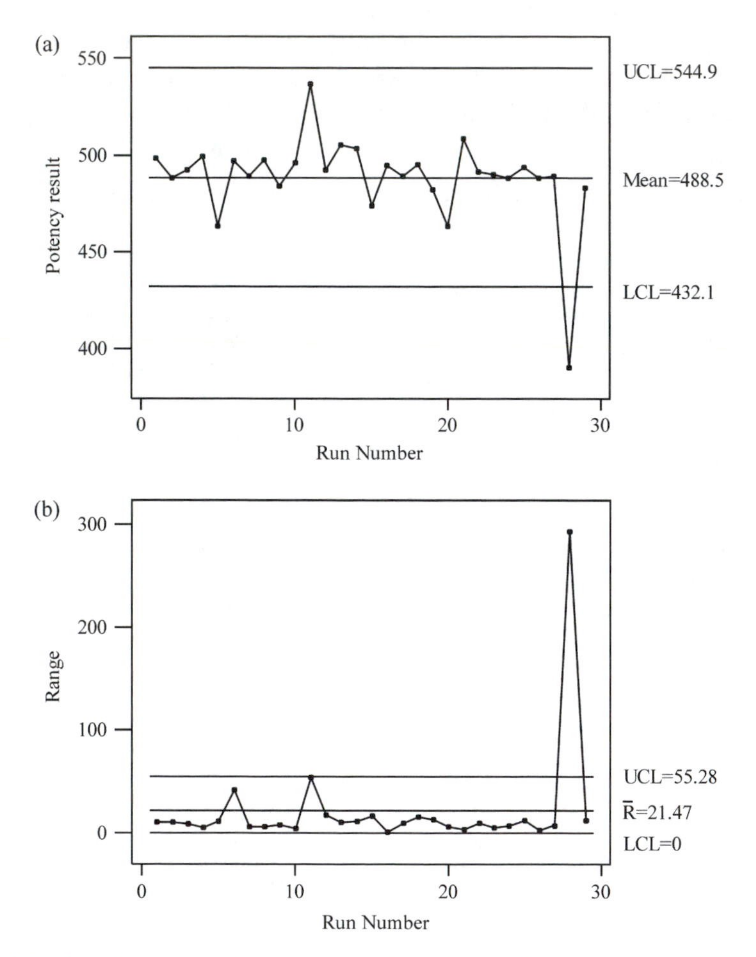

Figure 2.6 *Full dataset (a) X-bar chart and (b) Range chart*
(Reprinted from the *Analyst*, 1994, Vol. 124, pp 433. Copyright, 1999, Royal Society of Chemistry.)

differences between successive means); the control limits for Figure 2.6(b) are based on the average range determined from the within-run ranges, following the methods of calculation described above.

Chart Analysis. The control limits in Figure 2.6 are required to act as yardsticks by which the future stability of the analytical process will be monitored. It seems clear that they will provide useful signals of future instability only if they themselves are based on data that reflect in-control conditions. It is important, therefore, before accepting the control limits and centre lines as aids for routine monitoring of future assays, to assess the stability of the analytical process that generated the data on which they are based.

Examination of Figure 2.6 shows that run 28 is a very clear outlier in both charts. Inspection of the data in Table 2.7 shows one extremely low point (198.92) in run 28. Investigation of this point led the laboratory manager to believe that the analyst had omitted to vibrate the extraction vessel as required by the protocol. This would account for the very low value, as, without vibration, the active ingredient would not be extracted fully from the tablets. At this stage set number 28 was excluded from the dataset, as it was considered to reflect unstable conditions, and the two charts were redrawn as Figure 2.7.

Note that the average range has almost halved from 21.5 in Figure 2.6(b) to 11.8 in Figure 2.7(b), after exclusion of the large outlier. Also, in Figure 2.7(b) the upper control limit for the range chart has dropped from 55.3 to 30.3. This is a dramatic illustration of the impact outliers can have on a statistical analysis and emphasises the need for careful data scrutiny when setting up control charts. Exclusion of set 28 has also had an important impact on the X-bar chart. The centre line has shifted from 488.5 units in Figure 2.6(a) to 492 units in Figure 2.7(a). The control limits are $\pm$ 40.2 units from the centre line in Figure 2.7(a), whereas they were $\pm$ 56.4 units from the centre line in Figure 2.6(a).

Figure 2.7(b) shows runs numbers 6 and 11 as outliers. Figure 2.7(a), the corresponding X-bar chart, also shows run 11 as an outlier. Investigation of the two sets of results showed that they were produced by the same analyst. These results, together with other results on routine test samples, suggested problems with this analyst's technique. For example, when analyzing a batch of test samples this analyst tended to filter all samples in the batch before completing the dilution stages. This meant that open beakers of filtered material were left lying on the bench for varying amounts of time. This would allow the methanol to evaporate and thus concentrate the remaining material, leading to high results. Modification of the analyst's technique (including ensuring that the analysis of one test

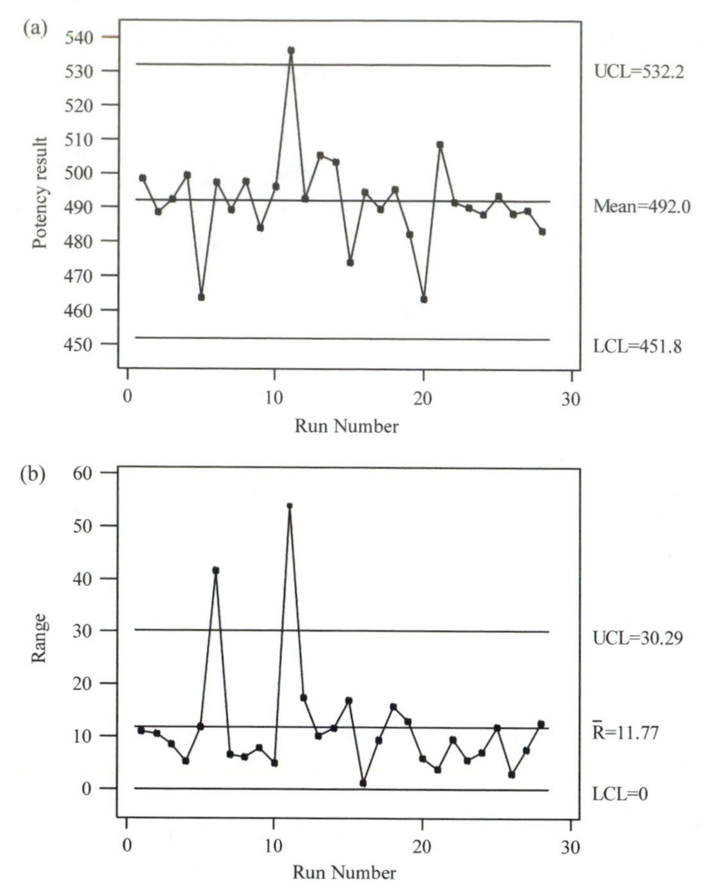

Figure 2.7 *After exclusion of set 28 (a) X-bar chart and (b) Range chart* (Reprinted from the *Analyst*, 1994, Vol. 124, pp 433. Copyright, 1999, Royal Society of Chemistry.)

sample is completed before filtration of a second) gave results in line with those of the other analysts. The investigation also called into question the ruggedness of the assay and led to further investigation of the analytical operating procedure.

At this stage sets 6 and 11 were also excluded from the dataset and the charts redrawn as Figure 2.8. The average range has decreased substantially again and is now 9.0 in Figure 2.8(b). Both charts appear to reflect a measurement process that is in control: there are no points outside the control limits and no extensive runs in the data. Despite this, the laboratory manager had doubts about runs 5 and 20 in the X-bar chart (because of the exclusion of sets 6 and 11, set 20 is the 18th value in Figure 2.8). Excluding these low points from the calculations of the control limits

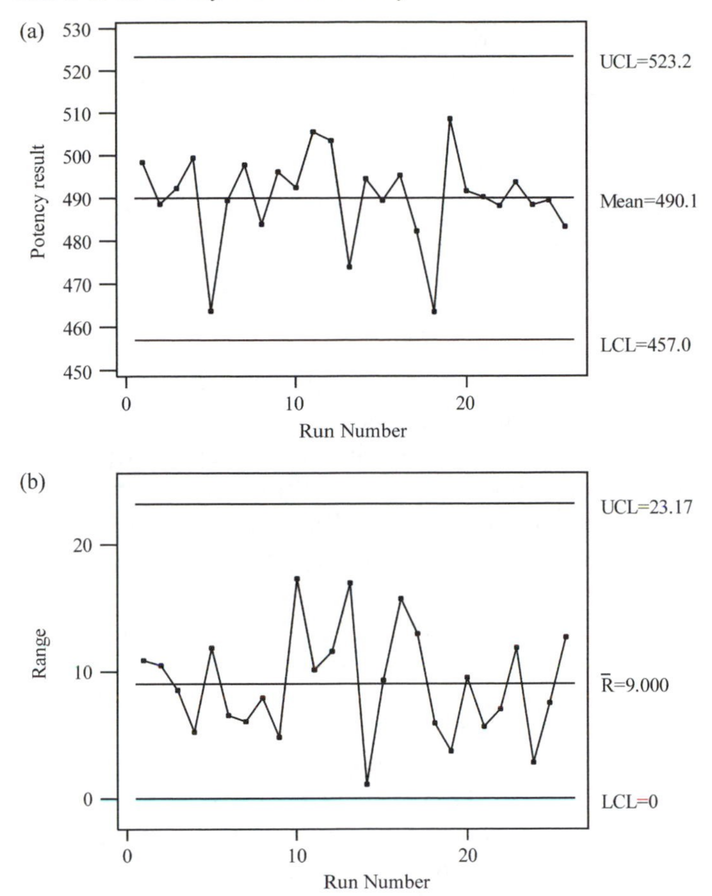

Figure 2.8 *After exclusion of sets 6, 11, 28 (a) X-bar chart and (b) Range chart* (Reprinted from the *Analyst*, 1994, Vol. 124, pp 433. Copyright, 1999, Royal Society of Chemistry.)

would give narrower limits, which would signal the two points as being out-of-control. However, there were no obvious explanations for either of these points representing out-of-control conditions. Also, Figure 2.8(a) indicates that the deviations of these points from the centre line are consistent with chance variation, as judged by its limits. The laboratory manager felt that it was important not to start with control limits that would be too narrow, since this could lead to disruptive false alarm signals. Accordingly, she decided not to exclude them from the dataset. If indeed these points do reflect unusual behaviour then, as more data accumulate, the control limits can be redrawn to reflect the better analytical performance that is being achieved.

A general word of caution is in order. In using historical data care must be taken not to be over-zealous in deleting what appear to be unusual observations. Wholesale trimming of the data runs the risk of producing estimates of analytical variability that are biased downwards, resulting in optimistic measures of precision, which could be misleading.

Appropriate Measures of Variability. This preliminary data collection exercise gave initial control limits (those of Figure 2.8) for use in the routine monitoring of the analytical system. Recall that different estimates of variability were used in constructing the X-bar and range charts. In the range chart the variability that is being monitored is that which arises under essentially repeatability conditions, *i.e.* one analyst makes three independent measurements on the same material within a short time interval under the same conditions. In the X-bar chart, the control limits are based on the variation between the averages of the three replicates. This variation contains two components: it includes the repeatability variation, but also contains the chance run-to-run variation which arises from many sources, including day-to-day, analyst-to-analyst and HPLC system-to-system variation.

In discrete manufacturing applications of control charts, the usual practice is to select from the process at regular time intervals a small number of products or components, and to measure some critical parameter of the selected items. The range chart is based on the within-sample ranges, just as the range chart in analytical applications is based on the within-run ranges. In the manufacturing context the within-sample range is, in most cases, also used as a basis for drawing the X-bar chart. The assumption behind this practice is that all the random variation is captured in the within-sample variation and that differences between sample averages either reflect this variability, or else are due to 'assignable causes', *i.e.*, disturbances which reflect out-of-control production conditions. This assumption works well in very many cases, but not universally, and needs to be investigated.[12,13] In the analytical context, however, experience suggests that superimposed on the within-run chance variation, there is also further random variation which is associated with chance perturbations that affect separate analytical runs differently. Some of the possible causes are listed above. This second component of chance variation is referred to as between-run variation, and can be more important in determining the precision of analytical results than is the within-run variation.

Figure 2.9 shows the results of incorrectly using the within-run variability as a basis for control limits for the X-bar chart. The correct limits, shown in Figure 2.8(a), are given by the centre line ± 33.1 units, while those in Figure 2.9 are ± 9.2 units. The limits shown in Figure 2.9 are too narrow and, if used routinely, would lead to many false alarm

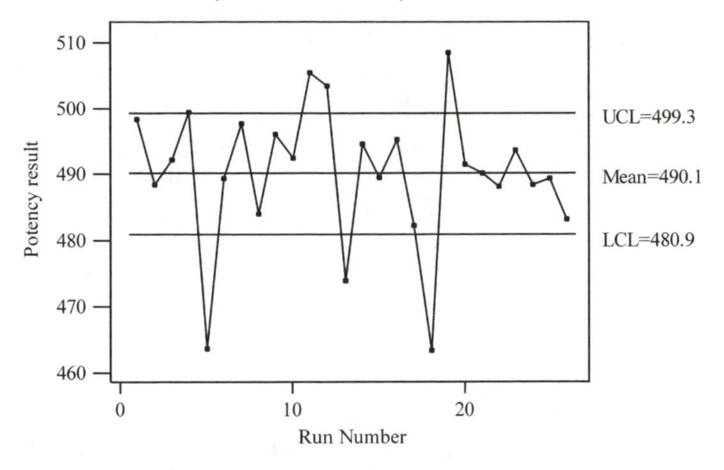

Figure 2.9 *X-bar chart with control limits incorrectly based on within-run variability* (Reprinted from the *Analyst*, 1994, Vol. 124, pp 433. Copyright, 1999, Royal Society of Chemistry.)

signals which would cause severe disruption to the analytical process. This shows the dangers of following the recipes for control chart construction that appear in most engineering quality control textbooks (probably the major source of such information) and which are inappropriate in the analytical context.

Exercise

2.6 Exclude the two runs with the low values in the X-bar chart (Figure 2.8(a)) and redraw both the X-bar and range charts. Compare your charts to those in the case study.

2.7 CONTROL CHART PERFORMANCE

The ability of X-bar charts to detect changes in an analytical system depends on various factors, including the inherent variability of the analytical process, the number of replicate measurements made and the rules used to signal an out-of-control state. In this section the effects of these factors on the time to detection of a shift in the analytical system are examined and illustrated using the case study data.

2.7.1 Average Run Length Analysis

Various performance measures are used to characterize the ability of control charts to detect problems with the systems they are used to monitor. Perhaps the most intuitively appealing of these is the average run length (ARL), *i.e.*, the average number of points plotted until an out-of-control signal is found. Here the word 'average' carries the same long-run interpretation as, for example, the statement that if a fair coin is tossed ten times, then, on average, we expect five heads. We do not interpret the coin tossing statement to mean that if it is done once, the result will be five heads and five tails. So also, in any particular use of a control chart, the actual run length, *i.e.*, the number of points plotted before an out-of-control signal occurs, may be either less than or greater than the ARL. Because the distribution of run lengths is skewed, the individual run lengths will be less than the ARL more than 50% of the time, but occasionally they will greatly exceed the ARL.

Table 2.8 is based on the pharmaceutical case study data and shows ARLs for different shifts in the mean level (*i.e.*, biases) under three rules for signalling out-of-control conditions for the X-bar chart:

- rule (a) a single point outside the action limits;
- rule (b) includes rule (a) but also recognizes as an out-of-control signal a run of 8 points above or below the centre line;
- rule (c) includes rule (a) but also recognizes two points in a row between the warning limits and the action limits as an out-of-control

Table 2.8 *Average run lengths for three rules*
(Reprinted from the *Analyst*, 1999, Vol. 124, pp 433. Copyright, 1999, Royal Society of Chemistry.)

Bias (mg)	*Action limits*	*Action lim. + runs rule*	*Action lim. + warning lim.*	*Bias (std err)*
0.0	370	150	280	0.0
2.2	310	110	220	0.2
4.4	200	60	130	0.4
6.6	120	34	75	0.6
8.8	72	21	43	0.8
11.0	44	15	26	1.0
13.3	28	11	16	1.2
15.5	18	8.6	11	1.4
17.7	12	7.0	7.4	1.6
19.9	8.7	5.9	5.4	1.8
22.1	6.3	4.9	4.1	2.0
24.3	4.7	4.1	3.2	2.2
26.5	3.6	3.4	2.6	2.4
28.7	2.9	2.8	2.2	2.6
30.9	2.4	2.4	1.9	2.8
33.1	2.0	2.0	1.7	3.0

signal. In principle, the two points could be on opposite sides of the centre line (this might happen if variability increased substantially), but, in practice, the signal will usually be triggered by two points between the warning and action limits on one side of the centre line.

The ARLs are tabulated in bias steps from zero up to a bias of 33.1 mg.

Inspection of Table 2.8 shows that the average run length is 370 while the system remains in-control and action limits only are used, *i.e.* on average, not until 370 analytical runs are carried out will an out-of-control signal be observed. This is one of the main reasons for choosing three standard error action limits – a very low false alarm rate is experienced. However, the consequence of ensuring that the false alarm rate is low is that the chart is not sensitive to small to moderate biases. Thus, the second column of Table 2.8 shows long ARLs for small biases: for a bias of 11 mg the ARL is 44, meaning that, on average, not until 44 analytical runs have been carried out will the analyst be given clear evidence of a shift in the analytical system.

The effects of adding supplementary rules to the basic action limits rule can be seen in the ARLs displayed in columns three and four of the table. Using the rule of a run of 8 in a row above or below the centre line, in addition to the action limits rule, has a dramatic effect on the ARL: at a bias of 11 mg the ARL is reduced to about a third of its previous value, *i.e.* from 44 to 15. However, when the system is in-control (*i.e.* no bias) the average run length between false alarms is now 150, as opposed to 370 when using action limits alone. The disadvantage of this increase in the false alarm rate must be set against the reduction in the ARL when biases occur. Comparison of columns two and three shows very substantial improvements in detection rates for small biases. Addition of the runs rule has little impact for large biases, as these are quickly detected by the action limits rule on its own. Column four shows the effect of supplementing action limits with a rule that considers two points in a row between the warning and action limits as a signal of bias. This column shows ARLs intermediate between those of columns two and three, for small to moderate biases.

It is clear that choosing a set of rules involves a trade-off between having long ARLs for zero biases, together with correspondingly long ARLs for moderate biases, and having short ARLs when there are moderate biases in the analytical system, together with the inconvenience of more frequent false alarm signals. The more rules in the set the shorter the ARL for zero bias, *i.e.* the more frequent are the false alarms, but the detection of biases will be correspondingly faster.

The implications of the possible biases and how quickly they are likely to be detected may now be judged in the light of the specification limits for the product, the analysis of which is to be monitored by the control chart. If product potency is well within the specification limits, then an assay which is subject to relatively large biases can still be 'fit for purpose'. Accordingly, potentially long ARLs will be acceptable when assessing the performance of the control chart. If, on the other hand, the manufacturing process is producing material with a potency close to a specification limit, then it will be important that measurement error does not carry the measured value over the limit, in either direction. In such circumstances, the measurement process requires tight control. It may even be necessary to reassess the assay protocol to achieve both good precision for test results and good performance characteristics for the control chart. Some ideas which are relevant to such an assessment are discussed below.

One general lesson that can be learned from the ARL analysis is that out-of-control signals do not necessarily appear immediately an assignable cause begins to affect the analytical system. For large biases, the chart may be expected to produce a signal quickly, but small to moderate biases can take some time to be detected. This means that while the search for assignable causes should focus initially on the immediate neighbourhood of the signal, it should be extended backwards in time if no obvious assignable cause is found.

ARL Analysis: General Application. To allow the ARL analysis to be applied in different practical contexts, the last column of Table 2.8 shows the ARLs tabulated in bias steps that are a multiple of the standard error of the plotted points. To adapt Table 2.8 to any given analytical system all that is required is to multiply column 5 by the appropriate standard error. Thus, for our case study, column 1 of Table 2.8 was obtained by multiplying column 5 by 11.046, which was the standard error used in calculating control limits for Figure 2.8(a). Champ and Woodall[14] present a table of ARLs for many different combinations of rules.

Note that the ARLs in Table 2.8 were obtained on the assumption that the data follow a Normal distribution. Since this is at best only an approximate model for real data, the ARL values should not be interpreted as precise results. Rather, Table 2.8 should be read as showing the broad qualitative implications of using different rules.

2.7.2 How Many Control Samples?

The principal considerations governing the number and frequency of control samples are cost (either financial cost of analyses or, more frequently, the analyst's time) and the stability of the measurement system. If a system is very stable and not prone to going out of control, a low level of monitoring is adequate. If, on the other hand, the system is not very stable and is known to go out of control a high level of monitoring is required if accuracy is to be assured. A general guideline on frequency would be to include at least one set of control samples in every analytical run and more than one set if the run is large or the stability of the system is in any way suspect.

The larger the number of samples in the control set the more likely it is that a measurement bias, if present, will be detected. The minimum level is obviously one. Many laboratories, however, use two, which, while improving the chances of detecting bias, also provide a check on precision and on gross errors: if the two measured values are unacceptably far apart it indicates a problem with the measurement system. The replicates can be used to draw a range chart, as discussed above, or simply to provide a rule that the measurements are unacceptable if more than a certain distance apart. A suitable distance would be $3\sqrt{2\sigma^2}$, where σ is the within-run or repeatability standard deviation; this rule is equivalent to operating a range chart using only the action limits rule.

The 'Harmonized Guidelines' document[6] contains a set of general recommendations which are intended to be adapted to the needs of individual laboratories: "In each of the following the order in the run in which the various materials are analysed should be randomised if possible. A failure to randomise may result in an underestimation of various components of error.

(i) *Short (e.g., $n<20$) frequent runs of similar materials.* Here the concentration range of the analyte in the run is relatively small, so a common value of standard deviation can be assumed. Insert a control material at least once per run. Plot either the individual values obtained, or the mean value, on an appropriate control chart. Analyse in duplicate at least half of the test materials, selected at random. Insert at least one blank determination.

(ii) *Longer (i.e., >20) frequent runs of similar materials.* Again a common level of standard deviation is assumed. Insert the control material at an approximate frequency of one per ten test materials. If the run size is likely to vary from run to run it is easier to standardise on a fixed number of insertions per run and plot the mean value on a control chart of means. Otherwise plot individual values. Analyse in duplicate a minimum of five test materials selected at random. Insert one blank determination per ten test materials.

(iii) *Frequent runs containing similar materials but with a wide range of analyte concentration.* Here we cannot assume that a single value of standard deviation is applicable. Insert control materials in total numbers approximately as recommended above. However, there should be at least two levels of analyte represented, one close to the median level of typical test materials, and the other approximately at the upper or lower decile as appropriate. Enter values for the two control materials on separate control charts. Duplicate a minimum of five test materials, and insert one procedural blank per ten test materials.

(iv) *Ad hoc analysis.* Here the concept of statistical control is not applicable. It is assumed, however, that the materials in the run are of a single type, *i.e.*, sufficiently similar for general conclusions on errors to be made. Carry out duplicate analysis on all of the test materials. Carry out spiking or recovery tests or use a formulated control material, with an appropriate number of insertions (see above), and with different concentrations of analyte if appropriate. Carry out blank determinations. As no control limits are available, compare the bias and precision with fitness for purpose limits or other established criteria."

2.8 LEARNING FROM CONTROL CHARTS

The technical details of how to set up and use control charts have now been discussed. We have seen that control charts capture and store information on analytical performance in a very simple and easily accessible form. Consequently, they lend themselves to retrospective analysis and can therefore act as a powerful tool in on-going quality improvement efforts in the laboratory. The remaining sections of this chapter will discuss using control charts in this way.

2.8.1 Improving Precision by Replication: Revisited

Chapter 1 discussed how obtaining the mean of several replicate measurements improves the precision of the final test result. It was noted, however, that the full benefits of averaging would not normally be realized from within-run replicates. That discussion will now be extended using the data from our case study to illustrate in detail the implications of different sources of random error. As will be seen below, requiring, for example, three analysts each to make a single determination of a measurand, as opposed to the usual practice of one analyst making all three determinations, may seriously improve the precision of the final test result. The reason for this is that the variability of the final determination has two components, *viz.*, within-run and between-run variability, and the way averaging affects the final result is determined by which way the measurements are made.

When combining variability from different sources it is normal to work in terms of variances (*i.e.* squares of standard deviations), as the components of the variation will be additive in this scale. Each of the means of the three replicates plotted in the case study X-bar chart is subject to a variance component due to the within-run variability; this is σ_W^2/n, where n is the number of replicates on which X-bar is based (here $n = 3$) and σ_W is the within-run standard deviation.* Each mean is also subject to a variance component, (σ_B^2), due to between-run variability. When the two components are combined we obtain the variance of the plotted means:

$$\sigma_{X\text{-}bar}^2 = \frac{\sigma_W^2}{n} + \sigma_B^2 \tag{2.8}$$

The form of Equation (2.8) expresses the fact that the averaging process affects only the within-run random error. Since all three observations on which each final test result (X-bar) is based come from only one run, they all experience those same random errors which affect that run as a whole (but which vary from run to run). Accordingly, there is no opportunity for averaging to reduce this aspect of the random influences on test results. The implications of this are illustrated and discussed below. The use of control chart data to estimate these components of analytical variability will be discussed first.

Estimating the Components of Analytical Variability. Given the data on which the control charts are based, the two variance components can be estimated in different ways. The standard statistical approach would be to use a technique known as 'Analysis of Variance' (ANOVA). This will be discussed in some detail in Chapter 7. Here, we will use a simple approach based on the summary statistics calculated for setting up the range and X-bar charts. The within-run standard deviation (σ_W) may be estimated directly from the range chart, which is based on the average of the within-run ranges. Factors for converting from ranges to standard deviations are given in Table A2. Using the average range from the final

*Note that σ_w is the repeatability standard deviation (labelled σ_r in Chapter 1). The label '*W*' is used here to emphasize the 'within'–'between' distinction in the discussion of variance components. Similarly, the between-run standard deviation is labelled with a '*B*' here, but with '*R*' in later chapters.

range chart (Figure 2.8(b)) an estimate of the within-run standard deviation is:

$$\hat{\sigma}_W = \frac{\bar{R}_2}{1.693} = \frac{9.00}{1.693} = 5.316 \tag{2.9}$$

The between-run standard deviation, σ_B, is estimated indirectly, using Equation (2.8) above re-expressed as:

$$\hat{\sigma}_B^2 = \hat{\sigma}_{X-bar}^2 - \frac{\hat{\sigma}_W}{n} \tag{2.10}$$

Recall that the standard error of the sample means (σ_{X-bar}) was estimated using Equation (2.5). For the data on which Figure 2.8(a) is based, this gives:

$$S_{X-bar} = \hat{\sigma}_{X-bar} = \frac{\bar{R}_1}{1.128} = \frac{12.46}{1.128} = 11.046 \tag{2.11}$$

Using this value and the $\hat{\sigma}_W$ value calculated above, the between-run variance $\hat{\sigma}_B^2$ is estimated as follows:

$$\hat{\sigma}_B^2 = \hat{\sigma}_{X-bar}^2 - \frac{\hat{\sigma}_W^2}{n}$$

$$= (11.046)^2 - \frac{(5.316)^2}{3} = 122.014 - 9.420 = 112.594$$

so that the between-run standard deviation is:

$$\hat{\sigma}_B = 10.611$$

We are now in a position to determine the impact of the number of replicates (n) on the standard error (σ_{X-bar}) of the means plotted in the X-bar chart. This quantity determines the precision of test sample results, also, since routine measurements of the product are made in the same way as those on which the control chart is based. We have already seen its role in determining the average run-length performance characteristics of the X-bar chart, as described by Table 2.8. Hence, it is a key parameter in determining both the precision of the analytical system and also our ability to monitor system stability.

Effects of Within-run Replicates on Test Result Precision. In very many cases the number of replicate measurements made will be decided by traditional practice rather than by analysis of the characteristics of the analytical system. Thus, in the case study described above the test results had been routinely based on three replicate measurements and so, when control charts were initiated, three control samples were included in each analytical run. Using the estimates of within-run and between-run variances, the implications for the magnitude of $\hat{\sigma}_{X-bar}$ of different numbers of within-run replicates can now be calculated as follows:

$$\hat{\sigma}_{X-bar} = \sqrt{\frac{\hat{\sigma}_W^2}{n} + \hat{\sigma}_B^2} \quad (2.12)$$

$$\hat{\sigma}_{X-bar} = \sqrt{\frac{(5.316)^2}{n} + (10.611)^2}$$

$$\hat{\sigma}_{X-bar} = \sqrt{\frac{28.260}{n} + 112.593}$$

The results are shown in Table 2.9.

Table 2.9 shows that reducing the number of control sample replicates to two per analytical run would increase the standard error of their mean, ($\hat{\sigma}_{X-bar}$), by approximately 2%. The reason for this is, of course, that the calculation is dominated by the between-run variance: $\hat{\sigma}_B^2 = 112.593$ is much larger than $\hat{\sigma}_W^2/n = 28.2601/n$, for all values of n. Thus, the analytical work involved in maintaining an X-bar chart for monitoring bias could be reduced by one third with little impact on the quality of the chart. Measuring the control material only once would reduce the work even further, but would mean an increase of approximately 7.5% in the standard error and also that a range chart to monitor precision could no longer be based on the control material measurements. This, however, could be overcome by using test sample replicates as a basis for monitoring assay precision (see discussion of standard deviation charts earlier). Moving in the opposite direction in the table shows that there is little advantage to be gained from increasing the number of within-run replicates to five.

Since test samples are measured using the same protocol as for control samples, Table 2.9 gives the precision with which test samples are measured, again for different numbers of within-run replicates. If a change of 2% in the precision of test results has little impact on their fitness for purpose, then the number of replicate analyses on test samples can also be reduced. This will have a major impact on laboratory workload, which, for this assay, would be reduced by one third. While this is highly desirable, a laboratory manager would need strong evidence of assay stability, particularly in a regulated environment, before reducing the number of replicates. Control charts provide this evidence. Thus, control charts have a dual role to play. Used on a

Table 2.9 *The effect of number of within-run replicates, n, on* $\hat{\sigma}_{X-bar}$

n	*1*	*2*	*3*	*4*	*5*
$\hat{\sigma}_{X-bar}$	11.87	11.26	11.05	10.94	10.87

day-to-day basis a control chart is a tool for monitoring the on-going stability of the analytical system and for signalling when problems occur. Analyzed retrospectively, it provides documentary evidence of system stability and provides measures of the quality of routine analyses. This is discussed further below. Note that while the discussion has been illustrated by the case study data, which were collected as a training set to establish control limits, in practice it would be desirable that such calculations be based on a long series of routine control chart data.

An Alternative Approach: Between-run Replicates. If either the precision of test results or the ARL characteristics of the control chart is considered unsatisfactory, then action is required. The natural response is to carry out a robustness/ruggedness study to try to identify and then reduce the sources of variability affecting the assay (experimental designs for such studies will be discussed in Chapter 5). A detailed ruggedness study will take time to carry out (and there is no guarantee of success in reducing variability). In the meantime, the analysis set out below may provide a short-term improvement or, where the variability cannot be reduced satisfactorily, an alternative longer-term solution.

The structure of the formula for the standard error of the plotted values carries implications for the way analysis might be carried out and suggests a second approach to reducing assay variability. The current protocol involves making three replicate determinations of potency within one analytical run. This applies to both test samples and the control material. The resulting standard error is:

$$\hat{\sigma}_{X\text{-}bar} = \sqrt{\frac{\hat{\sigma}_W^2}{n} + \hat{\sigma}_B^2}$$

where n is 3. If, instead, the protocol required a single determination of potency in each of three analytical runs, then the standard error of the mean of three such replicates would be:

$$\hat{\sigma}_{X\text{-}bar} = \sqrt{\frac{\hat{\sigma}_W^2 + \hat{\sigma}_B^2}{n}} \tag{2.13}$$

where n is 3. Under the current protocol only the within-run variation is averaged, since all measurements are made within one run. Under the alternative protocol averaging would take place over the between-run variation, also. The numerical implications for the case study data can be seen in Table 2.10. In the first row of Table 2.10 only the within-row variance, $\hat{\sigma}_W^2 = 28.260$, is divided by n. The second row of the table shows smaller standard errors, because in Equation (2.13) the between-run variance, $\hat{\sigma}_B^2 = 112.593$, is also

Table 2.10 *Estimated standard errors of means for two different protocols*

Protocol	*n*				
	1	*2*	*3*	*4*	*5*
n Replicates in 1 run	11.87	11.26	11.05	10.94	10.87
1 Replicate in n runs	11.87	8.39	6.85	5.93	5.31

divided by n. Since $\hat{\sigma}_B^2$ is much larger than $\hat{\sigma}_W^2$ the difference is substantial.

It is clear from Table 2.10 that the nature of the replicates has a major impact on the precision of the final measured value. The second protocol gives better precision for only two replicates than does the first for three replicates. There is nearly a 40% reduction in the standard error in switching from the first to the second protocol while keeping n at 3, and slightly more than a 50% reduction, when n is 5. From a practical point of view, although the same number of measurements may be involved in both cases, there are obvious logistical considerations involved in choosing between the two protocols. Nevertheless, the alternative approach is worth considering in situations where there is large run-to-run variation in results (in some biochemical assays, for example), as it may be the only way to achieve acceptable precision.

Exercises

2.7 Table 2.11 shows 42 sets of duplicate measurements on a control material. The data come from a pharmaceutical QC laboratory. The pairs of replicates were randomly placed in each analytical run. The response is potency expressed as a percentage.

Use the data to set up both an X-bar and a range chart – determine if the analytical system is in control. Using summary statistics from the control charts, estimate the within-run and between-run standard deviations. Using these, set up a table similar to Table 2.10 which will allow comparisons between different possible assay protocols. Examine the table you compile and assess the different protocols that might be adopted.

2.8 Table 2.10 is based on the case study data of Table 2.7. The analysis of these data led to runs 28, 6 and 11 being excluded, before the final control chart limits were adopted. The laboratory manager had doubts about runs number 5 and 20, also, but she decided to retain them in the analysis, nevertheless.

Exclude these two runs from the final dataset and recompute Table 2.10. Compare the new and original versions. What do you conclude?

Table 2.11 *Duplicate measurements of potency (%) on a QC control material*

Run	*Potency*		*Mean*	*Run*	*Potency*		*Mean*
1	94.68	94.82	94.750	22	94.20	94.75	94.475
2	93.07	93.96	93.515	23	93.50	94.06	93.780
3	94.61	94.53	94.570	24	94.13	94.30	94.215
4	93.23	93.76	93.495	25	94.31	93.75	94.030
5	93.67	93.36	93.515	26	94.44	93.96	94.200
6	94.70	94.88	94.790	27	94.10	95.28	94.690
7	93.32	92.97	93.145	28	93.78	94.11	93.945
8	95.21	94.95	95.080	29	94.91	94.26	94.585
9	94.27	94.03	94.150	30	94.84	94.97	94.905
10	93.95	92.90	93.425	31	93.88	94.20	94.040
11	93.48	92.87	93.175	32	93.29	92.55	92.920
12	94.04	93.19	93.615	33	94.66	94.43	94.545
13	94.56	94.15	94.355	34	94.20	94.65	94.425
14	94.32	93.69	94.005	35	95.39	95.63	95.510
15	93.48	94.67	94.075	36	94.97	94.55	94.760
16	93.15	93.74	93.445	37	94.10	95.45	94.775
17	93.90	94.16	94.030	38	94.13	93.29	93.710
18	95.12	94.45	94.785	39	95.27	94.05	94.660
19	93.96	93.74	93.850	40	94.15	94.33	94.240
20	94.70	94.45	94.575	41	94.02	94.01	94.015
21	93.74	93.88	93.810	42	94.79	94.96	94.875

2.8.2 Obtaining Measures of Precision from Control Charts

Measures of the capability of an analytical system are very often based on special studies. See, for example, the ICH guideline[15] referred to in Chapter 1 or the US National Committee for Clinical Laboratory Standards (NCCLS) guideline 'Evaluation of Precision Performance of Clinical Chemistry Devices'.[16] An alternative approach, especially appropriate after initial validation studies are completed, is to base precision performance measures on the data generated for control charting purposes, when the method is in routine operation.

Repeatability. In Chapter 1 the repeatability limit was defined as:

$$\text{Repeatability limit} = 1.96\sqrt{2\sigma_r^2}$$

where σ_r is the repeatability standard deviation, *i.e.*, the standard deviation of test results obtained under repeatability conditions. It was shown there how this parameter could be estimated from replicate measurements of test samples. An equivalent estimate is given directly by the within-run standard deviation (σ_W) on which the range chart is based: this is exactly the same estimator (assuming an identical matrix) except that it is based on the replicate measurements of control material rather than those on test samples.

Arguably, such an estimator is better than that which will result from a once-off special method validation study. Firstly, it encompasses the work of all the analysts who will be involved routinely in using the method. Once-off studies will frequently represent the work of a single analyst, often one who has more than average experience and who carries out special duties such as method validation studies. Secondly, the estimate it provides is based on a record of assay stability, without which any estimate is meaningless. Consequently, it represents a measure of precision based on a large number of replicate analyses, carried out over a relatively long time period, which makes it more broadly representative, while still fulfilling the requirement of replicates being close together in time.

Intermediate Precision. The reproducibility standard deviation was defined in Chapter 1 as $\sqrt{\sigma_r^2 + \sigma_L^2}$, where σ_r is the repeatability standard deviation and σ_L is a measure of the laboratory-to-laboratory variation. It was noted there that intermediate precision is similar to reproducibility in that the various factors which are maintained constant for the definition of repeatability may be allowed to vary, but all measurements take place in only one laboratory. The principal factors which may vary are time, operator, instrument and calibration. Replacing σ_L^2 in the reproducibility standard deviation formula by a variance component associated with varying conditions within one laboratory gives an intermediate measure of precision.

The measure of between-run variability, σ_B, which can be derived from the X-bar chart, is a global measure of the effects of all the factors that contribute to intermediate precision. As such it could be combined directly with the repeatability standard deviation to give a pragmatic measure of intermediate precision, resulting in an intermediate precision standard deviation of $\sqrt{\sigma_r^2 + \sigma_B^2}$. Alternatively, if the control chart data are dominated by one or two analyst/instrument combinations, it may be appropriate to delete (randomly) some of these data points to achieve a more balanced representation of different analytical conditions. Where a very long record of analyses is available it may even be possible to select a completely balanced dataset, *i.e.*, one which includes, for example, equal representation of all analyst/instrument combinations. This would mimic the conditions for a designed experiment, except that it would not involve the randomization procedures recommended in experimental studies. On the other hand, it would involve analyses carried out over a much longer time frame than would be normal for method validation studies.

Method validation studies provide initial evidence that a method is capable of performing the required analytical task and give estimates of the analytical precision that is achievable, as measured by repeatability, intermediate precision measures, or reproducibility. Control charts are complementary in that they provide evidence of stability, *i.e.*, assurance that the analytical system continues to be capable of generating data fit for purpose. As discussed above, control chart data can also provide updated estimates of two of these method validation parameters, *viz.*, repeatability and intermediate precision standard deviation, based on current data. In order that such calculations can be carried out without time-consuming searches of laboratory records, it is desirable that ancillary information, such as identity of analyst and instruments used, be recorded together with the test results in a readily accessible form.

2.8.3 Using Control Charts

Using control charts for monitoring system stability, once the charts are properly set up, is simply a matter of plotting the points and applying whatever set of rules has been decided upon. There are, however, a few practical points worth considering.

- Apart from acting as a working tool to monitor measurement stability, a control chart is a very useful historical record of the measurement process. As such, the more information it contains the better. A control chart should not be pretty! A pristine control chart, such as one produced by computer which includes nothing but plotted points and control lines, is infinitely less useful than a grubby piece of paper with annotations indicating, *e.g.*, changes of chromatographic column, mobile phase, new standards, problems with the automatic integrator requiring manual integration of peaks, *etc.* A very busy analyst will rarely have time to reflect for long on problems encountered in a particular assay; however, if the problems are noted on the chart, recurring problems will eventually be identified and action to rectify them should follow.

- Where the control chart contains extra information, such as a code for the instrument used and for the analyst who performed the assay, it is worthwhile examining these records periodically. Even if instruments or analysts give acceptably close results when the charts are first set up, it does not mean that they will continue to do so forever afterwards. Control charts provide a basis for learning more about the measurement process – used in this way, they can be important tools for analytical system improvement. If the control chart data are maintained as a computer record, as well as on the paper chart in the laboratory, it makes such retrospective investigations much easier.

- If a measurement system has poor precision a control chart does not, in itself, improve this state of affairs. The poor precision will be reflected in a large standard deviation, which will result in wide control limits. Provided the system is in statistical control the routine measurements will stay within these limits and no action will be signalled. Control charts are, however, a useful source of information for any quality improvement programme which attempts to improve the precision of an assay, in that they provide an historical record of measurement quality and show where problems have occurred. Close scrutiny of the possible causes of such problems is a very good starting point in process improvement studies.

- In planning experimental studies of any kind, the standard deviation is a key parameter in determining the appropriate sample size (see the discussion of sample size issues in Chapters 3 and 4). Control charts contain all the information required for determining the standard deviation and are therefore a kind of laboratory database. It is important, however, to use the right control chart in any given situation.

  Control charts based on a series of measurements on certified or house reference materials contain information on measurement error only. As such they contain the key information for planning method comparison studies or for studies of the performance of trainee analysts. They are not, however, the place to look for information on the likely variability that will be experienced in a study of the effects of, for example, a production process change. A control chart based on production samples will give this information: it will contain batch-to-batch product variability as well as measurement error.

- The psychological aspects of the use of control charts should not be neglected. It is important that the analyst views charts as tools with which to monitor and improve her or his own work. If charts are seen as a management device for monitoring and controlling the analyst, rather than the analytical process, then they are unlikely to be used creatively. Experience from manufacturing industry suggests that charts are powerful process improvement tools when used in a 'Total Quality' environment.

2.8.4 Concluding Remarks

The theory and practice of control charts were introduced by focusing on just two charts; these, however, are the charts most commonly used in analytical laboratories. Many other control charts are available.[4,10] For example, Cusum (Cumulative Sum) charts plot the accumulated difference between a target value and the measured value; they pick up small biases faster than the Shewhart chart described above. The Analytical Methods Committee of the Royal Society of Chemistry[5] gives an example which illustrates the use of Cusum charts. In Shewhart charts every observation is plotted independently of previous observations. The quantity plotted in Cusum charts includes all previous observations; all of which are given equal weight in calculating the plotted value. Exponentially Weighted Moving Average (EWMA) charts are a kind of compromise between these two extremes: the plotted quantity is a weighted average of all the observations to date, but the weights decrease very quickly backwards in time, so that the most recent observations are the main determinants of the current plotting point.

Some QC activities involve monitoring many responses simultaneously: monitoring whiskey production, for example, involves the analysis of more than twenty flavour compounds, ranging in magnitude from single figure to hundreds of ppm levels. Multivariate control charts were developed for such situations; they are, however, not as yet widely used and are still very much a research topic.[9,17] For an introduction to multivariate methods in analytical chemistry, see the review article by Brereton.[18] The charts discussed in this chapter are usually based on the assumption of at least an approximate Normal distribution for the measurement errors. Where the procedure involves counting (*e.g.* the membrane filter method for measuring airborne asbestos concentrations), assuming that the data have a Poisson frequency distribution would be more appropriate in many cases, and charts based on this assumption can be constructed. The concepts underlying the use of all such charts are, however, fundamentally the same and a thorough grasp of the foregoing material should enable the reader to evaluate their potential for use in his or her own laboratory.

2.9 PROFICIENCY TESTING

2.9.1 Overview

The use of control charts is, perhaps, the simplest and cheapest method of *internal* quality control for ensuring the continuing stability of an analytical system. Proficiency testing is a form of *external* quality control. It involves many laboratories measuring test portions of the same material and comparing the results. It provides assurance to laboratory management, accreditation bodies and customers that the results produced by the laboratory are consistent with those produced by other laboratories and, where a reliable assigned value is available for the test material, good performance in the proficiency test gives assurance of the trueness, *i.e.*, lack of bias, of the laboratory's analytical systems. Since there is an 'International Harmonized Protocol for Proficiency Testing of (Chemical) Analytical Laboratories'[19] (the 'IHP' in what follows) and since the issues are not, in the main, statistical, only a short introduction will be given here. Several excellent accounts of how to set up and run proficiency testing programmes are available in the literature.[20–22]

A common format for proficiency tests is for samples of a control material to be circulated to laboratories several times each year. The laboratories analyze the material and return a test result to the organizing laboratory. The results are then processed and each laboratory is supplied with a summary of the results for all laboratories, against which it can compare its own results. A typical graphical summary is shown in Figure 2.10 below (for larger schemes hundreds of laboratories could be involved): the results, x, are standardized by subtraction of an assigned value, μ, and division by an appropriate standard deviation, σ. This yields a performance score, z.

Assuming Normality, perhaps after deletion of outlying laboratories, the resulting z scores are interpreted as standard Normal scores, *i.e.*, approximately 68% would be expected to fall within one standard deviation of the mean, 95% within two standard deviations, while less

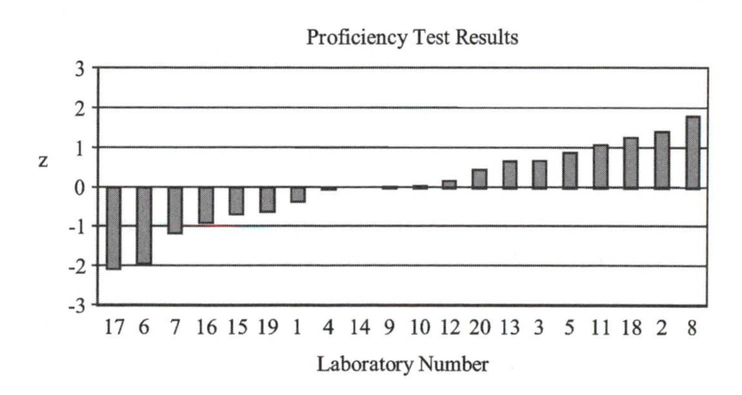

Figure 2.10 *Laboratory z scores in ascending order*

than three per thousand would be expected to lie further than three standard deviations from the assigned value, which will be the mean result if the population of laboratories produces unbiased results, on average. Thus scores less in magnitude (*i.e.*, ignoring sign) than 2 would be considered satisfactory, those between 2 and 3 questionable, while those greater than 3 would be regarded as unsatisfactory.

The summary results are coded so that each laboratory can identify its own result but cannot identify any other laboratory's result. The bar chart shows the laboratory how its own test result compares to the assigned value, *i.e.*, it gives it an objective measure of how accurate its results are likely to be. It also gives it a picture of how its performance compares to peer laboratories. Finally, if it monitors its z score from round to round, the proficiency testing programme can provide each laboratory with a view of its performance over time. If, for example, a laboratory is consistently above the assigned value, *i.e.*, the z scores are always positive, then the data suggest a positive bias, even if the z values are always less than 2.

2.9.2 Technical Issues

The Test Material. It is obvious that the test material that is used in the proficiency test must be stable, should be as close as possible in composition to the materials that are routinely tested, and be sufficiently homogeneous that the test portions supplied to different laboratories have effectively the same composition. Broadly speaking these are the same properties as are required for the control materials used in control charting within any one laboratory.

To provide evidence of the homogeneity of the bulk material prepared for the proficiency test, the IHP suggests that at least ten portions of the bulk material be analyzed in duplicate in a randomised order under repeatability conditions. Using a technique known as Analysis of Variance (see Chapter 7) two standard deviations can be calculated from the resulting data. One of these is the repeatability standard deviation, which measures the short-run measurement error. The second standard deviation measures the test portion to test portion variation. The analysis is essentially the same as that carried out earlier to calculate within-run and between-run standard deviations for control charting data. If the test portion to test portion standard deviation is less than 30% of the target standard deviation for laboratory-to-laboratory variability, then the material is considered by the IHP to be sufficiently homogeneous.

The Assigned Value. Several different methods may be used to arrive at the assigned value for the test material. According to the IHP, the best procedure in most circumstances is to use the consensus of a group of expert laboratories that achieve agreement by the careful execution of recognized reference methods. Where feasible, a good approach is by formulation, *i.e.*, by the addition of a known amount or concentration of analyte to a base material containing none. Another approach involves direct comparison with certified reference materials under repeatability conditions. Finally, the consensus of the results of all the participants in the round of a test is often used as the assigned value. The merits of the various approaches are discussed in the IHP and in the other references cited above.

The Standard Deviation. Wood and Thompson[20] suggest that the value of σ used in calculating the z scores should represent a fitness for purpose criterion, *i.e.*, a range for inter-laboratory variation that is consistent with the intended use of the resultant data. This may be based on professional judgement, it could be informed by data from collaborative studies or in certain areas there may be research guidance available, such as use of the Horwitz function[23] in the food industry. They strongly recommend against use of the standard deviation calculated from the results achieved within the round of the proficiency test that is being analyzed, since using such a standard deviation will mean, by definition, that 95% of participants will be identified as satisfactory, automatically, even if their results are objectively unacceptable.

2.9.3 Concluding Remarks

While the operation of a proficiency test is conceptually very simple, the interpretation of the resulting data will be strongly dependent on the validity of the parameters (assigned value and standard deviation) chosen for calculation of the z scores. It follows that laboratories intending to become involved in any such scheme should look carefully at the methods adopted by the scheme organizers. They should also consult the references cited above for further discussion of these issues, as well as the many logistical questions that have not been touched on in this short introduction.

Proficiency testing provides a check on results produced by a laboratory at selected infrequent time points. It is not a substitute for control charts based on the analysis of control materials within each batch of test materials. These provide assurance that the analytical system is functioning correctly in the short term and that the test results are likely to be valid and can be released to the customer. The primary purpose of proficiency tests, according to Wood and Thompson is "to allow participating laboratories to become aware of unsuspected errors in their work and to take remedial action". The two aspects of laboratory quality control are complementary rather than being alternative methods to achieve results that are fit for purpose. Indeed, it has been demonstrated that having good internal quality control can lead to better performance in proficiency tests.[24]

2.10 CONCLUSION

The main focus of this chapter has been on the use of control charts in the laboratory. Control charts are, perhaps, the simplest statistical method available to the analytical chemist. They are also, arguably, the most powerful. Their simplicity is particularly appealing: it requires little statistical sophistication to appreciate that if a batch of analyses includes a control material and if the result for this material is satisfactory, then the results for the test materials are likely to be satisfactory, also. Huge amounts of effort may be expended in developing and validating sophisticated analytical methods, but, without a concurrent check on the operation of the analytical system, it requires an act of faith to reach the conclusion that any given batch of results may be considered fit for purpose. Control charts provide empirical evidence to justify such a decision.

2.11 REVIEW EXERCISES

2.9 Exercise 1.11 refers to data reported by Moffat *et al.* on the percentage of paracetamol in batches of Sterwin 500 mg tablets. The tablets were assayed by both the BP1993 UV method and by an alternative NIR reflectance spectrophotometric method. Duplicate measurements on 25 production batches are reproduced in Table 1.6, p 31. For the purposes of this exercise we assume that the 25 batches were produced consecutively, though this may not have been the case.

Calculate the means of the duplicate UV measurements for the 25 batches. Use the means to construct an individuals chart for the data. Since the individual means refer to different batches of tablets, the chart you draw is suitable for assessing the stability of the *production* system (the charts discussed in the chapter were based on control materials and hence the variation monitored was generated by the *measurement* systems). Does the production system appear to be stable?

Repeat the exercise for the NIR data.

2.10 The data shown in Table 2.12 are the results obtained by one laboratory in a proficiency testing scheme concerned with the analysis of waste waters. The results, which were obtained between late 1993 and early 2002, are reported in time order. The data are the z scores calculated by subtraction of the assigned values from the laboratory's reported results and subsequent division by the assigned standard deviations. Three parameters were analyzed:

Table 2.12 *Proficiency test results*

Round	*Solids*	*BOD*	*COD*
1	−0.42	−1.73	0.23
2	−0.08	−1.41	0.17
3	−0.06	−1.93	−0.08
4	−0.20	0.09	−0.03
5	−0.28	−1.27	0.53
6	−0.15	0.99	0.52
7	−0.15	−1.68	0.09
8	−1.86	−0.78	0.35
9	−0.77	−0.08	0.28
10	−0.19	−0.45	−0.39
11	−0.21	0.51	−0.03
12	0.00	0.22	0.15
13	0.00	−0.42	0.29
14	0.09	−0.03	0.06
15	−0.84	−0.19	0.09
16	−0.57	−0.16	0.05
17	0.00	0.75	0.15
18	0.18	−0.25	0.41
19	0.25	−0.62	0.48
20	0.00	−0.33	−0.15
21	0.17	−0.08	−0.21
22	0.05	−0.65	0.39
23	−0.09	0.00	−0.22
24	0.23	−0.42	−0.03
25	−0.33	0.55	0.18
26	0.54	0.35	0.11
27	0.41	0.45	−0.04
28	−0.25	−0.12	1.00
29	−0.33	−0.04	−0.14
30	−0.49	0.46	0.79
31	0.25	*	0.01
32	0.28	−0.11	0.53
33	0.19	−0.41	0.38
34	0.00	−0.03	−0.17
35	0.34	−0.04	0.21
36	0.00	−0.47	−0.24
37	0.24	−0.59	−0.29
38	0.13	−0.12	−0.16
39	−0.10	−0.07	−0.44
40	0.07	0.35	0.20
41	0.11	−0.15	−0.31
42	−1.41	0.10	0.22
43	−0.03	−0.37	0.46

suspended solids, biological oxygen demand (BOD) and chemical oxygen demand (COD).

For all three datasets:

- Calculate the mean and standard deviation
- Plot the data in time order on a run chart
- Examine the scatter of the points and assess the performance of the laboratory

Notes on run charts:

- There are two relatively low values for the suspended solids scores. The second last point was found to be low due to sample losses from the sample container. The earlier low value is suspected to have been due to this problem also, but this is not known for sure.
- Note the instability of the BOD values for the early datapoints. This was due to an inadequate method which was replaced (the discovery of such inadequacies is one of the benefits of involvement in proficiency testing schemes). Assess the system performance from datapoint 9 onwards.

Table 2.13 *Duplicate measurements of fructose (g per 100 mL) in orange juices*

Run	*Fructose*		*Run*	*Fructose*	
1	2.52	2.46	23	2.06	2.08
2	2.56	2.46	24	2.07	2.06
3	2.45	2.46	25	2.10	2.08
4	2.44	2.53	26	2.09	2.09
5	2.51	2.46	27	2.13	2.19
6	2.48	2.48	28	2.18	2.13
7	2.52	2.51	29	2.14	2.21
8	2.55	2.49	30	2.10	2.08
9	2.50	2.46	31	2.10	2.16
10	2.43	2.44	32	2.56	2.51
11	2.50	2.54	33	2.42	2.47
12	2.57	2.54	34	2.50	2.50
13	1.69	1.73	35	2.54	2.49
14	1.69	1.67	36	2.25	2.30
15	1.70	1.73	37	2.22	2.20
16	1.67	1.72	38	2.60	2.63
17	1.90	1.90	39	2.34	2.42
18	1.82	1.79	40	2.48	2.46
19	1.86	1.82	41	2.47	2.45
20	1.98	2.00	42	2.54	2.59
21	1.99	1.93	43	1.62	1.66
22	2.10	2.09	44	1.70	1.70

2.11 Several exercises in this chapter have been based on data produced in a Public Analyst's laboratory in order to monitor the sugar content of orange juice. Tables 2.13 and 2.14 give data on the fructose obtained simultaneously with the glucose and sucrose measurements previously considered. Table 2.13 gives duplicate measurements of every tenth test sample. The data are presented in the order of measurement; the units are g per 100 mL. Table 2.14 gives percentage recovery results for a control sample, which is either a soft drink or clarified orange juice, spiked with fructose. The two data sets refer to the same time period, though there is not a one to one correspondence between the runs (some of the spiked samples are extra samples within the same runs, but this detail can be ignored).

Assess the stability of the analytical system with regard to both precision and bias.

2.12 A Public Analyst's laboratory routinely measures potable water samples by flame atomic absorption spectrometry to ensure compliance with the EU drinking water directive. One of the parameters that is measured is the zinc concentration. Every time a batch of test samples is measured the batch includes

Table 2.14 *Fructose recoveries from spiked controls*

Run	*Recovery(%)*	*Run*	*Recovery (%)*
1	96.00	26	101.08
2	94.50	27	99.53
3	100.00	28	98.97
4	97.50	29	100.28
5	96.00	30	100.18
6	100.50	31	97.13
7	99.50	32	98.45
8	101.35	33	102.83
9	99.13	34	95.75
10	99.85	35	98.75
11	97.35	36	98.13
12	99.90	37	97.95
13	98.35	38	98.65
14	102.40	39	99.47
15	95.55	40	96.50
16	99.70	41	101.70
17	100.30	42	97.70
18	98.45	43	101.70
19	99.75	44	98.20
20	99.55	45	98.45
21	98.95	46	102.20
22	99.95	47	97.25
23	105.28	48	98.00
24	98.88	49	101.65
25	101.75	50	100.50

Table 2.15 *Zinc recoveries from spiked controls*

Run	*Zinc*		*Run*	*Zinc*	
1	194	197	16	201	200
2	195	195	17	191	194
3	202	203	18	201	207
4	204	209	19	193	192
5	202	200	20	196	200
6	195	195	21	198	195
7	192	195	22	192	199
8	194	193	23	191	192
9	202	200	24	192	195
10	192	191	25	185	185
11	200	204	26	188	192
12	190	188	27	188	194
13	184	184	28	191	188
14	197	196	29	200	200
15	198	197	30	203	206

two replicates of a QC control sample spiked with 200 μg L^{-1} of zinc. Sample data are given Table 2.15 (units are μg L^{-1}).

Draw an individuals and range chart to assess the stability of the analytical system. Note that the laboratory uses two rules for signalling out of control conditions, *viz.*, a three-sigma rule and a run of nine points at one side of the centre line.

2.12 REFERENCES

1. W.A. Shewhart, *Economic Control of the Quality of Manufactured Products*, Macmillan, London, 1931.
2. E. Mullins, *Analyst*, 1994, **119**, 369.
3. Ford Motor Company, *Continuing Process Control and Process Capability Improvement*, Ford Motor Company, Dearborn, Michigan, 1983.
4. G.B. Wetherill and D.W. Brown, *Statistical Process Control: Theory and Practice*, Chapman and Hall, London, 1991.
5. Analytical Methods Committee, *Analyst*, 1989, **114**, 1497.
6. M. Thompson and R. Wood, *Pure Appl. Chem.*, 1995, **67**, 649.
7. *E-178-94: Standard Practice for Dealing with Outlying Observations*, American Society for Testing and Materials, West Conshohocken, PA, 1994.
8. *MINITAB Statistical Software, Release 13*, Minitab Inc., Pennsylvania, 2000.
9. R.J. Howarth, *Analyst*, 1995, **120**, 1851.

10. D.C. Montgomery, *Introduction to Statistical Quality Control*, Wiley, New York, 3rd edn, 1997.
11. E. Mullins, *Analyst*, 1999, **124**, 433.
12. R. Caulcutt, *Appl. Statist.*, 1995, **44**, 279.
13. L.C. Alwan and H.V. Roberts, *Appl. Statist.*, 1995, **44**, 269.
14. C. Champ and W.H. Woodall, *Technometrics*, 1987, **29**, 393.
15. International Conference on Harmonization of Technical Requirements for Registration of Pharmaceuticals for Human Use (ICH), *Guidelines for Industry: Validation of Analytical Procedures*, ICH Q2A, ICH, Geneva, 1994.
16. *Evaluation of Precision Performance of Clinical Chemistry Devices; Approved Guideline*, NCCLS-EP5-A, National Committee for Clinical Laboratory Standards, Wayne, PA, 1999.
17. R.J. Howarth, B.J. Coles and M.H. Ramsey, *Analyst*, 2000, **125**, 2032.
18. R.G. Brereton, *Analyst*, 2000, **125**, 2125.
19. M. Thompson and R. Wood, *Pure Appl. Chem.*, 1993, **65**, 2123.
20. R. Wood and M. Thompson, *Accred. Qual. Assur.*, 1996, **1**, 49.
21. R.E. Lawn, M. Thompson and R.F. Walker, *Proficiency Testing in Analytical Chemistry*, Royal Society of Chemistry, Cambridge, UK, 1997.
22. Analytical Methods Committee, *Analyst*, 1992, **117**, 97.
23. W. Horwitz, *Anal. Chem.*, 1982, **54**, 67A.
24. M. Thompson and P.J. Lowthian, *Analyst*, 1993, **118**, 1495.

CHAPTER 3

Some Important Statistical Ideas

3.1 INTRODUCTION

This chapter introduces the fundamental statistical ideas that underlie the various methods discussed later. These methods involve two closely related sets of ideas. One is concerned with deciding whether an observed outcome is consistent with some preconceived view of the way the system under study should behave; this set of ideas relates to 'statistical significance tests'. The second set is concerned with estimating the size of things and, in particular, with placing error bounds on the measured quantities – 'confidence intervals' are the relevant statistical tool for such purposes.

Recall the process of using an X-bar chart on a routine basis. For simplicity, consider only the use of the rule that a point outside the action limits is a signal of a system problem. Each time we plot a point on the chart and decide whether or not a problem exists, essentially, we test the hypothesis that the analytical system is producing results that fluctuate randomly about the centre line. This is a simple example of what is called a statistical 'significance test': the observed result (X-bar) is compared to what would be expected if the hypothesis of system stability were true. This hypothesis is rejected when the observed result is inconsistent with, *i.e.*, very unlikely under, such an hypothesis. This is judged to be the case when the plotted point falls outside the action limits. Several different significance tests will be introduced later. All such tests are conceptually the same and the basic ideas will be introduced in this chapter.

Whenever a study is planned, the question arises as to how many observations are needed. This question is inherently statistical, since it requires balancing risks: the risk of incorrectly deciding that a systematic effect (*e.g.*, a bias) exists, when it does not, against the risk of failing to detect an effect which does exist. Larger sample sizes will tend to reduce the influence of random variability, but at a cost. A structured approach to deciding on suitable study sizes will be introduced in this chapter in the context of single sample studies; it will be extended to comparative experimental studies in Chapter 4.

When replicate measurements are made on a batch of industrial product the test results will vary for two reasons, *viz*., non-homogeneity of the product and measurement error. If the results are averaged, an estimate is obtained of the batch average for the measured property. However, if a second sample of test portions were measured, a different estimate would, almost certainly, be obtained. It makes sense, therefore, to report the result not as a single point estimate, but rather in the form of an interval within which the true batch average is expected to lie. The most commonly used method for obtaining such an interval is that which gives a 'confidence interval'. Confidence intervals will be introduced in this chapter and discussed further in subsequent chapters.

The final section of the chapter is concerned with assessing the assumption of data Normality, which underlies so many statistical methods. Both a graphical method – a Normal probability plot – and a statistical significance test will be described.

3.2 STATISTICAL SIGNIFICANCE TESTS

Statistical significance tests are widely used (and abused!) in scientific and industrial work. The concepts involved in all such tests are the same irrespective of the specific purpose of the test. These ideas will be introduced here in the context of tests on sample means.

3.2.1 Example 1: A Method Validation Study

Pendl *et al.*[1] report a method for determining total fat in foods and feeds (Caviezel method based on GC). Several standard reference materials (SRMs) were measured by two laboratories, the submitting laboratory (A) and a peer laboratory (B). One of these was SRM 1846 (NIST), which is a baby food powder with a certified value of $\mu_o = 27.1\%$ fat. Each laboratory analyzed the SRM on ten different days and reported the results shown in Table 3.1. An important question for the laboratory managers is whether their analytical systems are unbiased, *i.e.*, are their results varying at random around a long-run value of $\mu_o = 27.1\%$?

For laboratory A, the average result is 26.9% fat. While this is lower than the certified value of $\mu_o = 27.1\%$ the deviation might simply be a consequence of the chance day-to-day variation that is evidently present in both measurement systems. To assess whether or not the deviation, $\bar{x} - \mu_o$, is large compared to the likely chance variation in the average

Table 3.1 *Test results (% fat) and summary statistics for the baby food study* (Reprinted from the Journal of AOAC INTERNATIONAL, 1998, Vol. 81, pp 907. Copyright, 1998, by AOAC INTERNATIONAL.)

	Laboratory			
	A		*B*	
	26.75	27.11	26.33	25.29
	26.73	27.18	26.68	25.88
	26.40	27.21	26.31	26.08
	26.68	27.20	26.59	25.85
	26.90	27.27	25.61	25.78
Mean	26.943		26.040	
SD	0.294		0.441	

result, it is compared to the estimated standard error of the sample mean, $s/\sqrt{n}$. The standard error is a measure of the variability of $\bar{x}$ values, just as the standard deviation is a measure of the variability in individual x values. The ratio:

$$t = \frac{\bar{x} - \mu_o}{s/\sqrt{n}} \tag{3.1}$$

has as a sampling distribution a Student's t-distribution with $n - 1$ degrees of freedom, provided the laboratory is unbiased, *i.e.*, provided the measurements are varying at random around the certified value of $\mu_o = 27.1\%$. If many thousands of samples of size n were obtained under the same conditions and $\bar{x}$ and s calculated in each case and then used to calculate the t-statistic using Equation (3.1), the histogram of the resulting t-values would approximate the smooth curve shown in Figure 3.1. This curve, which can be derived mathematically, is known as Student's t-distribution. The concept of a sampling distribution is important: it

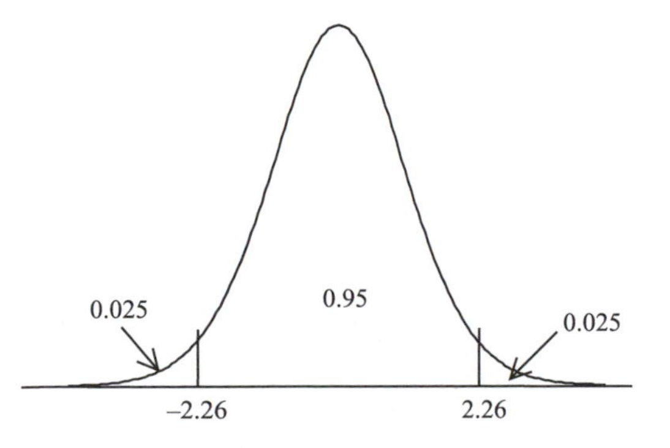

Figure 3.1 *Student's t-distribution with 9 degrees of freedom*

describes the distribution of values of the test statistic (here the t-ratio, Equation (3.1)) that would be obtained if very many samples were obtained from the same system. Accordingly, it describes the expected distribution of summary measures (such as $\bar{x}$, s or t) and allows us to identify unusual values.

If the resulting t-value is exceptional, by reference to this distribution, *i.e.*, if it lies far out in the tails of the sampling distribution, this would call into question the hypothesis that the measurements are varying around 27.1%. A value in the body of the sampling distribution would be likely if the analytical system is unbiased. Accordingly, such a result would support the assumption of an unbiased measurement system.

In order to carry out the test some value (called a 'critical value'), which will define what is considered exceptional, must be decided upon. If we consider as exceptional a value whose magnitude would have only a 5% chance of being exceeded if the laboratory is unbiased, then Figure 3.1 shows that 2.26 is the critical value in this case. A t-value either less than −2.26 or greater than +2.26 would be considered exceptional. The calculated t-statistic for Laboratory A is:

$$t = \frac{26.943 - 27.1}{0.294/\sqrt{10}} = -1.69$$

This lies close to the centre of the t-distribution, so we do not reject the hypothesis that the analytical system in Laboratory A is unbiased.*

Student's t-distribution is similar to the Standard Normal distribution (in that it is a symmetrical bell-shaped distribution centred on zero), but it gives somewhat larger cut-off values than the Normal, when bracketing a given percentage of the area under the curve. Thus, for the Normal curve the values ±1.96 bracket 95% of the area, while ±2.26 were the cut-off values above. The required t-value can be found in a table (see Table A3 in the Appendix) which gives values that enclose 95%, 99%, *etc.* in the centre of the distribution. The table is indexed by the number of degrees of freedom, which is a measure of how much information is available in computing s, the estimate of σ. The number of degrees of freedom is the denominator used in calculating s, $n - 1$ in this case.

Table 3.2 gives examples of cut-off values for enclosing 95% of the t-curve for various sample sizes. Note that, as the sample size increases,

*Note that failure to reject the hypothesis that the analytical system is producing results that vary at random around $\mu_o = 27.1\%$ fat is not the same as *proving* that this is the case. All that can be asserted is that the observed result is *consistent with* such an average. Other long-run values close to 27.1% would also be consistent with the data.

Table 3.2 *Selected t-values for central 95% of curve area*

Sample size	*Degrees of freedom*	*Central 95% t-value*
3	2	4.30
5	4	2.78
10	9	2.26
20	19	2.09
30	29	2.05
120	119	1.98

the t-distribution becomes more and more like a Normal distribution and the 95% cut-off value gets closer and closer to 1.96.

The Significance Testing Procedure. As we have seen, there are a number of steps involved in carrying out a significance test. These, together with the usual statistical terminology, will now be discussed, using Laboratory B as an illustration.

- *Specify the hypothesis to be tested and the alternative that will be decided upon if this is rejected*

  The hypothesis to be tested is referred to as the 'null hypothesis' and is labelled H_o. The 'alternative hypothesis' is labelled H_1. Here we have:

$$H_0 : \mu = 27.1\%$$

$$H_1 : \mu \neq 27.1\%$$

  The null hypothesis is given special status: it is assumed to be true unless the measurement data clearly demonstrate otherwise. Thus, the significance testing procedure parallels the legal principle that the accused is considered innocent until proven guilty: for conviction, the evidence must be such that the assumption of innocence is called into question.

- *Specify a statistic which measures departure from the null hypothesis* $\mu = \mu_o$ *(here* $\mu_o = 27.1$*)*

  The test statistic

$$t = \frac{\bar{x} - \mu_o}{s/\sqrt{n}}$$

  measures by how many standard errors $\bar{x}$ deviates from μ_o, *i.e.*, it scales the deviation of $\bar{x}$ from μ_o in terms of standard error units.

The t-distribution with $n - 1$ degrees of freedom, see Figure 3.1, describes the frequency distribution of values that might be expected to occur if, in fact, the null hypothesis is true.

- *Define what will be considered an exceptional outcome*

We consider a value to be exceptional if it has only a small chance of occurring when the null hypothesis is true. The probability chosen to define an exceptional outcome is called the 'significance level' of the test and is usually labelled α; we choose $\alpha = 0.05$, here. The cut-off values on the sampling distribution, defined by the significance level, are called 'critical values' and the regions of the curve beyond them are 'critical' or 'rejection' regions. Figure 3.1 shows that values either smaller than -2.26 or greater than $+2.26$ occur with probability 0.05.

Note that when we decide to reject the null hypothesis if the test statistic is in the rejection region, we automatically run a risk of $\alpha = 0.05$ of rejecting the null hypothesis *even though it is correct*. This is a price we must pay when using the statistical significance testing procedure.

- *Calculate the test statistic and compare it to the critical values*

For Laboratory B the test statistic is:

$$t = \frac{26.04 - 27.10}{0.441/\sqrt{10}} = -7.60$$

This is smaller than -2.26. Accordingly, we reject the null hypothesis that the laboratory measurements were varying around 27.1% and conclude that the analytical system in Laboratory B is biased downwards.

Note that the entire procedure and the criterion for rejecting the null hypothesis are defined before the calculations are carried out, and often before any measurements are made. This ensures that the decision criterion is not influenced by the data generated.

For Laboratory B the difference between the observed mean of 26.04 and the certified value of 27.10 is said to be 'statistically significant'. The difference between the mean for Laboratory A and the certified value would be declared to be 'not statistically significant'. 'Statistically significant' simply means 'unlikely to be due to chance variation' – it is evidence that a reproducible effect exists. However, 'statistically significant' does *not* mean 'important'. Whether or not the observed difference is of any practical importance is an entirely different question. The answer will be context specific and will require an informed judgement on the part of someone with an understanding of the context.

Exercises

3.1 In Chapters 1 and 2, several examples from a Public Analyst's laboratory of the analysis by HPLC of sugars in orange juice were presented. Exercise 2.2, p 49, gave 50 values of the recoveries of glucose from control samples using in monitoring the stability of the analytical system. The controls were either soft drinks or clarified orange juice, spiked with glucose.

The summary statistics for the 50 results are: mean = 98.443 and SD = 2.451. Carry out a *t*-test, using a significance level of 0.05, of the hypothesis that the method is unbiased, *i.e.*, that the long-run average recovery rate is 100%.

3.2 The Public Analyst's laboratory referred to above, routinely measures total oxidized nitrogen in water by continuous flow analysis (an automated colorimetric method). The internal QC system involves making duplicate measurements on water spiked with 5.05 mg L^{-1} of nitrogen. The 35 most recent in-control points (averages of two replicates) on the X-bar chart gave a mean of 5.066 and a SD of 0.051. If we can assume that before spiking the water is free of nitrogen, do the summary statistics suggest that the analytical system is biased or unbiased, *i.e.*, does the long-run mean equal 5.05 mg L^{-1} of nitrogen? Use a significance level of 0.05.

Note that the duplicates were averaged before the summary statistics were calculated. Accordingly the fact that the 35 values are averages is irrelevant – they are treated as 35 final results. No allowance has to be made, therefore, in your calculations for the averaging of the duplicates.

One-tail Tests. The tests described above are known as two-tail tests. They are so-called because the null hypothesis is rejected if either an unusually large or an unusually small value of the test statistic is obtained: the critical or rejection region is divided between the two tails. In some applications, we would want to reject the hypothesis only if the observed average were large. This might occur, for example, if we were assaying raw materials for some impurity. In others, we would reject the hypothesis if the observed average were small, for example if we were assaying a raw material for purity. In both cases the critical region would be entirely in one tail, so the tests would be one-tailed.

3.2.2 Example 2: Acceptance Sampling

Test portions were sampled from each of five randomly selected drums of a consignment of raw material and assayed for purity. The contract specified an average purity of at least 90%. If the average of the five results is $\bar{x} = 87.9\%$ and the standard deviation is $s = 2.4\%$, should the material be accepted? Note that the standard deviation in this case includes both sampling variability and measurement error.

To carry out a test we first specify null and alternative hypotheses:

$$H_0 : \mu \geqslant 90$$

$$H_1 : \mu < 90.$$

Here the hypotheses are directional. A result in the right-hand tail of the sampling distribution curve would support the null hypothesis while one in the left-hand tail would support the alternative hypothesis. We specify the significance level to be $\alpha = 0.05$. When the t-statistic:

$$t = \frac{\bar{x} - 90}{s/\sqrt{n}}$$

is calculated, it will be compared to the left-hand tail of a t-distribution with 4 degrees of freedom (see Figure 3.2). The null hypothesis will be rejected if the calculated t-statistic is less than the critical value. Table A3 gives a value of -2.13 as the cut-off value which has an area of 0.05 below it, for a curve with 4 degrees of freedom.

The test statistic is:

$$t = \frac{87.9 - 90}{2.4/\sqrt{5}} = -1.96$$

and since this does not lie in the critical region we do not reject the null hypothesis that the average purity is at least 90%. Given the large variability involved in sampling and measurement error ($s = 2.4$), a sample average value of 87.9, based on five measurements, is not sufficiently far below 90 to lead us to conclude that the average batch

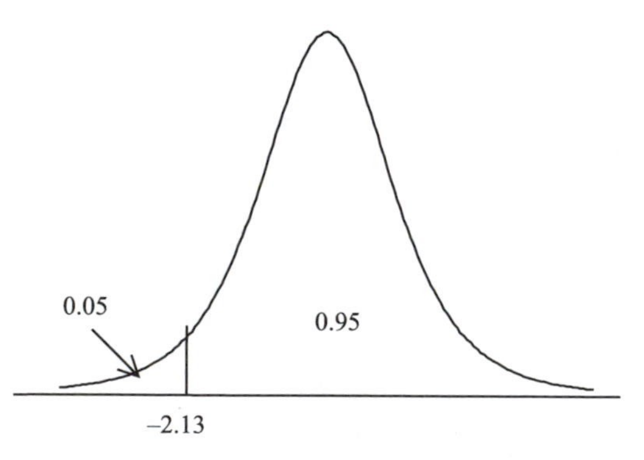

Figure 3.2 *One-tail critical region for the t-distribution with 4 degrees of freedom*

purity is below specification. Such a deficit could have arisen by chance in the sampling and measurement processes.

The null hypothesis being tested by the significance test is whether the results are varying at random around the lower specification bound of 90%. This might not be the case for two reasons, *viz.*, the material could be out of specification or the analytical system could be biased. In carrying out the test as we did above, we implicitly assumed that the analytical system is unbiased. For some types of analysis (*e.g.*, some biological assays) there can be very substantial analyst-to-analyst or run-to-run variation in test results. If this is likely it may be appropriate that the replicate analyses be carried by different analysts or in different analytical runs, so that the inherent biases will tend to cancel.

3.2.3 Summary

The statistical significance testing procedure is really very simple. It may be summarized in the following steps:

- Frame the question of interest in terms of a numerical hypothesis
- Calculate a statistic which measures departure from this hypothesis
- Compare the calculated value of the statistic to the sampling distribution of values that might have been found if the hypothesis were true
- If the calculated value of the test statistic is exceptional in relation to this distribution, reject the hypothesis
- Exceptional values of a test statistic are those that have only a small chance of being observed. The small probability that defines 'exceptional' is called the significance level of the test.

This structure underlies all the significance tests presented in this text (and hundreds of other tests also). The technical details of most of these tests are mathematically complicated, and only a very tiny fraction of analysts would have the least interest in delving into such matters. However, someone who has absorbed this simple logical structure and who understands the statistical assumptions required for the relevant test, could, using validated statistical software, apply effectively any of the tests without knowledge of the technical details.

Exercise

3.3 Incoming raw material has a specification that, on average, it should contain no more than 3% of an impurity. Test portions are taken from six randomly selected drums of the material. You assay them and get an average result of 3.44% and a standard deviation of 0.52%. Would you recommend acceptance or rejection of the consignment?

3.3 DETERMINING SAMPLE SIZE

A major problem facing an analyst about to engage in a study is how many observations are required to draw valid conclusions. Making too few measurements can jeopardize the objectives of the study; making too many is a waste of resources. This question is essentially statistical in nature: it involves balancing the risk of failing to detect important systematic effects, when they are present, with the risk of falsely concluding that effects are present when, in fact, they are not. A simple approach to deciding on study sizes is outlined below.

3.3.1 The Nature of the Problem

To provide a focus for the discussion we will consider again the problem of assaying incoming raw material to determine its purity; we will assume that the contract has specified that the average purity should be at least 90%. The acceptance sampling procedure requires that a small sample of test portions is selected, assayed and the consignment is accepted if the mean purity of the test portions is consistent with the required mean batch purity of 90% or better, otherwise it is rejected.

In making this accept/reject decision there are two ways in which we might be wrong:

- Incorrect rejection: if the mean batch purity really is 90% or better, we might be unlucky in our sampling and measurement errors and our mean test result might be sufficiently low to cause us to reject the batch when, in fact, it meets the quality requirement and should be accepted. Such decisions are not likely to endear us to our suppliers! The statistical terminology for incorrect rejection is 'Type I error'.

- Incorrect acceptance: if, on the other hand, the batch is below the agreed quality, we may get an exceptionally high mean result from our assays and accept the batch, when it should be rejected. Again, this might happen because of sampling and measurement errors. Such decisions may cause problems in production and will generally result in lower yields. The statistical terminology for incorrect acceptance is 'Type II error'.

In deciding on an appropriate sample size, we must protect ourselves against the possibility of making either error.

The effects of random measurement and sampling error can be reduced by averaging them out. The averaging process will, of course, be more effective the larger the sample size taken. There is, however, a cost involved in doing this. In selecting a study size, the cost of a large study must be balanced against the costs that would follow on making either of the incorrect decisions indicated above.

Specifying the Parameters of the Problem. Since chance variation lies at the heart of the problem, some estimate of its magnitude is needed before the required number of measurements can be determined. We need, therefore, to specify the size of the standard deviation (σ) which we consider describes the expected random variation. Given this, there are three other aspects of the problem that require specification:

- We need to specify whether the test will be one or two-tailed and the significance level (α) that will be used. The significance level implicitly defines the probability of incorrectly rejecting a good batch,* since there is a probability α of obtaining a test statistic in the rejection region even when the null hypothesis is true. This is called the 'Producer's Risk' in the quality control literature.

- While we would, naturally, like to detect any shortfall in batch purity, we must recognize that our ability to do so will depend on the amount of chance variability present. If this is large it could take an inordinate amount of analytical effort to detect a very small drop below a mean batch purity of 90%, *i.e.*, a very large number of test portions would need to be selected and measured. We must be willing, therefore, to tolerate the possibility of not detecting very small deviations below 90%. On the other hand, there is some deviation below 90% which we will want to be very sure of detecting. It is necessary, therefore, to specify the smallest deviation below 90% (call it δ) we consider important to detect.

- Having specified δ, we must specify the risk (which we label β) we are willing to take of accepting a batch whose mean purity level is

*The way the significance test is carried out allows for a probability α of rejecting the null hypothesis, *even when this is true* (see Figure 3.2).

$(90 - \delta)\%$. This is called the 'Consumer's Risk' in the quality control literature.

Some comments are appropriate:

- The size of the standard deviation σ is the key to the problem. A large standard deviation indicates large chance variation and immediately makes a large sample size necessary in order to detect small to moderate departures from the required mean purity level. The importance of having a good estimate of the likely order of magnitude of σ at the planning stage of a sampling exercise or experimental study cannot be overstated. This points to the need to retain information on measurement and product variability in an accessible form in the laboratory.

- The choice of a significance level is arbitrary, determined by whether one considers a one in twenty or one in a hundred chance as small (the most commonly used values for the significance level are 0.05 and 0.01). However, since the significance level is the probability of rejecting a good batch, the consequences, financial or other, of such a decision may help in choosing an appropriate value.

- The size of the drop below 90% that we consider important to detect (δ) and the risk (β) we are willing to take of missing such a shortfall, are obviously interconnected. While in principle we might like to detect the fact that the mean purity level was only 89.9%, in practice, we would be unlikely to worry about failing to notice this. Accordingly, we can tolerate a high risk of failing to detect such a shortfall. On the other hand, a mean batch purity of only 85% might have very serious financial implications and we might be unwilling to tolerate even a 1% chance of failing to detect such a shortfall in purity. If it is not obvious how to specify values for δ and β, the best strategy is to carry out a series of calculations using a range of plausible values and then assess the implied sample sizes in the light of the risks being specified.

3.3.2 Using the Sample Size Table

Suppose we require that if the mean batch purity is only 89% this will be detected with a high probability, say 0.95. Such a specification implies that we are willing to take a risk $\beta = 0.05$ of failing to detect a shortfall in purity of $\delta = 1$. We decide that if the mean purity matches the contract value of 90% we will take a risk $\alpha = 0.05$ of rejecting the batch.

Suppose, also, that experience in sampling and measuring this material suggests a standard deviation σ of about 0.9 units. Table A4 gives sample sizes that will ensure that the t-test will have operating characteristics equal to these specifications.

To use Table A4 we calculate:

$$D = \frac{\delta}{\sigma} = \frac{1.0}{0.9} = 1.1$$

which determines which row of the table is entered. The column entry is determined by $\alpha = 0.05$ (one-tail test) and $\beta = 0.05$. The resulting sample size value is:

$$n = 11.$$

Thus, a sample size of 11 test portions will ensure that good batches (*i.e.*, ones with mean purity levels of at least 90%) will be rejected with a probability of no more than 0.05, while non-conforming batches, whose mean purity level is 89% or less, will be rejected with a probability of at least 0.95.

The effects on sample size of varying our specifications can now be examined. For example:

- varying σ and holding the other specifications fixed (*i.e.*, $\alpha = 0.05$, $\beta = 0.05$ and $\delta = 1$):

$$\sigma = 1 \text{ implies } D = 1 \text{ and requires } n = 13$$

$$\sigma = 1.2 \text{ implies } D = 0.83 \text{ and requires } n = 18$$

$$\sigma = 1.5 \text{ implies } D = 0.67 \text{ and requires } n = 26$$

- holding $\sigma = 0.9$ and $\delta = 1$, but choosing:

$$\alpha = 0.01 \text{ and } \beta = 0.01 \text{ requires a sample size of } n = 21.$$

A series of such calculations will show the sensitivity of the required sample size to the various specifications. Note, for example, the importance of the standard deviation σ in the calculations above: a 50% change from $\sigma = 1$ to $\sigma = 1.5$ doubles the sample size that will be required to detect a shortfall of one percentage point in mean purity level. This underlines the importance of maintaining good records regarding the variability of measurements in different situations: this is key information in planning future studies.

The sample size table can also be used to investigate the implications of using a sample size different from that determined by the original set of parameter specifications. Suppose, having calculated the sample size $n = 11$ for our illustrative example above, the laboratory supervisor decided that due to staff shortages a sample size of $n = 5$ would be used routinely. What would be the implications of such a decision in terms of the shortfall in purity that might be detected with high probability?

If the specifications of $\alpha = \beta = 0.05$ are maintained, then in Table A4 a sample of size 5 corresponds to a D value of 1.9 for a one-tail test. Since we estimated σ to be 0.9 this corresponds to:

$$\delta = D\sigma = 1.9(0.9) = 1.7$$

A sample of size 5 would give a probability of 0.95 of detecting the fact that the mean purity level was only $(90 - 1.7)\% = 88.3\%$. Expressed differently, the laboratory would run a risk of 0.05 of accepting material which had a mean purity as low as 88.3%.

Two-tailed Tests. No new issues arise in determining sample sizes when the test that will be carried out is two-tailed. It is simply a matter of using the appropriate columns of the sample size table. Exercise 3.4, below, requires determining the appropriate sample sizes for two-tailed tests.

3.3.3 Discussion

Replicate measurements are made in order to average out chance variation. This raises the questions as to what variation is being averaged, and how the measurements should be made in order to ensure that the appropriate averaging takes place.

Where the test portions to be measured are sampled from a non-homogeneous material, *e.g.*, coal, grain, geochemical or environmental material or waste water samples, the sampling variability is likely to be dominant and the purely analytical variability is likely to be much less important. In such cases the purpose of replication is to average out the sampling/analyte variability and all test portions may be analyzed in a single analytical run.

In method validation studies within a single laboratory where a certified material is measured, or in other instances where homogeneous material is measured, it will often be inappropriate to carry out all the measurements within a single analytical run. If the dominant variability originates in the analytical system itself then, as we saw in the last chapter, between-run variation can be as important, or indeed more important, than within-run variation. If all measurements are made within a single analytical run, then the within-run variation will tend be averaged out. However, averaging will have no effect whatsoever on the single run bias. Accordingly, when the within-run/between-run distinction is important and the magnitude of possible run biases are such as to affect the fitness for purpose of the final result, it will usually be the case that the test portions should be analyzed in different analytical runs. When this is done the run biases will also tend to be averaged out. These ideas are illustrated and discussed further in Chapter 4, Section 4.5.

Exercises

3.4 Refer back to the Pendl *et al.* study of the determination of total fat in baby food in Section 3.2. Suppose that in planning the study the authors had decided to carry out a test with a significance level of 0.05 and that they required a probability of 0.95 of detecting a difference of (a) 0.5 (b) 0.4 (c) 0.25 (d) 0.1 (units are % fat) between the long-run average result obtained by their laboratory and the certified value of 27.1%. Assume that from their method development work they estimated that the standard deviation for replicates would be about 0.25%. For each of (a) to (d) obtain the required sample size.

3.5 Consider the problem of assaying an incoming raw material whose mean impurity level should be no more than 3%. If, on the basis of experience with similar material, it is expected that the standard deviation of measurements (including sampling variation) will be about 0.5%, how many samples would you suggest should be taken?

3.3.4 Some Useful Graphs: Power Curves

The sample size tables allow only limited exploration of the implications of using different sample sizes for a given set of risk specifications. Widely available general-purpose statistical software packages provide powerful facilities both for calculating sample sizes (*i.e.*, they would allow the user to compute Table A4 easily) and for analyzing the properties of any particular sampling plan. For example, Figures 3.3 and 3.4 were drawn using the sample size determination and graphing facilities of Minitab.

Figure 3.3 is a plot of the power of the significance test, against various sample sizes. The power of the test is the probability of rejecting the null hypothesis (here $\mu \geqslant 90\%$) when it is false. Thus, Figure 3.3 shows the probability of rejecting $\mu \geqslant 90\%$, when the mean purity is 89%, for a range of sample sizes up to $n = 30$. We can see that for $n = 11$

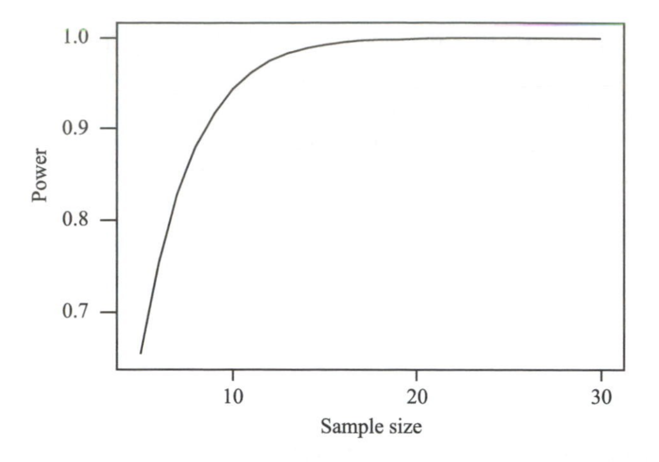

Figure 3.3 *Probability of rejecting the batch when $\mu = 89\%$ ($\alpha = 0.05$ and $\sigma = 0.9$)*

the probability of rejecting the null hypothesis is around 0.95: this was one of our risk specifications in arriving at $n = 11$, when introducing the sample size table. The graph makes clear that almost nothing would be gained by having a sample size greater than $n = 15$.

Figure 3.4 shows power curves for three different sample sizes plotted against the mean purity of the material being sampled; the shortfall in purity (δ) increases from left to right. It shows the influence of sample size when seeking to detect small departures from the hypothesis of $\mu \geqslant 90\%$ – compare the probabilities of detecting the shortfalls in purity (*i.e.*, the power values) for purities between 90 and 89%. For large

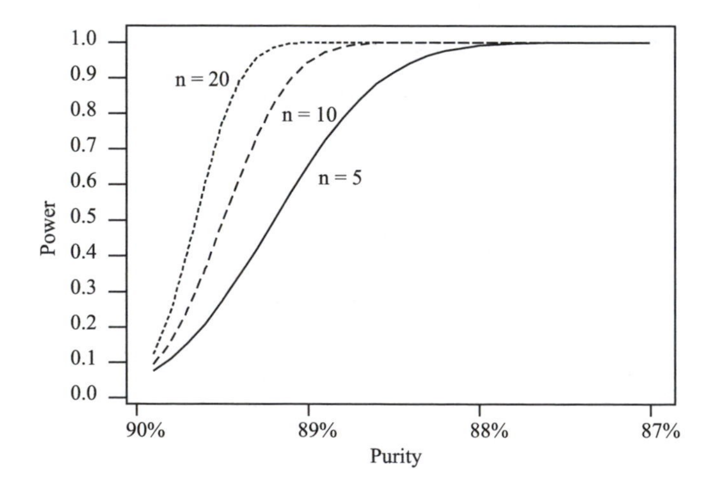

Figure 3.4 *Power curves for three sample sizes ($\alpha = 0.05$ and $\sigma = 0.9$)*

departures from 90% (*e.g.*, 88% or lower) a small sample size is just as useful as larger ones: $n = 5$ gives virtually the same probability of rejecting the raw material as does $n = 20$.

The power curve shows the probability of *rejecting* the batch of material plotted against the true batch purity. An alternative curve, commonly used in the acceptance sampling and quality control literature, is the 'operating characteristics' or OC curve – this is a plot of the probability of *accepting* the material against the true batch purity. Since the probability of acceptance is equal to one minus the probability of rejection, the information contained in the power and OC curves is the same. OC curves and corresponding sample size tables are given in, for example, ANSI/ASQC Z1.4-1993 'Sampling Procedures and Tables for Inspection by Attributes'[2] (equivalent to the 'Military Standards Tables'), which are widely used as a basis for acceptance sampling in industry.

3.4 CONFIDENCE INTERVALS FOR MEANS

The analysis of a randomly selected set of test portions from incoming raw material is a typical exercise carried out in an industrial QC laboratory. When a summary measure, such as the sample mean, is calculated from the results, it will be subject to chance variation, since the data generated by the study will contain both sampling and measurement error. Here, we address the problem of specifying an interval around the calculated mean which will allow for this chance variation and, thus, provide a more realistic estimate of the long-run quantity, in this case, the batch average, that is being measured.

In Chapter 1 we saw that 95% of the area under all Normal curves lies within ± 1.96 standard deviations* of the long-run average or true value, μ. This means that for an unbiased stable analytical system 95% of individual measurements fall within $\pm 1.96\sigma$ of the true value, μ, where σ is the standard deviation of individual values.

We saw also that the effect of averaging is to reduce the variability of final results. Figure 3.5 illustrates this effect: if the standard deviation of single measurements is σ, then the standard error for $\bar{x}$, the average of n measurements, is $\sigma/\sqrt{n}$. This means that if n measurements are made and the sample mean, $\bar{x}$, is calculated, and this process is repeated a very large number of times, 95% of the sample means will lie within a distance $\pm 1.96\sigma/\sqrt{n}$ of μ. We can express this result equivalently by saying that μ

*Very often the ± 1.96 is rounded to ± 2.

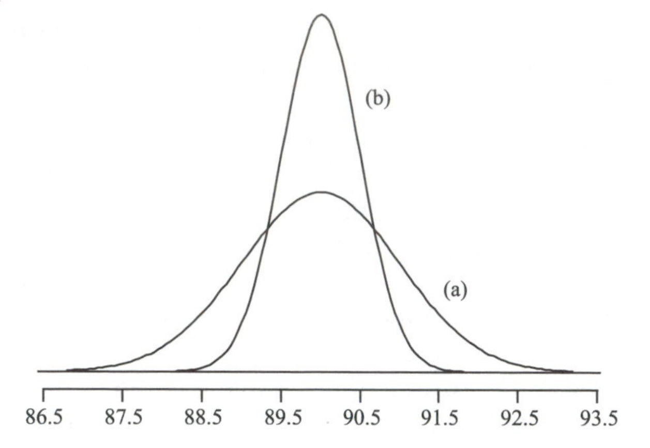

Figure 3.5 *Idealized histograms of the distributions of individual measurements (a) and of the averages of $n = 4$ measurements (b).*†

lies within this same distance of $\bar{x}$. Therefore, if intervals are calculated repeatedly, using the formula:

$$\bar{x} \pm 1.96\frac{\sigma}{\sqrt{n}} \tag{3.2}$$

then 95% of the resulting intervals will contain μ. Thus, we can have a high degree of confidence that any one interval calculated in this way will contain μ. Such an interval is known as a *95% confidence interval* for μ. This is the most frequently used method for placing error bounds around a sample mean, $\bar{x}$, when using it to estimate an unknown process or population mean, μ.

3.4.1 Example 1: Estimating the Average Potency of a Pharmaceutical Material

A quantity of pharmaceutical product was set aside for use as a control material when analyzing the potency of production test samples of the product. One vial of control material is analyzed per run and the result is plotted on a control chart. The material has been in use in the QC laboratory for several months, during which 100 results have accumulated and the control chart has shown that the analytical system is stable. The average, based on the 100 results, was found to be 93.40% of the label claim and the standard deviation was 0.82%. How well determined is the value 93.40% as an estimate of the average potency of the material?

†This picture is a reproduction of Figure 1.17 in Chapter 1; the underlying ideas are discussed there.

We can answer this question by calculating a confidence interval. This is obtained by replacing the symbols in the formula:

$$\bar{x} \pm 1.96\frac{\sigma}{\sqrt{n}}$$

by their calculated values. Thus for the control material we obtain:

$$93.40 \pm 1.96\frac{0.82}{\sqrt{100}}$$

that is,

$$93.40 \pm 0.16$$

The confidence interval estimates that the average potency is somewhere between 93.24 and 93.56% of the label claim. This calculation essentially assumes that the standard deviation is known to be 0.82%. Strictly speaking, this is not the case but with such a large number of measurements it is a reasonable assumption.

A Simulation Exercise. The properties of this method of placing error bounds on test results, based as it is on the idea of repeated sampling, can be illustrated usefully by computer simulation of repeated sampling. The simulation described below is designed to help the reader to understand the implications of the confidence bounds and of the associated confidence level.

Fifty samples of size $n = 4$ test results were generated randomly, as if from a measurement process with known true value, μ, and standard deviation, σ. From each sample of four test results the sample mean, $\bar{x}$, and the corresponding interval:

$$\bar{x} \pm 1.96\frac{\sigma}{\sqrt{n}}$$

were then calculated; the intervals are shown as horizontal lines in Figure 3.6.

The centre of each horizontal line is the observed mean, $\bar{x}$, its endpoints are the bounds given by adding or subtracting $1.96\sigma/\sqrt{n}$ from $\bar{x}$. The vertical line represents the known true value, μ. The flat curve represents the frequency distribution of individual test results; the narrow curve the sampling distribution of averages of four results. Out of the fifty intervals only two do not contain the true value, μ. Some intervals have the true value almost in the centre of the interval (*i.e.*, $\bar{x}$ is very close to μ), but some just about cover the true value, which means $\bar{x}$ is relatively far from μ. Theory suggests that 5% of all such intervals

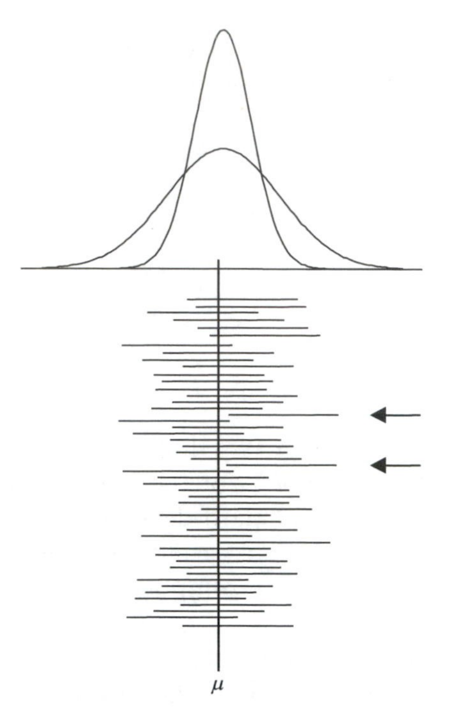

Figure 3.6 *50 simulated confidence intervals based on samples of size n = 4*

would fail to cover the true value; the results of the simulation exercise are consistent with this.

In practice, of course, only a single set of n measurements will be made and a single interval calculated, and we will not know whether or not the calculated interval contains the unknown true value, μ. However, the theory guarantees that if intervals are calculated repeatedly in this way, 95% of them will contain the true value. This gives us confidence in the particular interval we have calculated. Accordingly, the calculated interval is described as *a 95% confidence interval* for μ. Note that the 95% confidence level refers to the method we use, rather than the particular interval we have calculated.

The Confidence Level. The interval discussed above is said to have a '95% confidence level'. If a different level of confidence is required, the value of 1.96, which corresponds to 95%, is replaced by the appropriate value from the standard Normal curve, *e.g.*, 1.645 for 90% confidence and 2.58 for 99% confidence. To see the implications of the different values we return to the example above. Three confidence intervals corresponding to different confidence levels are given by:

90% confidence interval:

$$93.40 \pm 1.645\frac{0.82}{\sqrt{100}} = 93.40 \pm 0.13 \Rightarrow (93.27, 93.53)$$

95% confidence interval:

$$93.40 \pm 1.96\frac{0.82}{\sqrt{100}} = 93.40 \pm 0.16 \Rightarrow (93.24, 93.56)$$

99% confidence interval:

$$93.40 \pm 2.58\frac{0.82}{\sqrt{100}} = 93.40 \pm 0.21 \Rightarrow (93.19, 93.61)$$

Figure 3.7 illustrates the effect of the confidence level on the width of the calculated interval. Note that the higher the confidence level the wider (and perhaps less useful) the confidence interval. Accordingly, there is a trade-off between the level of confidence and the width of the interval. The 95% confidence level is most commonly used.

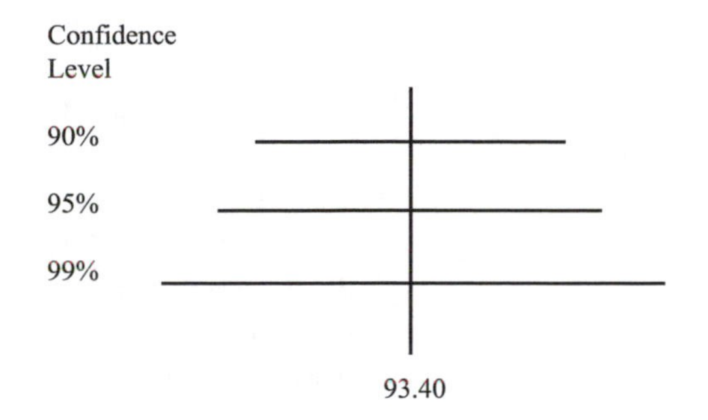

Figure 3.7 *Confidence intervals with different confidence levels*

Sample Size. It is clear from the confidence interval formula (3.2) that the number of measurements controls the width of the interval for a given standard deviation and a given confidence level. Suppose the QC manager wants an estimate of the potency of this control material with error bounds of no more than ±0.1. How many measurements would be required?

Assuming a measurement standard deviation of 0.82% and a 95% confidence level, the formula for the interval gives:

$$\bar{x} \pm 1.96\frac{0.82}{\sqrt{n}}$$

Equating the bounds to ±0.1 gives:

$$1.96\frac{0.82}{\sqrt{n}} = 0.1$$

which may be rearranged to give the sample size:

$$n = \left(\frac{(1.96)(0.82)}{0.1}\right)^2 = 258.3$$

In order to get a confidence interval of width 0.2 (*i.e.*, ± 0.1) 259 measurements will be required; the manager will have to wait!

Small Sample Sizes. In many, if not most, situations the standard deviation will not be known and the number of measurements available from which an estimate may be obtained will not be large. Consequently, the chance variation associated with the estimate of the standard deviation must also be taken into account when calculating confidence intervals around sample means.

In practice, all that is required is to use the sample estimate of the standard deviation, s, instead of σ, and a Student t-multiplier instead of the corresponding Standard Normal value (*e.g.*, 1.96). The interval then becomes:

$$\bar{x} \pm t\frac{s}{\sqrt{n}} \tag{3.3}$$

As we saw when discussing significance tests, Student's t-distribution is similar to the Standard Normal distribution (in that it is a symmetrical bell-shaped distribution), but it gives somewhat larger multipliers than the Normal, when bracketing a given percentage of the area under the curve. As the sample size increases the t-distribution becomes more and more like a Standard Normal distribution and the multiplier for a 95% confidence interval gets closer and closer to 1.96.

3.4.2 Example 2: The Method Validation Study Revisited – Estimating Bias

In Example 1 of Section 3.2 we discussed data reported by Pendl *et al.* concerning a method for determining total fat in foods and feeds. Several reference materials were measured by two laboratories, the submitting laboratory (A) and a peer laboratory (B). One material measured was SRM 1846 (NIST), which is a baby food powder. Each laboratory analyzed the SRM on ten different days and reported the test results reproduced in Table 3.3.

In Section 3.2 we used statistical significance tests to examine the data for evidence of biases on the parts of the two laboratories. Here, we address the same question but we approach it from a different point of view. Instead of asking if the data are consistent with a long-run or true value of 27.1% fat, we ask 'with what long-run values are the data consistent?' We answer this question by calculating confidence intervals.

Table 3.3 *Test results (% fat) and summary statistics for the baby food study* (Reprinted from the Journal of AOAC INTERNATIONAL, 1998, Vol. 81, pp 907. Copyright, 1998, by AOAC INTERNATIONAL.)

	Laboratory			
	A		*B*	
	26.75	27.11	26.33	25.29
	26.73	27.18	26.68	25.88
	26.40	27.21	26.31	26.08
	26.68	27.20	26.59	25.85
	26.90	27.27	25.61	25.78
Mean	26.943		26.040	
SD	0.294		0.441	

For Laboratory A the mean result was 26.943% and the standard deviation was 0.294%. Since these summary statistics are based on ten values, the degrees of freedom of the t-multiplier appropriate for use with equation (3.3) is $n - 1 = 9$; the corresponding multiplier is $t = 2.26$ (see Figure 3.1, page 89). Substituting the numerical values into the confidence interval formula:

$$\bar{x} \pm t\frac{s}{\sqrt{n}}$$

gives

$$26.943 \pm 2.26\frac{0.294}{\sqrt{10}}$$

that is

$$26.94 \pm 0.21$$

This result suggests that the long-run value around which the laboratory's results are varying is somewhere between 26.73 and 27.15% fat. Since the certified value of 27.1% lies within this interval there is no basis for claiming that the laboratory produces biased results; the results are consistent with the claim that the laboratory's analytical system is unbiased. The average of the measurements is 26.94%, which is a little below the certified value. However, given the magnitude of chance day-to-day variation in the measurements, this observed difference from the certified value is consistent with chance variation and does not necessarily indicate an inherent bias in the laboratory's analyses.

For Laboratory B the mean result was 26.040% and the standard deviation was 0.441%, again based on ten values. The confidence interval in this case is:

$$26.040 \pm 2.26\frac{0.441}{\sqrt{10}}$$

that is

$$26.040 \pm 0.315$$

The long-run value around which the measurements from Laboratory B are varying is estimated to be somewhere between 25.73 and 26.36% fat. Since the certified value is above this interval, we conclude that the analytical system is biased downwards. The deviation of $\bar{x} = 26.04\%$ from the certified value of $\mu_o = 27.10\%$ fat is larger than would be expected from purely chance variation.

If the certified value is essentially free of measurement error (*i.e.*, the error can be considered negligible in this case) then subtracting the bounds of the confidence interval from 27.1% gives a confidence interval for the laboratory bias. Thus, we estimate that Laboratory B is biased downwards by between 0.5 and 1.4 percentage fat units.

Our interpretation of the results of the two laboratory analyses are the same irrespective of whether our approach is based on significance tests or on confidence intervals. This is because there is a one-to-one correspondence between the two statistical methods. If a confidence interval contains the null hypothesis value, as happened for Laboratory A, then the significance test will *not* give a statistically significant result. If, on the other hand, the confidence interval does not contain the hypothesized value μ_o, the significance test *will* produce a statistically significant result. This was the outcome for Laboratory B.

Exercises

3.6 In Exercise 3.1 we carried out a *t*-test to assess the presence of bias for a routine recovery check for glucose in spiked control materials and found that the system was biased. Obtain a 95% confidence interval for the magnitude of the bias (*i.e.*, for the difference between the long-run mean recovery and 100%), using the summary information given in Exercise 3.1, p 93. Does the interval confirm the existence of a bias?

3.7 In Exercise 3.2 we tested the mean result for total oxidized nitrogen in water for a bias away from the spiked value of 5.05 mg g^{-1}. Obtain a 95% confidence interval for the long-run mean result produced by the analytical system. Compare this to the outcome of the *t*-test, carried out in Exercise 3.2, p 93. Are the interpretations of the two analyses consistent with each other?

3.4.3 Example 3: Estimating the Potency of a Pharmaceutical Material: Revisited

Figure 2.8 of Chapter 2 is a control chart based on 26 measurements of the potency of a single batch of tablets, a quantity of which is used as a

Table 3.4 *Potency results for the control material*
(Reprinted from the *Analyst*, 1999, Vol. 124, pp 433. Copyright, 1999, Royal Society of Chemistry.)

Potency (mg per tablet)					
498.38	489.31	505.50	495.24	490.17	483.20
488.47	497.71	503.41	482.25	488.14	
492.22	483.96	473.93	463.40	493.65	
499.48	496.05	494.54	508.58	488.42	
463.68	492.46	489.45	491.57	489.33	

control material for monitoring the analysis of tablets of this type. The reported values (mg per tablet) are given in Table 3.4.

We can obtain a confidence interval for the average potency of the batch of tablets being used as control material from these results. We already did this for control chart data in Example 1; there, we assumed the standard deviation was essentially a known constant, since it was based on 100 test results. Here, we take into account the chance variation embodied in the sample standard deviation, s, by using a t-multiplier instead of the Standard Normal value of 1.96 used in Example 1. The interval is calculated by substituting the numerical values into the formula:

$$\bar{x} \pm t\frac{s}{\sqrt{n}}$$

which gives

$$490.096 \pm 2.06\frac{10.783}{\sqrt{26}}$$

that is

$$490.096 \pm 4.356$$

Thus, the batch average potency is estimated to be somewhere between 485.74 and 494.45 mg per tablet for the control material. Note that $t = 2.06$ is the multiplier for a 95% confidence interval based on $n - 1 = 25$ degrees of freedom.*

Exercise

3.8 We saw in Chapter 2 that two of these results were considered suspiciously low by the laboratory manager. These were 463.68 and 463.40. Exclude these values and recompute the confidence interval for the batch mean. Compare the result to that given above.

*Statistics packages such as Minitab, or spreadsheet packages such as Excel, will calculate t-multipliers for degrees of freedom not contained in Table A3.

3.4.4 Example 4: Error Bounds for Routine Test Results

Samples of tablets from batches of these tablets are routinely measured in triplicate in the same way as the control material (see description of the analytical procedure in Chapter 2, page 55) and the three results are averaged to give the final reported result. How should we calculate a confidence interval around the reported result?

Note that the three replicate measurements are made by a single analyst within a single analytical run. Consequently, they do not contain any information about the run-to-run component of the overall measurement error affecting the analytical system. It would be inappropriate, therefore, to use the standard deviation of the three replicates as a basis for error bounds on the reported result. The other sources of variability that impinge on different runs (different analysts, instruments, calibrations, glassware, reagents, *etc.*) need to be accounted for, if realistic error bounds are to be obtained.

The 26 control material results reported above are measured in exactly the same way as production material and the control material is itself production material set aside for control purposes, and so its characteristics are identical to those of the test samples. Accordingly, the variation between the 26 results, as measured by their standard deviation, provides a direct estimate of the likely chance analytical variation to which routine production results are subject, including run-to-run variation as well as within-run variation. Accordingly, we use the standard deviation in calculating a confidence interval around a single reported production result.

The confidence interval formula for the average potency of the test material, based on a reported value of x, is:

$$x \pm ts$$

since the sample size is $n=1$. The standard deviation estimate, $s=10.783$, calculated from the control chart data, has $26-1=25$ degrees of freedom; this means that the t-multiplier will correspond to a t-distribution with 25 degrees of freedom, also. Hence, given a reported value of $x=496.80$, say, the 95% confidence interval for the mean potency of the production material is:

$$496.80 \pm 2.06(10.783)$$

that is,

$$496.80 \pm 22.21 \text{ mg per tablet}$$

The control chart data show that when material of this type is measured repeatedly, by different analysts at different times within this laboratory, the reported results will vary substantially. Accordingly, when a test sample is measured once (even though this measurement is the average of three within-run replicates) the reported result cannot be taken at face value. The confidence interval, based on the information derived from the control chart, shows that error bounds of ±22.21 mg per tablet need to be attached to the reported result.

Discussion. The foregoing calculation is based on within-laboratory variability only. For industrial applications this will often be acceptable. In a wider context, other sources of variability may need to be included in the error bounds. Thompson[3] described a test result as arising from a 'ladder of errors', which is a useful conceptual model for the measurement process. He describes a single analytical result as the sum of a number of components:

$$\text{Result} = \text{true value} + \text{method bias} + \text{laboratory bias} + \text{run bias} + \text{repeatability error}$$

The interval calculated above essentially assumes the absence of both a method bias and a laboratory bias and takes account of the last two random components of the model in obtaining an estimate of the fixed, but unknown, first element, the true value. Inter-laboratory collaborative trials can provide information on the method and laboratory biases. The problem of estimating the various sources of error or uncertainty in test results will be considered again in Chapters 7 and 8.

One aspect of Thompson's model, the concept of a 'true value', though it appears at first sight to be obvious, is, however, quite subtle. It can be thought of in two ways. It is the result that would be obtained if there were no measurement errors. However, since there are *always* measurement errors, this is not an entirely satisfactory definition. An alternative one says that the true value is the average result that would be obtained if the analyte were measured a very large number of times by an unbiased analytical system. The assumption that the system is unbiased means that it is centred on the true value – positive errors will tend to cancel negative errors and the average will correspond to the true value. The statistical terminology is that the 'true value' is the 'expected value' of the system of measurements, *i.e.*, the long-run average. This alternative definition, of course, begs the question as to how we could know that the analytical system is unbiased! However, despite these definitional problems it is a useful practical concept, in that it gives us a simple conceptual model of the measurement process.

In summary, the most commonly used statistical approach to expressing the possible error inherent in an analytical result is to calculate a standard deviation which measures the variation between replicate measurements and then to use this to calculate a confidence interval. The method of calculating the interval is such that a given percentage of such intervals (usually 95%) will cover the 'true value' of the quantity being measured. This property is applied to an individual interval by describing it as a 95% confidence interval: we have confidence in the particular interval because of the long-run properties of the method we use in our calculations.

3.5 SAMPLING

The examples discussed above are all concerned with assessing the likely magnitudes of the measurement errors involved in measuring essentially homogeneous material (though some product variability would be likely in the case of the tablets). If the material being studied is not homogeneous further sources of variability will arise. In many cases the analytical variability will constitute only a very small part of the total. Massart *et al.*[4], for example, refer to a potassium determination carried out routinely in an agricultural laboratory: "It was found that 87.8% of the error was due to sampling errors (84% for sampling in the field and 3.8% due to inhomogeneity of the laboratory sample), 9.4% to between-laboratory error, 1.4% to the sample preparation, and only 1.4% to the precision of the measurement. It is clear that, in this instance, an increase in the precision of measurement is of little interest."

Representative sampling is clearly of major concern in achieving meaningful results when analyzing non-homogeneous materials. However, the practice of sampling is a subject in itself and since it is generally context-specific and usually takes place outside the laboratory only some brief comments about sampling will be made in this introduction to statistical methods within the laboratory.

Statistical methods assume random sampling from the population of interest. 'Random' sampling is not the same as choosing 'haphazardly' – technically it means that every possible sample of a given size is equally likely to be selected from the population. To see how this might be achieved in a relatively simple case, consider the problem of selecting a random sample of 80 vials from a consignment of 900 vials each containing a reagent. If the vials are numbered (from 001–900, say) then a sample may be selected using random numbers – tables of random digits are widely available or they can be generated by computer. Thus, Table 3.5 was generated in a spreadsheet package.

Table 3.5 *Random numbers*

8337706	8627780	5543560
6708984	5413303	6158922
1048135	8887517	6414572
4051167	9627741	6152833
5846237	9090684	1723352
8009640	7155387	5172820
4789447	4374526	4328907
9275657	6309732	9992322

Starting at the top left and selecting triplets of digits, we could work either across the rows or down the columns. Taking the latter approach, we might select the items numbered 833, 670, 104, 405, 584, 800, *etc*, until we have filled our sample size. Often, the items are not numbered. In such cases they might, for example, be unpacked from containers and spread out on a table with a numbered grid, which would allow each item to be uniquely identified. Sampling could then proceed as described above.

It is obvious that the logistics of drawing a sample randomly can be tedious and time consuming, even in such a relatively simple case. It is important to remember, though, that failure to sample randomly can defeat the purpose of the sampling plan – which is to make decisions about the properties of the consignment, or the process from which it is drawn, by examining a *representative* sample of items. Selecting, for example, a small number of boxes of vials from a consignment and examining all the vials in those boxes runs the risk of those boxes being unrepresentative. It is quite likely that all the vials in any one box contain material manufactured close together in time. Their properties may be quite different from those of vials in boxes that were manufactured either much earlier or later in the production sequence, due to process changes. If a sampling method that is not truly random is to be used, the possible sources of bias need to be considered carefully and the selection method should be such as to exclude them.

It is obvious that sampling in cases other than our acceptance sampling example will be even more difficult. For example, sampling air, water, geochemicals and consignments of bulk products, such as ores or feedstuffs, will all involve specialized techniques. Accounts of these are available in International Standards and the relevant research literature. A general introduction to sampling practice is given by Crosby and Patel[5] and information on bulk sampling may be found in Smith and James[6] and Gy[7]. Wetherill[8] gives a technical introduction to the statistical aspects of sampling inspection.

Table 3.6 *Potency results for batches of product*

243.11	239.64	236.93	238.52
242.39	237.75	243.83	243.42
245.65	238.48	242.52	240.50
242.90	236.11	246.18	240.04
245.04	238.74	235.51	242.40
244.99	239.85	*	235.20

Exercises

3.9 Table 1.3 of Chapter 1 shows data from an in-process potency assay of a pharmaceutical product. Three independent replicate measurements were made on each of 24 consecutive batches of product. Having excluded the outlier (batch 18), the remaining 23 final test results (averages of the three replicates) are shown in Table 3.6. From these obtain a 95% confidence interval for the production process average potency. Note that your interval contains batch-to-batch variability as well as measurement error.

3.10 Exercise 1.11 refers to data reported by Moffat *et al.* on the percentage of paracetamol in batches of Sterwin 500 mg tablets. The tablets were assayed by both the BP1993 UV method and by an alternative NIR reflectance spectrophotometric method. Duplicate measurements on 25 production batches are reproduced in Table 1.6, p 31. Calculate the average results for each batch and use the 25 results to obtain a confidence interval for the production process long-run mean. Do this for both the UV and NIR results.

3.11 Two exercises in this chapter have been based on data produced in a Public Analyst's laboratory in order to monitor the glucose content of orange juice. Table 2.14, p 83, gives recoveries for fructose; these are measured simultaneously with the glucose recoveries. Use these results to carry out a *t*-test for the presence of bias; use a significance level of 0.05. Calculate a 95% confidence interval for the bias. Compare the results of your two analyses.

3.6 CONFIDENCE INTERVALS FOR STANDARD DEVIATIONS

In the last section the idea of estimating a long-run mean or true value using a confidence interval based on sample data was introduced. The sample mean provides a point estimate, while the sample standard deviation is used as a basis for placing confidence bounds around this point estimate. Here, our concern will be to obtain an interval estimate of the long-run standard deviation, based on the sample value. The method involves obtaining a confidence interval for the variance, the square of the standard deviation, and then taking the square roots of its confidence bounds to obtain bounds for the standard deviation.

If a random sample of size n is selected from a process or population with standard deviation σ and the sample standard deviation s is calculated, then, assuming Normality, the sampling distribution of

$$\frac{(n-1)s^2}{\sigma^2} \tag{3.4}$$

is the chi-square (χ^2) distribution with $n - 1$ degrees of freedom. Just like the t-distribution, the chi-square distribution is a family of curves, indexed by the degrees of freedom. Table A5 (Appendix) shows critical values which can be used for calculating upper and lower confidence bounds.

Figure 3.8 is a typical chi-square curve (the shape varies with the degrees of freedom, this one has 9); it indicates that 95% of the area under the curve lies between $\chi^2_L = 2.70$ and $\chi^2_U = 19.02$. This means that the observed value of $\frac{(n-1)s^2}{\sigma^2}$ will lie between χ^2_L and χ^2_U 95% of the time. Equivalently, if a single s^2 is calculated the probability is 0.95 that the value of $\frac{(n-1)s^2}{\sigma^2}$ will be between χ^2_L and χ^2_U.

The bounds for $\frac{(n-1)s^2}{\sigma^2}$ shown in the figure can be written as expression (3.5):

$$\chi^2_L < \frac{(n-1)s^2}{\sigma^2} < \chi^2_U \tag{3.5}$$

If the two parts of this expression are considered separately, it can be rearranged, by cross-multiplication and cross-division, to give an alternative representation of the implications of Figure 3.8:

$$\frac{(n-1)s^2}{\sigma^2} < \chi^2_U \Rightarrow \frac{(n-1)s^2}{\chi^2_U} < \sigma^2$$

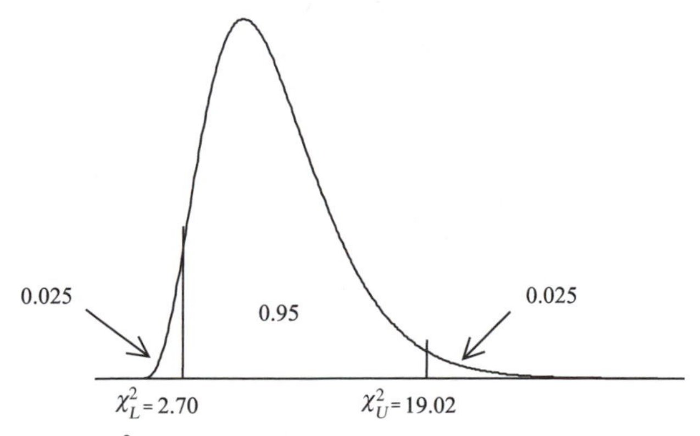

Figure 3.8 *A typical χ^2 distribution*

$$\chi_L^2 < \frac{(n-1)s^2}{\sigma^2} \Rightarrow \sigma^2 < \frac{(n-1)s^2}{\chi_L^2}$$

and we arrive at (3.6)

$$\frac{(n-1)s^2}{\chi_U^2} < \sigma^2 < \frac{(n-1)s^2}{\chi_L^2} \tag{3.6}$$

The end-points in expression (3.6) are random quantities, since the sample variance, s^2, is subject to chance variation. Accordingly, the two bounds define a random interval. The interpretation of this interval is that if n measurements are made and s^2 is calculated and the two end-points of the interval are obtained, then 95% of the random intervals calculated in this way will contain the true variance σ^2. When the numerical result is substituted for s^2, expression (3.6) becomes a 95% confidence interval for σ^2. A simulation study, which would result in a graph similar to Figure 3.6, could be produced in this case also; the distribution curve would, of course, be skewed. In order to obtain confidence bounds for σ, the square roots of the bounds given by (3.6) are calculated.

The Pendl *et al.* data for Laboratory A – ten replicate measurements of the percentage fat content of baby food powder – are reproduced in Table 3.7. Suppose an interval estimate of the intermediate precision of the laboratory is required (recall that the ten test results were obtained on ten different days).

The standard deviation of the ten results is 0.2938 which corresponds to a sample variance of $s^2 = 0.0863$. There are 9 degrees of freedom and so Table A5 gives 95% critical values as $\chi_L^2 = 2.70, \chi_U^2 = 19.02$. The confidence interval for σ^2 is

$$\frac{9(.0863)}{19.02}, \frac{9(.0863)}{2.70}$$

Table 3.7 *Replicate measurements of fat in baby food powder for Laboratory A* (Reprinted from the Journal of AOAC INTERNATIONAL, 1998, Vol. 81, pp 907. Copyright, 1998, by AOAC INTERNATIONAL.)

% Fat	
26.75	27.11
26.73	27.18
26.40	27.21
26.68	27.20
26.90	27.27

which reduces to

$$0.041, 0.288.$$

When the square roots are calculated we obtain 0.20 and 0.54 as 95% confidence bounds for σ. Note the skewness in the distances of the error bounds from the point estimate of the standard deviation, which was 0.29. This is a reflection of the skewness in the sampling distribution curve. It means that when the degrees of freedom are small the true standard deviation could be markedly larger than the point estimate.

Exercises

3.12 The second Laboratory, B, in the % fat measurement study, gave the test results shown in Table 3.8 for the baby food powder. Calculate a 95% confidence interval for the intermediate precision standard deviation and compare it to that obtained for the submitting laboratory, A.

Table 3.8 *Replicate measurements of fat in baby food powder for Laboratory B* (Reprinted form the Journal of AOAC INTERNATIONAL, 1998, Vol. 81, pp 907. Copyright, 1998, by AOAC INTERNATIONAL.)

% Fat	
26.33	25.29
26.68	25.88
26.31	26.08
26.59	25.85
25.61	25.78

3.13 Exercise 1.4 related to the determination of antibacterial drug residues in catfish muscles. The % recoveries of two drugs, which were added to the muscle at approximately 20 μg g^{-1} levels, are shown in Table 3.9. For each residue obtain 95 and 99% confidence intervals for the long-run standard deviations.

Table 3.9 *Recoveries of drug residues in catfish muscles (%)* (Reprinted from the Journal of AOAC INTERNATIONAL, 1995, Vol. **78**, pp 343. Copyright, 1995, by AOAC INTERNATIONAL.)

Oxolinic acid		*Flumequine*	
95.4	103.0	81.2	86.3
97.0	94.9	81.3	78.7
96.5	95.9	78.7	75.7

Repeatability Limits. Where the standard deviation estimate is based on a single set of n test results, the degrees of freedom will be $n-1$. Often, however, the standard deviation is estimated by combining the standard deviations obtained from several samples. Thus, the case study of Section 1.6 was concerned with combining the information from three replicate measurements on each of 24 batches of a pharmaceutical product. One batch was excluded, as an outlier, and the remaining 23 batches gave a combined standard deviation estimate:

$$\hat{\sigma} = s = \sqrt{(s_1^2 + s_2^2 + s_3^2 \cdot \cdot \cdot s_{23}^2)/23} = \sqrt{1.72} = 1.31$$

This estimate has 46 degrees of freedom (d.f.); each batch contributes $3 - 1 = 2$ d.f. to the total. A 95% confidence interval for σ^2 is:

$$\frac{46(1.72)}{66.62}, \frac{46(1.72)}{29.16}$$

which reduces to

$$1.188, 2.713$$

The chi-square values used in the calculations are based on 46 d.f.; they were calculated using Minitab. Substituting the confidence bounds for σ^2 into the expression* $1.96\sqrt{2\hat{\sigma}^2}$ results in a 95% confidence interval for the repeatability limit. The confidence bounds are given by:

$$1.96\sqrt{2(1.188)}, 1.96\sqrt{2(2.713)}$$

Thus, we estimate that the repeatability limit lies between 3.02 and 4.57, with a confidence level of 95%.

Exercises

3.14 In Exercise 1.8 a point estimate of the repeatability limit and repeatability critical difference (based on five replicates) was required for the % casein content of milk. Calculate interval estimates of these quantities. The data are shown in Table 3.10, p 121.

3.15 For the antibacterial drug residue data of Exercise 3.13 assume the replicates were obtained under repeatability conditions. Calculate 95% confidence intervals for the repeatability limits and for the repeatability critical differences, assuming final test results would be based on six replicates.

*Use of 1.96 in the repeatability formula strictly speaking assumes that σ is known. To allow for the uncertainty in $\hat{\sigma}$; a t-curve multiplier could be used. Alternatively, a rounded value of 2 will often be used.

Table 3.10 *Duplicate measurements of casein*
(Reprinted form the Journal of AOAC INTERNATIONAL, 1998, Vol. 81, pp 763. Copyright, 1998, by AOAC INTERNATIONAL.)

Milk	*% Casein*	
1	2.5797	2.5723
2	3.0398	3.0504
3	2.5771	2.5368
4	2.6462	2.6225
5	2.5026	2.4859
6	2.5380	2.5358
7	2.5497	2.5651
8	2.4211	2.4305
9	2.5039	2.4852

3.7 CHECKING NORMALITY

The assumption that data follow a Normal distribution underlies the calculation of confidence intervals for both means and standard deviations and the significance tests for means described earlier in this chapter. For the methods related to means the assumption is less important for larger sample sizes, as the averaging process results in sample means following a sampling distribution which will be well approximated by a Normal curve, unless the data are sampled from a very non-Normal (*e.g.*, highly skewed) distribution. For small sample sizes (and quite small samples are routinely used in many analytical laboratories) there is less opportunity for the averaging process to take effect in this way. Accordingly, it may be important that the assumption of Normality be validated before routinely applying statistical methods based upon it.

An obvious first step in doing this is to draw a histogram of the data. This will give a quick impression of the general shape of the distribution, though histograms may need to be 'read' with some caution. Figure 3.9 shows four histograms, all of which are based on the same data (119 measurements of the potency of a pharmaceutical product). Note that the shape depends strongly on the number of bars in the histogram. Comparing 3.9(a) with 3.9(d) gives quite different impressions of the data structure: 3.9(a) suggests a symmetrical, perhaps Normal, distribution, while 3.9(d) is a very spiky picture with bars detached from the main body of the data at each end of the distribution. In any particular context, it would be easy to suggest possible causal explanations for the various features apparent in 3.9(d). In fact, the spikiness is due to data sparsity – there are only four observations per bar, on average, and the

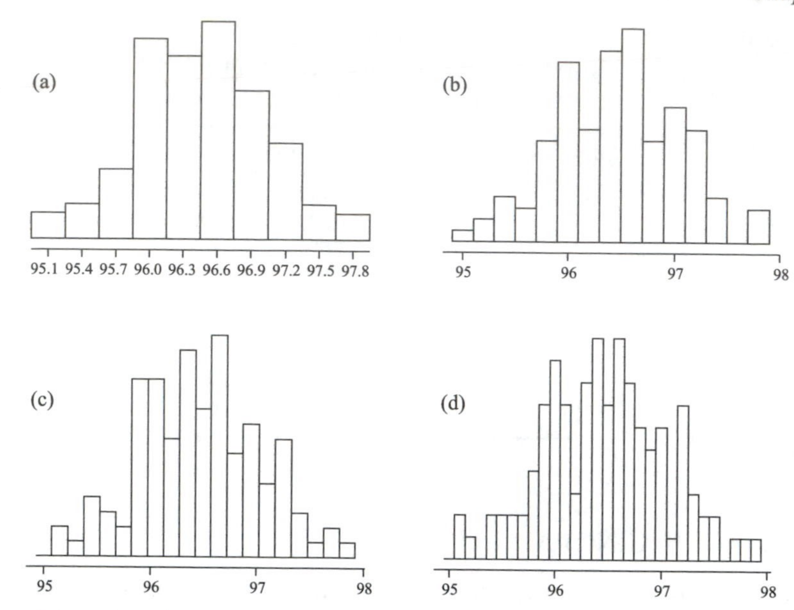

Figure 3.9 *The influence of number of categories on histogram shape*

three bars on the right hand side each represent a single observation. Here we have a relatively large dataset: the problem of shape instability in the histogram will increase as the number of data values decreases.

3.7.1 Normal Probability Plots

An alternative approach to investigating Normality is to draw a Normal (probability) plot. Unlike histograms, Normal plots do not involve grouping the data and are not affected, therefore, by arbitrary decisions as to how many grouping categories should be used. For the purposes of explaining the logic behind the plots a set of only ten numbers will be used: this should not be interpreted as suggesting that a sample size of ten is adequate for assessing Normality. As for histograms, the more data available the better – a small dataset is unlikely to contain much information on the shape of the tail areas, in particular.

Suppose the data in the first column of Table 3.11 represent a random sample of size ten from what may be a Normal distribution. If the sample really reflects an underlying Normal distribution, then for each x value in our sample there will be a corresponding z value in the Standard Normal distribution. To set up the correspondence, we first calculate for each x value the fraction of the sample that is equal to or less than x. With each x value we then associate the z value that has a corresponding fraction of the area of the Standard Normal curve below it. Thus, column 3 of

Table 3.11 *Calculating Normal Scores*

x	*Ranks i*	*Fraction* $\leq x$ (i/n)	z	$\frac{i-3/8}{n+1/4}$	*N-scores*
12.6	9	0.9	1.282	0.841	1.000
13.1	10	1.0	∞	0.939	1.547
9.8	2	0.2	−0.842	0.159	−1.000
11.7	7	0.7	0.524	0.646	0.375
10.6	4	0.4	−0.253	0.354	−0.375
11.2	6	0.6	0.253	0.549	0.123
12.2	8	0.8	0.842	0.744	0.655
11.1	5	0.5	0.000	0.451	−0.123
8.7	1	0.1	−1.282	0.061	−1.547
10.5	3	0.3	−0.524	0.256	−0.655

Table 3.11 shows that 20% of the sample values are equal to or less than 9.8; similarly 90% are less than or equal to 12.6. The corresponding z values can be found from Table A1 (Appendix). Thus, the z value with 20% of the area below it is −0.84 and the z value with 90% of the area below it is 1.28; these are shown in column 4.

Since the theoretical relationship between corresponding x and z values is

$$z = \frac{x - \mu}{\sigma} \quad \text{or} \quad x = \mu + \sigma z$$

if we plot x against z we should get a straight line with intercept μ and slope σ as shown in Figure 3.10.

In practice two issues need consideration:

- For any finite sample we would not expect exact correspondence between the sample value x and the theoretical z value, calculated as outlined above. For example, if instead of 12.6 the first test result in

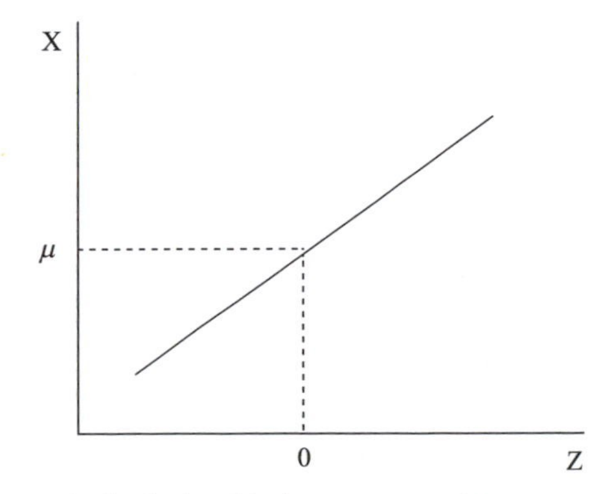

Figure 3.10 *The theoretical relationship between x and z*

the table had been 12.7 (or, indeed, anything between 12.2 and 13.1), this would still be the 9th largest value in the sample and would correspond to the same z value of 1.28. Due to sampling variation, therefore, the value of x which will correspond to a particular z value will be subject to chance variation. The observed values would, however, be expected to vary around a straight line and, in particular, they should not show any tendency to depart systematically from a straight line.

- Note that the largest x value corresponds to a z value of infinity. This leads to some difficulties in plotting! A simple solution to this problem is to perturb the calculation a little to avoid the infinity. This could be done by using $\frac{i-1/2}{n}$ or $\frac{i}{n+1/2}$ instead of $\frac{i}{n}$ when calculating the fraction equal to or less than x: for the purposes of drawing a graph it makes little difference which convention is used. Minitab uses $\frac{i-3/8}{n+1/4}$, which is justified on theoretical grounds that need not concern us here. These fractions are shown in column 5, while column 6 shows the corresponding z values – the estimated Normal scores.

Figure 3.11 shows column 1 plotted against column 6; an approximately straight line is obtained which suggests that the data are consistent with a Normal distribution. Note that some packages plot the Normal scores on the vertical axis and the data on the horizontal axis. For example, Figure 3.12 was generated using the automatic Normal plot menu option in Minitab. The labelling on the vertical axis gives the cumulative areas

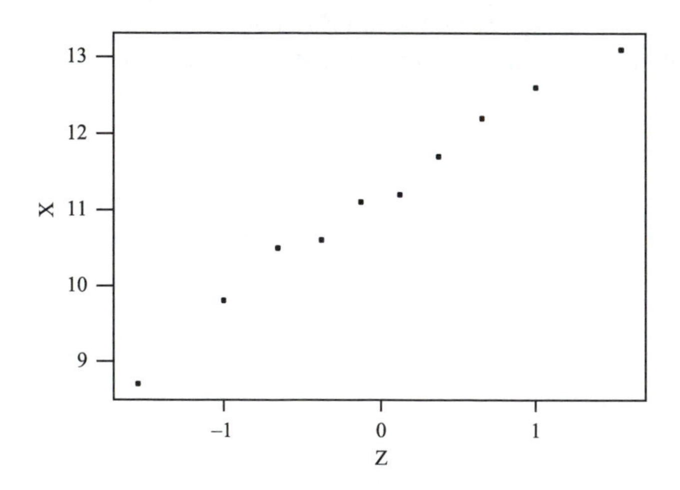

Figure 3.11 *A Normal plot for the sample data*

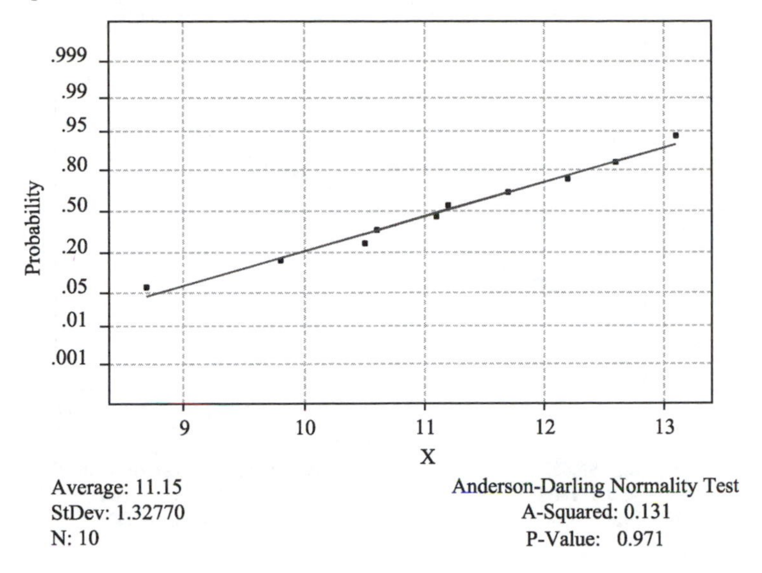

Figure 3.12 *Normal plot for the sample data**

(*i.e.*, the cumulative probabilities) under the Standard Normal curve corresponding to the estimated Normal scores, rather than the estimated Normal scores themselves. Thus, the smallest x value is 8.7; the fraction of the sample that is equal to or less than 8.7 is $\frac{1-3/8}{10+1/4} = 0.061$ (using the Minitab convention), which corresponds to a z value of -1.546. Changing the labelling of the vertical axis from $z = -1.546$ to a cumulative probability of 0.061 does not change the position of the plotted data point. The Minitab output also provides a formal test for departures from Normality (this is discussed below) – the test leads to the same conclusion as an informal inspection of the graph: the assumption of Normality cannot be rejected in this case.

3.7.2 A Significance Test for Normality

Figure 3.13 shows a Normal probability plot for the potency data of Figure 3.9; the data give an exceptionally straight line, thus confirming the impression given by Figure 3.9(a) which suggested a Normal distribution. The output also contains the results of a statistical significance test (the Anderson–Darling test) for Normality. This output is typical of the way packages report the results of significance tests – through a p-value – and will serve as an example of how the results of a test can be interpreted even without knowledge of the full details of the calculations.

*The line shown on the plot is a 'least squares' line – see Chapter 6 for a discussion of fitting lines by least squares.

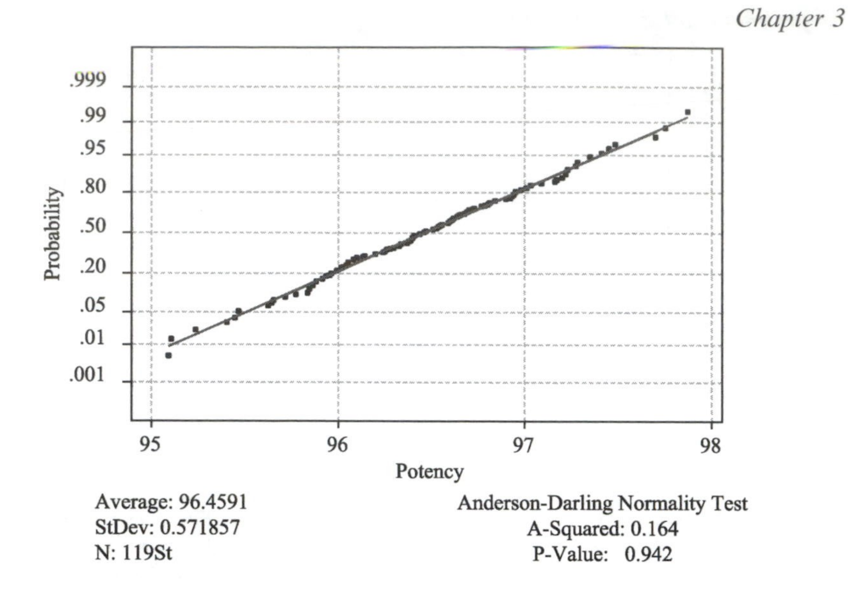

Figure 3.13 *Normal plot and significance test for Normality of potency data*

The null hypothesis is that the data come from a Normal distribution. The test statistic (A-squared) measures departure from this hypothesis. The null hypothesis is rejected if the test statistic is improbable under this hypothesis. The *p*-value is the probability of observing, in a repeat of the study, a more extreme value of the test statistic (A-squared) than the one observed in the study, if, in fact, the null hypothesis is true. If we require a test with a significance level of, say, $\alpha = 0.05$ the null hypothesis of Normality will be rejected if the p-value is less than 0.05. Here, the *p*-value is 0.94 which indicates that data randomly selected from a Normal distribution would show greater departures from a straight line in 94% of cases. In other words, the potency data are not at all unusual when considered as a possible sample from a Normal distribution. Hence there is no basis for rejecting the assumption of Normality: our data are consistent with such an assumption. If anything, the straight line is unusually good!

3.7.3 Departures from Normality

The schematic diagrams in Figure 3.14 indicate some patterns which may be expected when the underlying distribution does not conform to the Normal curve. Plots similar to Figure 3.14(a) will be encountered commonly, as outliers are often seen in real datasets – refer, for example, to the case study on setting up control charts in Chapter 2. Figure 3.14(c)

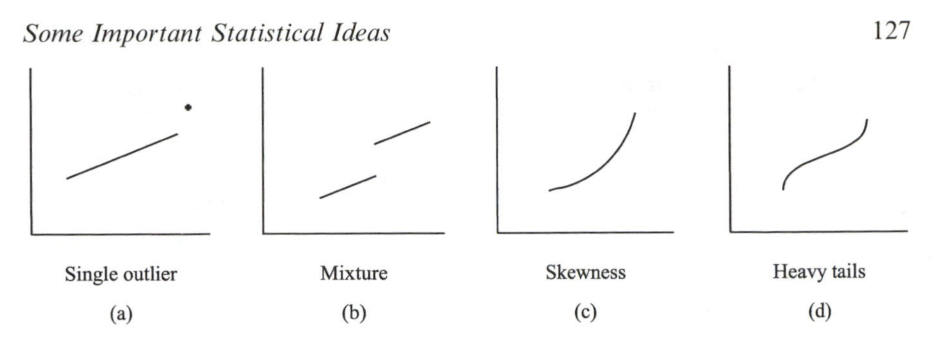

Figure 3.14 *Idealized patterns of departure from a straight line*

might, for example, arise where the data represented blood measurements on patients – many blood parameters have skewed distributions. Figures 3.14(b) and (d) are less common but (b) may be seen where the data are generated by two similar but different sources, *e.g.* if historical product yields were been studied and raw materials from two different suppliers had been used in production. The term 'heavy tails' refers to distributions which have more area in the tails than does the Normal distribution (the *t*-distribution is an example). Heavy tails mean that test results appear more frequently further from the mean than would be expected from a Normal distribution; Thompson[3] reports that proficiency test results distributions are often skewed or heavy-tailed.

Exercises

3.16 Exercises 3.1 and 3.6 refer to the percentage recoveries in a HPLC analysis of either a soft drink or clarified orange juice, spiked with glucose. The statistical tests and confidence intervals are based on the assumption of at least approximate Normality. So also is the individuals chart that was drawn in Exercise 2.2. Use the data given in Table 2.3, p 49, to draw a Normal probability plot and determine whether this assumption appears reasonable.

3.17 Several sugars are measured simultaneously by the Public Analyst's laboratory. Exercise 3.11 refers to the fructose measurements. These were used to draw an individuals chart in Exercise 2.11 (the data are in Table 2.14, p 83). Assess the Normality assumption for the fructose data, also.

3.18 The Public Analyst's laboratory referred to above, routinely measures total oxidized nitrogen in water by continuous flow analysis (an automated colorimetric method). The internal QC system involves making duplicate measurements on water spiked with 5.05 mg L^{-1} of nitrogen. The most recent data are shown in Table 3.12. Draw an individuals chart for the means of the duplicates and confirm that run 10 is an outlier. Draw a range chart for the duplicates and confirm that runs 18 and 21 are outliers. Exclude these runs

Table 3.12 *Total oxidized nitrogen (TON) in water spiked with nitrogen*

Run	*TON*		*Run*	*TON*	
1	5.044	5.093	20	5.022	5.071
2	5.032	5.144	21	5.134	4.970
3	5.067	5.070	22	5.051	5.002
4	5.064	5.056	23	5.072	5.113
5	5.048	5.067	24	5.056	5.110
6	5.129	5.129	25	5.094	5.089
7	5.054	5.103	26	4.983	4.989
8	4.952	5.010	27	5.140	5.129
9	5.062	5.068	28	5.076	5.018
10	5.247	5.275	29	5.011	4.967
11	5.115	5.092	30	4.990	5.087
12	5.137	5.104	31	5.134	5.232
13	5.208	5.125	32	5.022	5.040
14	5.006	4.998	33	5.006	5.006
15	4.965	4.985	34	5.030	5.049
16	5.106	5.089	35	5.046	5.068
17	5.166	5.098	36	5.048	5.084
18	5.036	5.273	37	5.038	5.046
19	5.077	5.117	38	5.044	5.052

and draw a Normal probability plot for the remaining 35 means. Is the assumption of Normality, required for the *t*-test of Exercise 3.2 and confidence interval of Exercise 3.7, supported by the data?

3.7.4 Transformations

In very many cases the Normal distribution will be an adequate model for the data generated in analytical laboratories. However, it will certainly not be appropriate for all analytes in all laboratories and it is important to establish when it is not applicable. When this is the case it may be possible to transform the test results (by taking logarithms, square roots, reciprocals, *etc.*) such that the transformed results are at least approximately modelled by a Normal curve. When this is done analyses may be carried out in the transformed scale and the results back-transformed to the original scale for interpretation.

We have already encountered an example of an analysis being carried out on a scale different from that where our real interest lay. Recall when we wanted a confidence interval for the long-run standard deviation of a set of data: our analysis was carried out in terms of the sample variance (the square of the standard deviation) and the resulting interval was back-transformed to give confidence bounds for the standard deviation. In that case the reason for the transformation was different – the sampling distribution of s^2 is mathematically simpler to work with than that of s – but the principle is the same. The Normal distribution is relatively easy to work with and it makes sense, therefore, to make life simpler by transforming data from more complex distributions so that they can be analyzed by the familiar Normal theory methods.

In many cases it will already be known within a laboratory what variables need transformation and how best to do so. For example, many variables encountered in clinical chemistry, environmental or geochemistry laboratories will be known to be skewed – see Figure 3.15(a). Where such prior knowledge is not available, there is no substitute for collecting data and trying out various transformations until the Normal plot gives acceptable results.

One aspect of non-normal data is worth mentioning. Repeated measurements on a control material will often produce an approximately Normal distribution, even where the measured parameter is not Normally distributed in the population being studied. Figure 3.15(a),

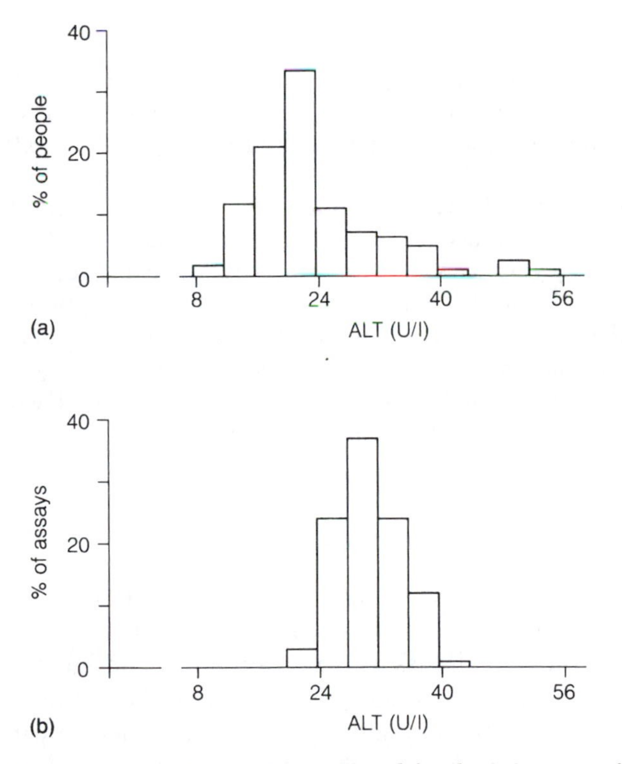

Figure 3.15 *Distributions of ALT results (a) 129 adults (b) 119 assays of one specimen* (Reprinted from *Statistics for the Life Sciences 2/E* by Samuels/Winter, © Reprinted by permission of Pearson Education, Inc., Upper Saddle River, NJ)

for example, shows serum ALT* concentrations for 129 adults. The distribution is right-skewed, that is, it has a long tail on the right-hand side, which means that some individuals have much higher ALT levels than the majority of the population.

Figure 3.15(b) shows the distribution of results obtained when a single serum specimen was analyzed 119 times. The shape of the distribution which describes the chance analytical variability is very different from that which describes the overall variation in ALT measurements. It suggests that a Normal distribution might well describe the analytical variability. If, therefore, we were interested in studying the analytical procedure – for, example, to quantify the precision, which is obviously poor – we might work directly in the arithmetic scale. On the other hand, if we wished to make statements about the distribution of ALT concentrations in the population from which the measurements derive, then we might choose to transform the data before proceeding with our analysis.

Exercise

3.19 The means of the three replicates for 26 runs of the data from the case study on setting up control charts in Chapter 2 are given in Table 3.4, p 111. The three means which were excluded were run 6: 497.24, run 11: 536.37, and run 28: 390.34.

Draw Normal probability plots for all 29 means and then the reduced set of 26. Note the effect of exclusion of outliers on the p-value of the Normality test, if one is provided by your package. Note that the p-value for the second plot is still statistically significant, thus rejecting the Normality assumption.

Confirm that the two smallest values in the plot correspond to the two results about which the laboratory manager was suspicious (she did not exclude them, nevertheless, as she had no explanation for their being small). Exclude these two values and redraw and interpret the Normal plot.

3.8 CONCLUDING REMARKS

This chapter introduced the fundamental ideas of statistical inference, *viz*., confidence intervals and significance tests, which will be used in the remaining chapters in discussing, for example, the design and analysis of comparative studies and in calibration studies. The implementation of these ideas will vary, but the underlying logical structure will remain the same. For example, in the next chapter we will see a confidence interval for the difference between two means. In the next chapter, also, we will

*Alanine aminotransferase (ALT) is an enzyme found in most human tissue.

discuss how to decide upon an appropriate sample size for a comparative experiment. Again, the details change from those presented earlier, but the logic remains unchanged: it is essentially a matter of balancing the risk of failing to detect real effects against the risk of deciding effects exist when, in fact, they do not.

Our discussion of Normal probability plots is a first example of 'model validation'. Just as analysts need to validate analytical methods, so also the statistical models that are used for data analysis need to be validated; this will be a recurring theme in the remainder of the book.

3.9 REVIEW EXERCISES

3.20 In a separate part of the study on determination of % total fat in food and feeds lyophilized pork meat (CRM 384: BCR) was measured by two laboratories on ten different days. The data are shown in Table 3.13, where A is the submitting laboratory and B is the peer laboratory.

For each laboratory carry out a t-test to determine if the laboratory's results are statistically significantly different from the certified value of 10.80%; compare the results of the tests. Use a significance level of 0.05 in the tests. Calculate 95% confidence limits for the long-run average result for each laboratory. Relate the interpretation of the confidence intervals to the results of the significance tests. Calculate 95 and 99% confidence intervals for the intermediate precision achieved by each of the two laboratories.

Table 3.13 *Total fat (%) in lyophilized pork meat*
(Reprinted from the Journal of AOAC INTERNATIONAL, 1998, Vol. 81, pp 907. Copyright, 1998, by AOAC INTERNATIONAL.)

Laboaratory			
A		*B*	
10.75	10.58	10.58	10.61
10.65	10.72	10.92	10.65
10.66	10.77	10.69	10.72
10.69	10.95	10.79	10.60
10.61	10.78	10.57	10.56

3.21 Suppose that in planning the above study the authors estimated from their development work that the standard deviation for replicates would be about 0.1%. Decide what sample sizes would be required to detect differences of (a) 0.1% and (b) 0.2% from the certified value of 10.80% under the four conditions specified by significance levels of 0.05 and 0.01, together with power values of 0.95 or 0.99.

3.22 Exercise 1.11 refers to data reported by Moffat *et al.* on the percentage of paracetamol in batches of Sterwin 500 mg tablets. The tablets were assayed by both the BP1993 UV method and by an alternative NIR reflectance spectrophotometric method. Duplicate measurements on 25 production batches are reproduced in Table 1.6, p 31.

Exercise 2.5 suggested drawing an individuals chart for the data and Exercise 3.10 involved calculating a confidence interval for the production process mean level. In both cases the methods are based on the assumption of at least approximate Normality – does this appear a reasonable assumption in this case? To answer this question, calculate the means of the duplicates for the 25 batches and use the means to draw a Normal probability plot. Do the means appear to come from a Normal distribution? Do the analysis for both the UV and NIR data.

3.23 Refer back to Exercise 2.10 in which proficiency testing results for suspended solids, biological oxygen demand (BOD) and chemical oxygen demand (COD) were reported as the z values obtained by a single laboratory over a seven year period. As indicated in the exercise, you should ignore two suspended solids and eight BOD results. Assess the Normality of the remaining data by drawing Normal probability plots. Carry out t-tests of the hypothesis that the z values vary around zero, *i.e.*, that the analytical systems are unbiased. Calculate the corresponding confidence intervals and interpret the results.

3.24 A Public Analyst's laboratory routinely measures potable water samples by flame atomic absorption spectrometry to ensure compliance with the EU drinking water directive. One of the parameters that is measured is the zinc concentration. Every time a batch of test samples is measured the batch includes two replicates of a QC control sample spiked with 200 μg L^{-1} of zinc. Exercise 2.12 involved drawing control charts to assess the stability of the analytical system. The data are presented in Table 2.15, p 84.

Draw a Normal plot of the duplicates means, to assess the Normality assumption underlying the charts. Carry out a t-test of the hypothesis that the data vary around a long-run average value of of 200 μg L^{-1}. If your test is statistically significant, calculate a 95% confidence interval for the bias.

3.10 REFERENCES

1. R. Pendl, M. Bauer, R. Caviezel and P. Schulthess, *J. AOAC Int.*, **81** (1998) 907.
2. ANSI/ASQC Z.14-1993, *Sampling Procedures and Tables for Inspection by Attributes*, American Society for Quality, Milwaukee, WI, 1993.
3. M. Thompson, *Analyst*, **125** (2000) 2020.

4. D.L. Massart, B.G.M. Vandeginste, S.N. Deming, Y. Michotte and L. Kaufman, *Chemometrics: a Textbook*, Elsevier, Amsterdam, 1998.
5. N.T. Crosby and I. Patel, *General Principles of Good Sampling Practice*, Royal Society of Chemistry, Cambridge, 1995.
6. R. Smith and G.V. James, *The Sampling of Bulk Materials*, Royal Society of Chemistry, London, 1981.
7. P.M. Gy, *Sampling for Analytical Purposes*, Wiley, Chichester, 1998.
8. G.B. Wetherill, *Sampling Inspection and Quality Control*, 2nd edn, Chapman and Hall, London, 1982.

CHAPTER 4

Simple Comparative Studies

4.1 INTRODUCTION

Comparative studies are carried out routinely in the analytical laboratory in method development, in troubleshooting and as part of method validation exercises. They are initiated in response to questions like 'would the assay be more stable or would the recovery rate be better if the HPLC column temperature were controlled?' A comparative study might involve several measurements of a test material with and without column temperature control, followed by a statistical comparison of the test results. Other examples might involve comparison of the results produced by two analysts, two instruments, or two laboratories, for the same analysis.

In this chapter the issues involved in the design and analysis of simple comparative studies, *i.e.*, ones designed to investigate the impact of changing a single system parameter, will be discussed in some detail. Studies involving changes to many system parameters will be considered in Chapters 5 and 7. The analysis of the resulting data typically involves formal comparison of the means and standard deviations of the two sets of test results, using statistical significance tests and confidence intervals. Design aspects, such as sample size determination, pairing, randomization of the study sequence and ensuring that the replication scheme is appropriate to the study objectives, will also be discussed in this chapter.

4.2 A TYPICAL COMPARATIVE STUDY

One stage in a HPLC method development programme involved a study of the effect of mobile phase composition on the percentage recovery of the nominal level of the active ingredients of pharmaceutical products. Two different sulfonic acid sodium salts (Methods A and B) were compared. The study was carried out in a randomized order and involved 24 separate analytical runs over a period of several days;

Table 4.1 *Percentage recoveries for 24 analyses*

	Recovery (%)			
	A		*B*	
	95.0	94.2	97.3	98.5
	97.3	93.0	97.2	95.9
	95.7	96.2	95.2	96.0
	95.7	95.9	95.6	98.0
	94.8	96.2	99.2	95.9
	95.8	94.9	96.2	96.8
Mean	95.39		96.82	
SD	1.11		1.25	

the reported results are the averages of a fixed number of within-run replicates. The results for one of the test materials are shown in Table 4.1, while Figure 4.1 displays the data as a pair of dot plots.

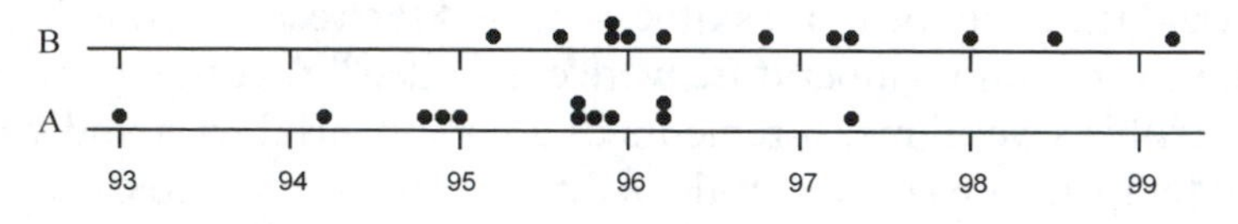

Figure 4.1 *Dotplots of the recoveries under the two methods*

It is clear from the dotplots and the summary statistics that in this study Method B gives a higher average percentage recovery. Several questions arise:

- Given that there is quite a lot of chance variability present in the data (thus, the recoveries for Method A vary from 93% to 97%), could the difference of 1.43% between the two averages be a consequence of the chance variation?

- If there is a recovery difference, what is the best estimate of its long-run average value?

- Does the variability of the results depend on which mobile phase is used, *i.e.*, is there a change in precision in switching between the two methods?

The first of these questions will be addressed by carrying out a *t*-test of the difference between the means. The second by obtaining a confidence interval for the long-run mean difference between results produced by the two system configurations, and the third by using a different significance test (the *F*-test) for comparing the standard deviations. Initially, the analyses will assume that the answer to the third question is that the precision is the same for both methods, but this assumption will be investigated later. All three analyses will be based on the assumption of data Normality; this assumption will also be investigated later.

4.2.1 A Statistical Significance Test for Comparing Method Means

A simple statistical model for the data is that all the observations may be regarded as coming from a Normal distribution with standard deviation σ, common to both system configurations. The values generated using Method A have a long-run mean μ_1, while those from method B have a long-run mean μ_2. The question as to whether there is a long-run difference between the recovery rates may then be posed in terms of the difference (if any) between μ_1 and μ_2. The question is addressed directly, using a statistical significance test.

In Chapter 3 we carried out a *t*-test of the statistical significance of the difference between a single sample mean $\bar{x}$ and a certified value μ_o by dividing by the estimated standard error of $\bar{x}$:

$$t = \frac{\bar{x} - \mu_o}{s/\sqrt{n}} = \frac{\bar{x} - \mu_o}{\sqrt{s^2/n}}$$

and comparing the resulting *t*-value to the appropriate *t*-distribution. We will do essentially the same here, but first we need to estimate the standard error of the difference between two sample means.

For a single sample mean, $\bar{x}$, the standard error is $\sqrt{\frac{\sigma^2}{n}}$ – the variance is $\frac{\sigma^2}{n}$. The variance of the difference between two independent sample means is the sum of their variances, *i.e.*, $\frac{\sigma_1^2}{n_1} + \frac{\sigma_2^2}{n_2} = \frac{2\sigma^2}{n}$ for common standard deviation σ and sample size n. Note that the variances add, even though the means are subtracted.* It follows from this result that the standard error of the difference between two independent sample means is $\sqrt{\frac{2\sigma^2}{n}}$. We have two estimates of σ, one from each method, and these can be combined into a single estimate of σ^2; this is done by averaging the two sample variances.

$$s^2 = \frac{s_1^2 + s_2^2}{2} = 1.40$$

This estimate has 22 degrees of freedom, since each of the sample standard deviations has 12–1 = 11 degrees of freedom.The means are

*If this seems odd, ask yourself if you would expect the combination of two uncertain quantities to be more or less uncertain than the two quantities being combined.

compared formally by specifying the null hypothesis of no difference and the alternative hypothesis that denies this:

$$H_o : \mu_2 - \mu_1 = 0$$
$$H_1 : \mu_2 - \mu_1 \neq 0$$

and then carrying out a t-test. An obvious estimator of the difference between the long-run means, $\mu_2 - \mu_1$ is the difference between the sample means, $\bar{y}_2 - \bar{y}_1$; this time we label the data as y, instead of x – just for a change! If this sample difference is very far from zero, it would be evidence that H_o is false. In order to assess the magnitude of the observed difference, given the presence of chance variation, the distance from $\bar{y}_2 - \bar{y}_1$ to zero is re-scaled by dividing by its estimated standard error, and the result, t, is compared to a Student's t-distribution with 22 degrees of freedom:

$$t = \frac{(\bar{y}_2 - \bar{y}_1) - 0}{\sqrt{\frac{2s^2}{n}}} \tag{4.1}$$

As in Chapter 3, a definition is required for what constitute exceptionally large or small values of the test statistic, t. This is specified by choosing the significance level for the test: the significance level is a small probability which is used to determine cut-off points (the critical values) on the tails of the distribution curve beyond which observed t-values are considered exceptional. Thus, if H_o is true, the t-value from the study will be expected to lie between – 2.07 and 2.07, since 95% of the area under the curve lies between these values, see Figure 4.2. If, in fact, there is no long-run recovery rate difference, a t-value beyond these critical values will only occur with probability 0.05.

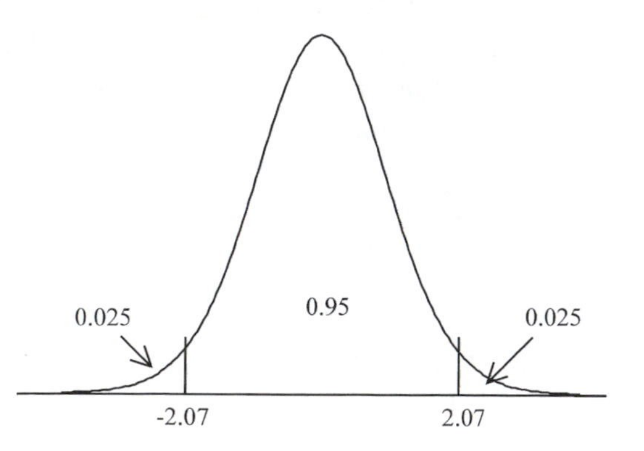

Figure 4.2 *Student's t-distribution with 22 degrees of freedom*

The test statistic can fall into the tails for one of two reasons: just by chance when there is no long-run difference, or because the two long-run means are not the same. In using statistical significance tests, the latter is always assumed to be the reason for an exceptional test-statistic. Accordingly, an unusually large or small test-statistic will result in the null hypothesis being rejected.

The calculated t-value is:

$$t = \frac{96.82 - 95.39}{\sqrt{\frac{2(1.40)}{12}}} = \frac{96.82 - 95.39}{0.48} = 2.96$$

and as this value falls outside the critical values (– 2.07, + 2.07) H_o is rejected. The two sample means are said to be statistically significantly different. Method B is considered to give a higher recovery rate.

If you compare the steps involved in carrying out the two-sample t-test described above with those discussed in Section 3.2 for a single-sample t-test (see summary on page 95 of Chapter 3), you will see that the two procedures are virtually identical. The differences are purely technical: they simply allow for the fact that in the two-sample case there are two sample means each subject to chance variation, whereas there is only one for the one-sample t-test.

Exercises

4.1 A laboratory supervisor felt that a particular potency assay of a pharmaceutical material was capable of better precision and began to explore various ways of reducing assay variability. As part of this process she reviewed the most recent control chart (for control material) and picked out the measurements of the two analysts who had carried out the assay most frequently. Her thinking was that if differences between analysts do exist, then by identifying and eliminating the root causes, the assay variability would be reduced. The data are shown in Table 4.2 (units are % potency).

Do these results suggest that the two analysts differ in their average assay results? Assume that the precision achieved by the two analysts is the same and carry out a t-test for a relative bias (% potency) between the two analysts.

Table 4.2 *Control chart summaries*

Analyst	*Mean*	*SD*	*Sample size*
A	86.72	0.29	6
B	87.35	0.23	6

4.2 Pendl *et al.*[1] describe a fast, easy and reliable quantitative method (GC) for determining total fat in foods and animal feeds. The data in Table 4.3 (% fat) represent the results of replicate measurements on coconut bars, which were made on ten different days by two laboratories A and B.

Assume the within-laboratory precision is the same for the two laboratories and obtain a combined estimate of the standard deviation. Use this to carry out a t-test to determine if there is a relative bias between the laboratories.

Table 4.3 *Replicate measurements(% fat) on coconut bars*
(Reprinted from the Journal of AOAC INTERNATIONAL, 1998, Vol. 81, pp 907. Copyright, 1998, by AOAC INTERNATIONAL.)

Laboratatory			
A		*B*	
25.89	25.44	25.40	25.51
25.69	25.65	25.50	25.63
25.99	25.73	25.45	25.53
25.76	26.05	25.25	25.70
25.99	25.79	25.83	25.49

4.2.2 Estimating the Difference in Recovery Rates

For the HPLC method development study, the difference between the two sample means, $\bar{y}_2 - \bar{y}_1$, is a point estimate of the long-run method difference, $\mu_2 - \mu_1$. This observed difference is obviously subject to chance variation, so we might ask what it tells us about the long-run difference. Following the approach of Chapter 3, where long-run means were estimated from sample means, a natural approach to answering this question is to calculate a confidence interval for $\mu_2 - \mu_1$.

In Chapter 3 a 95% confidence interval for a single mean was given by:

$$\bar{x} \pm t_{.025}\frac{s}{\sqrt{n}}$$

or equivalently:

$$\bar{x} \pm t_{.025}\sqrt{\frac{s^2}{n}}$$

where $\sqrt{\frac{s^2}{n}}$ is the estimated standard error of $\bar{x}$. By extension, a confidence interval for the difference between two means $\mu_2 - \mu_1$ is:

$$\bar{y}_2 - \bar{y}_1 \pm t_{.025}\sqrt{\frac{s^2}{n_2} + \frac{s^2}{n_1}}$$

which in this case reduces to:

$$\bar{y}_2 - \bar{y}_1 \pm t_{.025}\sqrt{\frac{2s^2}{n}} \tag{4.2}$$

Here, $\sqrt{\frac{2s^2}{n}}$ is the estimated standard error of the difference between the two sample means $\bar{y}_2 - \bar{y}_1$, each based on n values, and s is the combined estimate of the common standard deviation σ, which was calculated above.

The calculated confidence interval is:

$$(96.82 - 95.39) \pm 2.07\sqrt{\frac{2(1.40)}{12}}$$

$$1.43 \pm 1.0$$

Although the study showed an average difference of 1.43 units, the confidence interval estimates that the difference in long-run recovery rates is somewhere between 0.43 and 2.43, with 95% confidence. The long-run difference in recovery rates could be as small as half a unit or it might be almost two and a half units.

A Confidence Interval that Covers Zero. Suppose that the confidence interval had turned out as 1.43 ± 2.43, *i.e.*, ranging from – 1.0 to 3.86. How would this result be interpreted?

Such a result could mean that Method B gives a long-run recovery rate that is greater by as much as 3.86 units; it could, alternatively, mean that Method A gives results higher by one unit, on average. In other words the data cannot tell unambiguously which method gives higher results. In such a case the sample means are said to be not statistically significantly different from each other, *i.e.* the observed difference could have resulted from chance variation.

The relationship between confidence intervals and significance tests is the same for two-sample studies as it was for single means in Chapter 3. If the confidence interval covers zero the null hypothesis that $\mu_2 - \mu_1 = 0$ will not be rejected by the significance test. If the interval does not contain zero then the null hypothesis will be rejected and the two sample means will be declared to be 'statistically significantly different'.

Exercises

4.3. Refer back to Exercise 4.1 and calculate a 95% confidence interval for the relative bias (% potency) between the two analysts. Relate the confidence interval you obtain to the result of the t-test carried out earlier.

4.4 Refer back to Exercise 4.2 and obtain a 95% confidence interval for the relative bias between the laboratories. How should the calculated interval be interpreted? Does it indicate a relative bias? Relate the confidence interval you obtain to the result of the t-test carried out earlier.

A Typical Computer Analysis. The computer output shown in Table 4.4 comes from Minitab, but very similar output will be obtained from any general-purpose statistical package. The analysis was specified under the assumption of equal long-run standard deviations for both methods. The results are the same as those presented above. Note that the reason for the negative signs in the confidence interval is because the package has estimated $\mu_1 - \mu_2$ rather than $\mu_2 - \mu_1$.

Interpretation of p-Value. It is common that statistical packages show p-values as indicators of the statistical significance, or otherwise, of the results. This was discussed in Chapter 3 in the context of a Normality test. In the current context, the p-value is the answer to the question 'In a repeat of the study, what would be the probability of obtaining a more extreme t-value than that observed, if there really were no difference between the long-run means?' As Figure 4.3 shows, the probability of obtaining a t-value either less than -2.96 or greater than $+2.96$ is 0.0072.

Both tails are taken into account since a t-value of $+2.96$ would be considered as indicating a statistically significant difference, also. The test is called 'two-tailed' for this reason – one-tailed tests were encountered in Chapter 3.

If a significance level of 0.05 is required for the significance test, then a p-value less than 0.05 is taken as indicating that the observed difference between the means is statistically significant. If the p-value is greater than

Table 4.4 *A Minitab analysis of the method development data*

Two Sample T-Test and Confidence Interval

Two sample T for Method A *vs* Method B

	N	Mean	StDev	SE Mean
Method A	12	95.39	1.11	0.32
Method B	12	96.82	1.25	0.36

95% C.I. for mu A−mu B: (−2.42, −0.43)

T-Test mu A = mu B (*vs* not =): T = −2.96 P = 0.0072 DF = 22

Both use Pooled StDev = 1.18

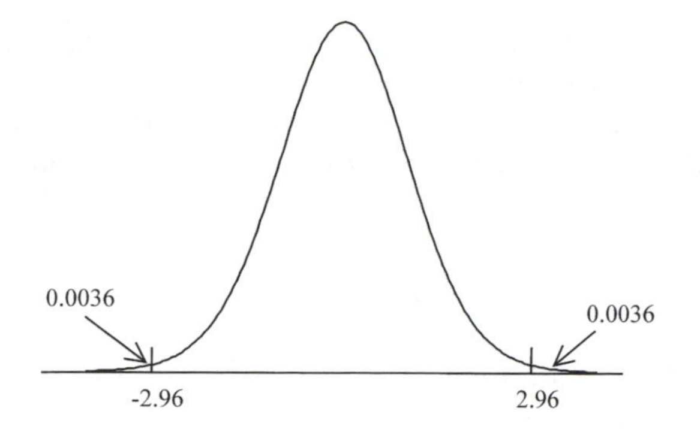

Figure 4.3 A *t-distribution with 22 degrees of freedom*

0.05 then the result is not statistically significant, *i.e.*, the observed difference is considered consistent with only chance variation away from zero. The advantage of quoting *p*-values is that it is immediately obvious how extreme the *t*-value is, *i.e.*, how unlikely such a value is under the hypothesis of no long-run difference.

4.2.3 Comparing Standard Deviations

In carrying out the *t*-test and in calculating the confidence interval it was assumed that the long-run standard deviations were the same for the two system configurations being compared. Clearly, the sample standard deviations are close, but a formal statistical test of the equality of the long-run values may be desired. Such a test (the *F*-test*) will be described below. However, the value of this widely used test is open to question on two grounds. Moore[2] argues against it on the grounds that it is highly sensitive to the assumption of data Normality. It is also known not to be powerful. While Normality is very often a justifiable assumption in an analytical context, it will be shown in Section 4.4 that very large sample sizes are required in order to be reasonably confident of detecting even moderately large differences between the standard deviations of the systems being compared. For the modest sample sizes typically encountered in laboratory studies the test will not be powerful. This means that there could well be a substantial difference between the standard deviations but the test will fail to detect this difference. Accordingly, careful study of the analytical protocols and professional

*The test is named *F* for Fisher, in honour of Sir Ronald Fisher who made major contributions to the theory and practice of statistics in the first half of the 20th century.

experience and judgement may well have to replace, or at least supplement, statistical significance testing in making such comparisons.

To carry out the test, the null hypothesis of equal standard deviations is specified; the alternative hypothesis denies their equality:

$$H_0 : \sigma_1 = \sigma_2$$
$$H_1 : \sigma_1 \neq \sigma_2$$

The test statistic is the ratio of the standard deviations squared, or equivalently, the ratio of the sample variances:

$$F = \left(\frac{s_2}{s_1}\right)^2 \tag{4.3}$$

If the two long-run standard deviations are equal, *i.e.*, $\sigma_1 = \sigma_2$, this test statistic follows an F-distribution with $n_2 - 1$ numerator and $n_1 - 1$ denominator degrees of freedom, associated with s_2 and s_1, respectively. Here, the F-distribution has 11 degrees of freedom for both numerator and denominator. Figure 4.4 shows that if a significance level of $\alpha = 0.05$ is selected the critical values are 0.29 and 3.47; this means that if the long-run standard deviations are indeed equal, there are tail probabilities of 0.025 of observing an F-ratio either greater than 3.47 or less than 0.29. Either very large or very small F-ratios result in the null hypothesis being rejected; large F ratios ($F > 3.47$) suggest that $\sigma_2 > \sigma_1$, while small F ratios ($F < 0.29$) suggest that $\sigma_2 < \sigma_1$.

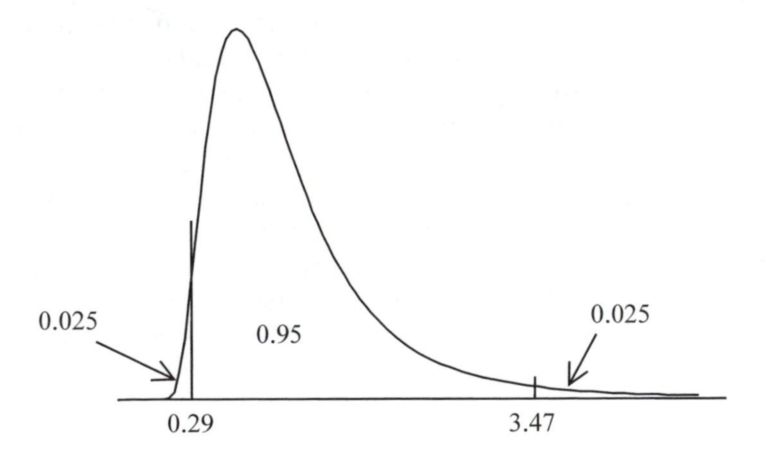

Figure 4.4 *The F-distribution with 11, 11 degrees of freedom*

For the method development study the F-ratio is:

$$F = \left(\frac{1.25}{1.11}\right)^2 = 1.27$$

Since the calculated F statistic lies in the body of the distribution the test provides no basis for rejecting the assumption of equal long-run standard deviations, *i.e.*, the precision appears to be about the same for the two methods.

Note that the F-distribution is not symmetrical and, therefore, for a two-tailed test critical values are required which are different in magnitude for the two tails. Statistical tables usually give the right-hand tail critical values only (see, for example, Table A6, Appendix).* To avoid having to calculate the left-hand critical value, it is conventional to use the larger sample standard deviation as the numerator of the F-ratio. If this is done, then the test asks if the sample ratio is too large to be consistent with random fluctuation away from $F = 1.0$, which is what the result would be if the long-run standard deviations were known and equal. In doing this, it is important that a table which gives the critical value for 0.025 in the right-hand tail should be consulted, even though a significance level of 0.05 is required.

Exercise

4.5 Carry out F-tests to check the equality of the long-run standard deviations for the data in Exercises 4.1, p 139, and 4.2, p 140.

4.2.4 Comparing Means when Standard Deviations are Unequal

The standard t-test (and related confidence interval) assumes equality of long-run standard deviations for the two sets of data being compared. An alternative approximate t-test allows comparison of the sample means where the standard deviations cannot be assumed to be equal. This might arise in situations where the work of a trainee was to be compared to that of an experienced analyst, or where two quite different analytical methods (*i.e.*, not just simple variants of one another) were being compared. The mobile phase comparison data will be used to illustrate the test, even though it is not necessary to relax the assumption of equal standard deviations in this case.

*The left-hand tail value $F_{0.025,a,b}$ (the value that leaves 0.025 in the left-hand tail where the degrees of freedom are 'a' for the numerator and 'b' for the denominator) is given by the reciprocal of $F_{0.975,b,a}$ (the value that leaves 0.025 in the right-hand tail; note the reversal of degrees of freedom).

Table 4.5 *Minitab output where the standard deviations are not assumed equal*

Two Sample T-Test and Confidence Interval

Two sample T for Method A *vs* Method B

	N	Mean	StDev	SE Mean
Method A	12	95.39	1.11	0.32
Method B	12	96.82	1.25	0.36

95% C.I. for mu A− mu B: (−2.43, −0.42)
T-Test mu A = mu B (*vs* not =): T = −2.96 P = 0.0075 DF = 21

Table 4.5 shows Minitab output for the test; the output is only marginally different from that of Table 4.4 and none of the practical conclusions from the study would change. Note that the degrees of freedom and the p-value have both changed slightly. The t-statistic was calculated using:

$$t = \frac{(\bar{y}_2 - \bar{y}_1) - 0}{\sqrt{\frac{s_1^2}{n_1} + \frac{s_2^2}{n_2}}} \tag{4.4}$$

and the degrees of freedom (df) were calculated using the following rather complicated expression:

$$\mathrm{df} = \frac{(var_1 + var_2)^2}{(Var_1)^2/(n_1 - 1) + (Var_2)^2/(n_2 - 1)} \tag{4.5}$$

where:

$$var_1 = \frac{s_1^2}{n_1} \text{ and } var_2 = \frac{s_2^2}{n_2}$$

and the resulting value for df was rounded down to an integer. For the mobile phase comparison data the t-value remains the same at $t = -2.96$ and the degrees of freedom come out as df = 21.69 (using $s_1 = 1.106$ and $s_2 = 1.247$), which is truncated to df = 21.

Exercise

4.6 Pendl *et al.*[1] describe a fast, easy and reliable quantitative method (GC) for determining total fat in foods and animal feeds. The data shown in Table 4.6 (% fat) represent the results of replicate measurements on margarine, which were made on ten different days by two laboratories A and B. Verify that an F-test, with a significance level of 0.05, will reject the hypothesis of equal precision in the two laboratories. Carry out a test of the significance of the difference between the average results.

Table 4.6 *Replicate measurements(% fat) on margarine*
(Reprinted from the Journal of AOAC INTERNATIONAL, 1998, Vol. 81, pp 907. Copyright, 1998, by AOAC INTERNATIONAL.)

Laboratory			
A		*B*	
79.63	79.19	74.96	79.17
79.64	79.66	74.81	79.82
78.86	79.37	76.91	79.31
78.63	79.42	78.41	77.65
78.92	79.60	77.95	78.36

4.2.5 Validating the Assumption of Normality

To assess whether the method development data are Normal, separate Normal plots might be drawn for the twelve test results from each method. However, the sample size would be small in both cases. If the long-run standard deviations can be assumed equal for the two groups, the data may be combined, having first subtracted the group means* from each observation in their respective groups. This adjusts the mean for each group to zero, and so the 24 observations will have the same distribution, *i.e.*, one with a common mean of zero and a common, but unknown, standard deviation. The resulting deviations are called 'residuals', as they are what remain after the group means are subtracted from the individual results.

Figure 4.5 shows a Normal plot of these residuals. This is close to a straight line, as would be expected if they come from a Normal distribution. The p-value indicates that 24 observations selected randomly from a truly Normal distribution would have a probability of $p = 0.83$ of showing stronger departure from a straight-line relationship than that observed in this study. Accordingly, there is no reason to reject the hypothesis that the data come from a Normal distribution.

Exercise

4.7 The data from Exercise 4.2, representing the results of replicate measurements on coconut bars by two laboratories A and B, are reproduced in Table 4.7. By subtracting the group means from the two sets of results, create a set of 20 residuals. Draw a Normal plot of these residuals and check

*Ideally this would be the long-run mean for each method, but, in practice, all we have available to us is the observed sample mean.

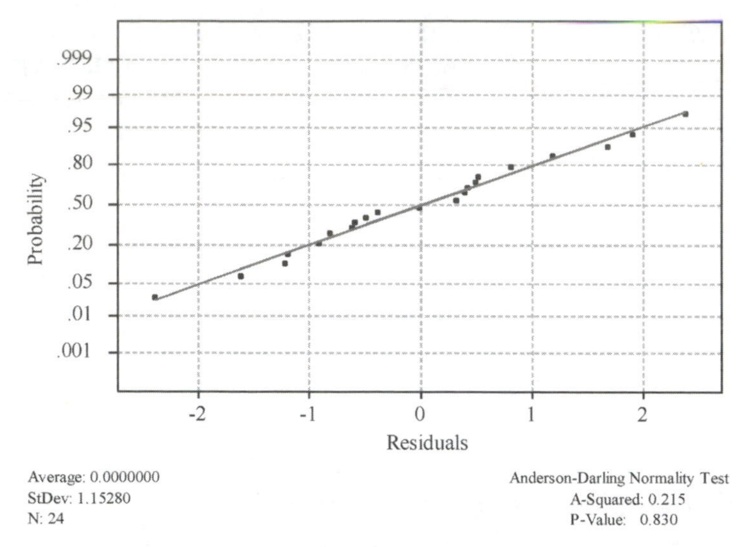

Figure 4.5 *A Normal plot and significance test for the Normality of the residuals*

the validity of the Normality assumption, which underlies both the confidence interval and the significance test computed earlier.

4.3 PAIRED COMPARISONS

In Section 4.2, the means of two groups were compared using t-tests and confidence intervals. In carrying out these analyses the individual data points made no direct contribution – all the information they had to contribute was captured in the summary statistics (the sample means and standard deviations). In some situations this approach would ignore an important aspect of the study design. Suppose a medical or psychological study involved comparing measurements on two groups of people who

Table 4.7 *Replicate measurements(% fat) on coconut bars*
(Reprinted from the Journal of AOAC INTERNATIONAL, 1998, Vol. 81, pp 907. Copyright, 1998, by AOAC INTERNATIONAL.)

Laboratory			
A		*B*	
25.89	25.44	25.40	25.51
25.69	25.65	25.50	25.63
25.99	25.73	25.45	25.53
25.76	26.05	25.25	25.70
25.99	25.79	25.83	25.49

had been treated differently, for example, with two different drug regimes. If the two groups had been randomly selected from some population (*e.g.*, students of the same sex in the same college) then the methods previously discussed might be applied to the responses of interest. If, however, the subjects in the study were sets of twins and one twin had been allocated to each treatment group, then a different analysis would be appropriate. The differences between twins, before any treatments are given, will tend to be smaller for most if not all response variables, than differences between any two randomly selected individuals that are not twins. Accordingly, if direct comparisons are made between the two members of each pair then the analysis will involve less variability than would be the case under random allocation of subjects. Hence, the comparison will be more sensitive and capable of detecting smaller systematic differences between the groups. Laboratory studies are sometimes designed in such a way that the resulting data may be considered paired; in the following sections both the design and analysis of such studies will be discussed.

4.3.1 A Trouble-shooting Exercise

Disagreements arose regarding the purity of material being supplied by one plant to a sister plant in a multinational corporation. The QC laboratories at the two plants routinely measured the purity of each batch of the material before and after shipping. The results (units are % purity) from the six most recent batches as measured by each laboratory are presented in Table 4.8.

The questions to be addressed in our analysis of these data are:

- Is there a statistically significant difference between the results produced by the two laboratories, *i.e.*, are they biased relative to each other?
- If yes, what is the magnitude of the bias?

Table 4.8 *Purity data for six batches, analyzed by two laboratories*

Batch	*Lab 1*	*Lab 2*
1	90.01	90.45
2	89.34	89.94
3	89.32	90.05
4	89.11	89.62
5	89.52	89.99
6	88.33	89.40

Table 4.9 *Purity data with differences and summary statistics*

Batch	*Lab 1*	*Lab 2*	*Difference*
1	90.01	90.45	0.44
2	89.34	89.94	0.60
3	89.32	90.05	0.73
4	89.11	89.62	0.51
5	89.52	89.99	0.47
6	88.33	89.40	1.07
Mean	89.272	89.908	0.637
SD	0.553	0.364	0.237

The questions of interest in this study relate to the measurement systems. However, the standard deviations of the six test results from each laboratory contain batch-to-batch product variability, as well as analytical variation. To exclude this, the analysis focuses on the differences between the test results obtained by the two laboratories for each of the six batches. Table 4.9 shows the test results, their differences and some summary statistics. By focusing on the differences, the analysis eliminates the effects of, for example, batch 1 showing a purity of 90.01% in Laboratory 1, while the batch 6 result is only 88.33%. Batch-to-batch variation in purity has no bearing on the objectives of the study and, hence, it is advantageous to eliminate it from the analysis.

A simple statistical model for the data asserts that the differences are Normally distributed about a mean, μ, with a standard deviation σ. The question of the existence of a relative bias can then be addressed by asking if the long-run mean difference between the results produced by the two laboratories, μ, is zero. If the significance test rejects this hypothesis, then the magnitude of the bias, μ may be estimated using a confidence interval.

Significance Test. If the proposed statistical model is appropriate for the data, then a one-sample *t*-test of the hypothesis that $\mu = 0$ may be carried out exactly as in Chapter 3. To carry out the test a significance level $\alpha = 0.05$ is chosen, the null and alternative hypotheses are specified as:

$$H_o : \mu = 0$$

$$H_1 : \mu \neq 0$$

and the test-statistic t:

$$t = \frac{\bar{d} - 0}{\frac{s}{\sqrt{n}}} \tag{4.6}$$

$$t = \frac{0.637 - 0}{\frac{0.237}{\sqrt{6}}} = 6.58$$

is calculated.* The resulting value is then compared to Student's t-distribution with $n - 1 = 5$ degrees of freedom (Figure 4.6).

This distribution has 95% of its area between + 2.57 and – 2.57. Since the calculated t-value lies far out in the right-hand tail, the data strongly suggest that the laboratories do not, on average, obtain the same purity results when measuring the same material (the p-value for the test is 0.0012).

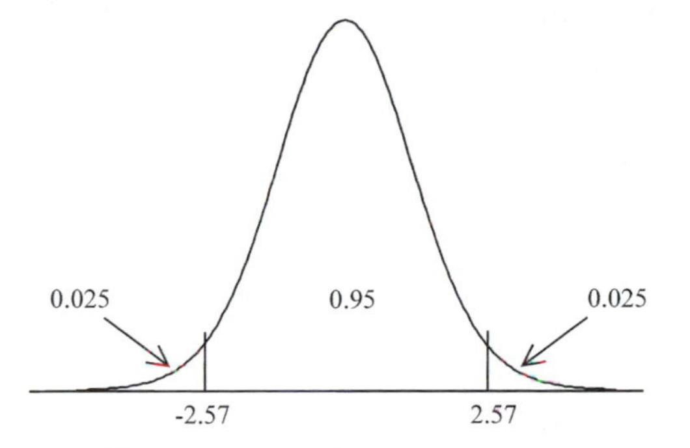

Figure 4.6 *Student's t-distribution with 5 degrees of freedom*

Confidence Interval. To measure the magnitude of the relative bias between the two laboratories, a confidence interval is calculated in exactly the same way as for single samples in Chapter 3, except that here the sample mean, $\bar{d}$, and the sample standard deviation, s, refer to differences between pairs of test results, rather than to individual measurements.

A 95% confidence interval for the long-run mean difference in purity results obtained by the two laboratories is:

$$\bar{d} \pm t\frac{s}{\sqrt{n}} \tag{4.7}$$

$$0.636 \pm 2.57\frac{0.237}{\sqrt{6}}$$

$$0.636 \pm 0.249$$

*Zero has been inserted explicitly in the formula to emphasize that what is being examined is the distance from $\bar{d}$ (the sample mean) to the hypothesized mean of zero.

The estimate of the relative bias between the laboratories indicates that it is somewhere between 0.39 and 0.89 units.

Exercises

4.8. The QC manager who carried out the analysis of the laboratory comparison data in the preceding example noticed that the difference between the results for the last batch was by far the largest difference in the dataset. Since the result for this batch from Laboratory 1 was the only value in the dataset which was less than 89.0, she wondered if it was correct and whether the large difference for batch 6 was responsible for the strong statistical significance of the difference between the laboratory means. Before returning to the laboratory notebooks to investigate this result, she decided to exclude this batch from the dataset and re-analyze the remaining data. The data with summaries are shown in Table 4.10. Carry out the calculations and draw the appropriate conclusions. Compare your results to those obtained from the full dataset.

Table 4.10 *Purity results for two laboratories*

Batch	*Lab 1*	*Lab 2*	*Difference*
1	90.01	90.45	0.44
2	89.34	89.94	0.60
3	89.32	90.05	0.73
4	89.11	89.62	0.51
5	89.52	89.99	0.47
Mean	89.460	90.010	0.550
SD	0.340	0.297	0.117

4.9 Several exercises in earlier chapters were based on recovery data for sugars from spiked control materials which are used to provide individuals charts to monitor routine analyses of sugars in orange juice. Thus, Table 2.3, p 49, presents 50 percentage recovery results for glucose, while Table 2.14, p 83, gives corresponding results for fructose. These parameters are measured simultaneously and so the 50 pairs of results come from the same analytical runs.

Calculate the differences between the percentage recoveries for the two sugars. Carry out a paired t-test to decide if the average recovery rate for the two sugars is the same or different. Calculate a confidence interval for the long-run mean difference. What do you conclude?

The Effect of Ignoring the Pairing. Suppose the person analyzing the data in Table 4.8 had not realized that the appropriate analysis was a paired t-test, and had carried out a two-sample t-test; what would be the effect on the analysis?

Two Minitab analyses of the data, one which does and one which does not take account of the paired structure are shown in Table 4.11. The differences between the two analyses can be expressed either in terms of the confidence intervals or the t-tests. The confidence interval for the paired analysis is narrower (± 0.25 as opposed to ± 0.61) and the corresponding t-value is much larger (equivalently, the p-value is much smaller, 0.001 instead of 0.040). In both cases the difference is due to the measure of standard deviation being used in the analysis. Where pairing is ignored, the variability is inflated by the batch-to-batch variation: the paired analysis eliminates this source of variability from the calculations, the unpaired analysis includes it.

In many cases ignoring the pairing can have an even more dramatic effect on the analysis, resulting in what should be a statistically significant difference being found to be non-significant. Accordingly, it is important to recognize the structure inherent in the data when carrying out the analysis.

Table 4.11 *Two different analyses of the laboratory bias data*

Paired Analysis:
T-Test of the Mean
Test of mu = 0.0000 *vs* mu not = 0.0000

Variable	N	Mean	StDev	SE Mean	T	P-Value
diff.	6	0.6367	0.2368	0.0967	6.59	0.0012

Confidence Interval

Variable	N	Mean	StDev	SE Mean	95.0% C.I.
diff.	6	0.6367	0.2368	0.0967	(0.3881, 0.8852)

Analysis ignoring pairing:

Two Sample T-Test and Confidence Interval

	N	Mean	StDev	SE Mean
Lab-2	6	89.908	0.364	0.15
Lab-1	6	89.272	0.553	0.23

95% C.I. for mu Lab-2 − mu Lab-1: (0.03, 1.24)
T-Test mu Lab-2 = mu Lab-1 (*vs* not =): T = 2.36 P = 0.040 DF = 10
Both use Pooled StDev = 0.468

Exercise

4.10 Repeat the analyses for Exercises 4.8 and 4.9 but this time ignore the pairing. Compare the results of the paired and independent group analyses.

4.3.2 Case Study

A pharmaceutical company supplies batches of tablets to a customer. The QC laboratories of both companies perform extensive analyses on samples from each batch. For many months the results from the two laboratories had been unsatisfactory in their lack of agreement. Results for one parameter, the one-hour dissolution rate, are shown in Table 4.12.

One possible display of the data is shown in Figure 4.7 below. This picture illustrates the batch-to-batch variation in the measurements. It shows the level about which the data vary – the mean is 37.1% and the fact that the data range from 11 to 55%. However, Figure 4.7 is not an effective way of displaying the differences between the analytical results produced in the two laboratories, as it is dominated visually by the variation between batches.

Table 4.12 *One-hour dissolution rates (%) for 112 batches of tablets as measured by laboratories A and B*

Batch	*A*	*B*	*Batch*	*A*	*B*	*Batch*	*A*	*B*	*Batch*	*A*	*B*
1	34.9	34.2	29	38.5	45.8	57	35.0	38.6	85	33.8	41.4
2	37.6	33.5	30	38.8	46.2	58	37.1	39.0	86	34.8	39.7
3	40.2	36.1	31	42.8	53.2	59	36.5	47.0	87	34.1	39.0
4	36.8	34.9	32	37.4	47.7	60	30.7	36.6	88	39.6	42.5
5	31.5	32.2	33	31.8	41.7	61	28.9	28.9	89	35.4	39.3
6	38.6	43.2	34	37.9	48.4	62	26.5	23.4	90	37.7	37.1
7	38.4	39.4	35	33.5	39.7	63	37.5	42.2	91	31.7	35.6
8	39.7	43.3	36	27.5	34.9	64	27.9	30.9	92	33.0	37.8
9	31.4	38.4	37	32.6	40.3	65	38.2	43.3	93	34.4	37.5
10	41.5	40.8	38	30.9	37.8	66	26.2	29.3	94	33.9	36.9
11	34.9	36.5	39	29.1	36.6	67	26.7	28.1	95	36.5	40.0
12	31.9	34.1	40	34.9	32.0	68	31.6	36.9	96	41.8	43.2
13	34.2	38.7	41	28.9	28.1	69	27.5	32.6	97	30.7	35.9
14	39.0	41.4	42	26.4	27.5	70	28.7	33.6	98	34.0	38.6
15	40.9	43.3	43	28.6	30.4	71	33.5	34.9	99	32.8	37.3
16	48.0	50.1	44	30.6	31.1	72	35.6	36.2	100	30.0	34.4
17	37.8	40.4	45	33.0	35.2	73	33.5	35.7	101	34.4	35.5
18	40.9	43.0	46	31.7	34.2	74	29.1	30.6	102	34.8	39.9
19	41.2	50.4	47	36.0	37.0	75	35.1	34.3	103	33.8	39.6
20	37.0	44.0	48	48.1	51.8	76	29.2	32.4	104	30.5	34.3
21	39.9	48.2	49	52.1	54.7	77	35.9	35.9	105	44.7	43.0
22	37.6	46.1	50	45.8	49.5	78	41.2	45.1	106	34.6	37.0
23	43.2	50.0	51	11.1	11.9	79	35.3	39.1	107	35.6	37.6
24	39.5	52.5	52	12.6	15.3	80	41.7	43.4	108	35.3	38.7
25	43.6	51.7	53	41.2	41.9	81	34.4	40.2	109	26.7	30.7
26	39.8	42.2	54	49.3	51.7	82	35.3	42.5	110	33.9	38.5
27	39.4	46.2	55	43.1	49.7	83	35.8	40.3	111	36.5	39.2
28	39.5	46.1	56	37.1	39.0	84	32.0	36.9	112	36.8	38.8

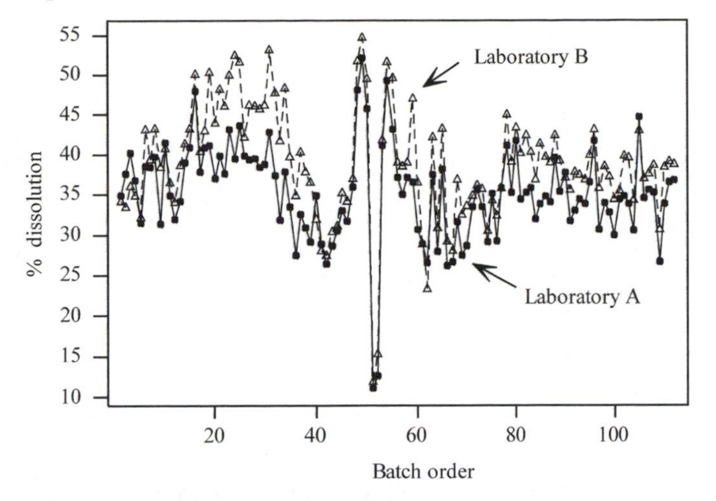

Figure 4.7 *The 112 batch results from the two laboratories*

Figure 4.8 shows the differences between the two sets of results plotted in time order of production. Because differences are plotted, the graph gives no indication of batch-to-batch variation in the one-hour dissolution rate. However, it focuses attention on the differences in analytical performance on the part of the two laboratories, which is what is of immediate interest. The plot is in the form of an 'individuals' control chart – the control limits are based on the moving ranges, calculated from successive plotted points.

The centre line of the control chart is the overall average difference between the two sets of 112 measurements. It shows a clear relative bias

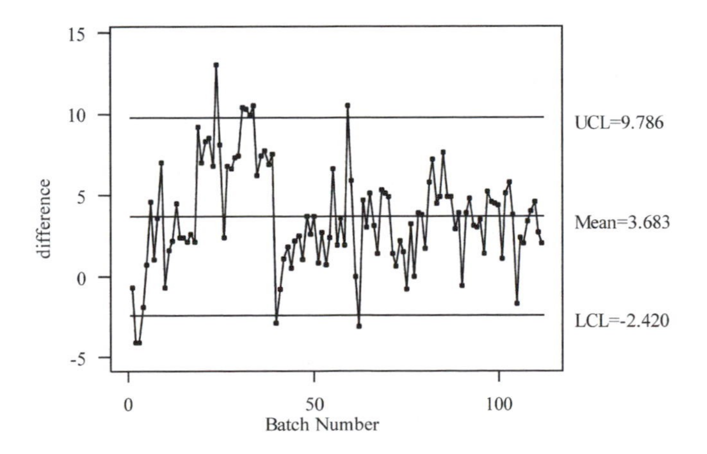

Figure 4.8 *An individuals chart for the differences between the pairs of results*

Table 4.13 *The next 40 batches (113–152)*

Batch	*A*	*B*	*Batch*	*A*	*B*	*Batch*	*A*	*B*	*Batch*	*A*	*B*
1	41.7	40.6	11	40.9	41.4	21	34.1	36.2	31	38.0	39.6
2	42.1	43.6	12	41.3	44.9	22	39.3	40.6	32	42.9	43.2
3	37.0	39.0	13	40.5	41.9	23	37.3	40.9	33	37.3	38.7
4	37.6	39.0	14	37.8	37.3	24	38.2	38.8	34	42.8	42.6
5	35.2	38.3	15	39.6	39.6	25	37.7	39.8	35	40.2	38.2
6	43.9	48.9	16	39.9	44.4	26	42.0	44.7	36	49.2	51.3
7	39.0	40.5	17	39.2	41.2	27	36.3	39.4	37	40.3	43.0
8	39.2	40.8	18	39.4	40.1	28	39.8	42.7	38	41.7	41.8
9	37.8	40.0	19	40.2	41.6	29	45.8	44.3	39	40.7	41.0
10	46.3	46.1	20	37.8	39.3	30	41.6	43.7	40	40.0	44.8

between the two laboratories. The differences were calculated as Lab B minus Lab A, so the chart indicates that, on average, Lab B is biased upwards relative to Lab A. This bias is not constant, however, because the system of differences is far from being in statistical control. Much work was carried out in an attempt to eliminate the disagreements between the laboratories. Table 4.13 and Figure 4.9 show the corresponding results for the next 40 batches.

It is clear that a major improvement has been achieved. The average difference is now only 1.56% compared to the previous average of 3.68%. More importantly, the system of differences is now stable. This means that, although there is still a relative bias between the two laboratories, the difference between them is predictable. The hope would be that with

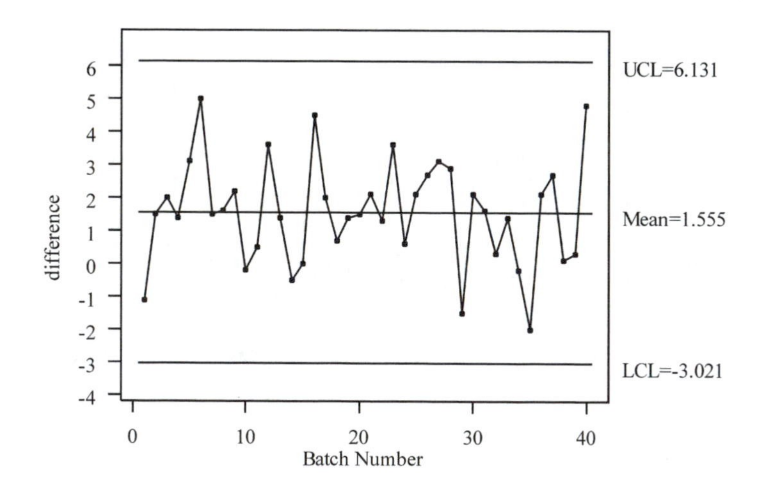

Figure 4.9 *Individuals chart for the laboratory differences for batches 113 – 152*

further work this difference could be reduced and, ideally, eliminated entirely. In the meantime, the control limits of Figure 4.9 can be used to monitor the system of differences and can help to maintain stability. For example, an out-of-control point for a new batch might be used as a signal to trigger re-analysis of samples of the product by both laboratories.

The centre line of Figure 4.9 provides a point estimate of the relative bias between the two laboratories. Before calculating a confidence interval for the long-run or true difference, we will check the assumptions underlying the analysis. Figure 4.10 shows a plot of the laboratory differences *versus* the means of the pairs of measurements. The paired t-test and corresponding confidence interval are based on the assumption that the set of differences vary at random around a fixed mean with a constant standard deviation. Figure 4.10 supports such an assumption. There is no tendency for the differences to either increase or decrease with the magnitudes of the measurements.

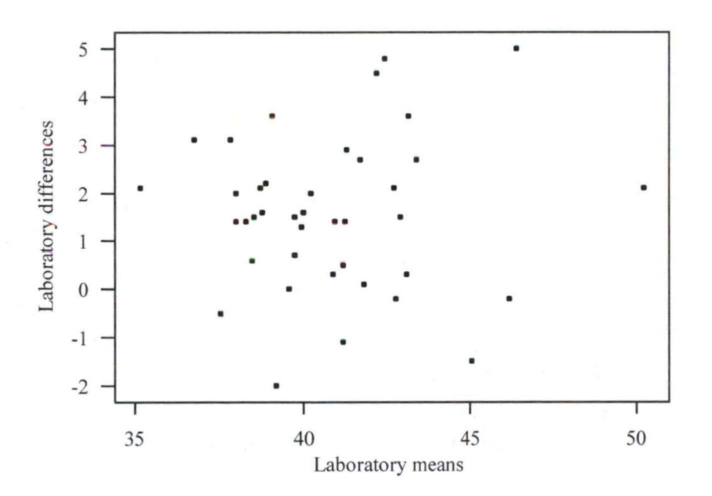

Figure 4.10 *Laboratory differences versus laboratory means for the 40 pairs of data, batches 113–152*

Figure 4.11 is a Normal probability plot of the 40 differences. The assumption of data Normality is clearly supported by the plot. The data are close to a straight line and the p-value is 0.558, both of which support the Normality assumption underlying the test and confidence interval.

Table 4.14 shows a Minitab analysis of the 40 laboratory differences. The null hypothesis for the t-test is that the long-run mean difference is zero – the very small p-value clearly rejects this hypothesis. It is extremely unlikely that the observed mean difference of 1.555 is a result of only the chance measurement error in the system. The conclusion is that

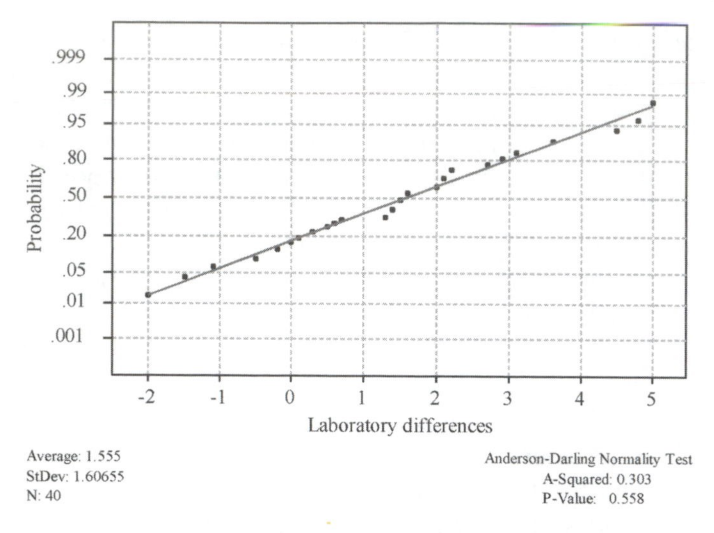

Figure 4.11 *Normal probability plot of laboratory differences*

the laboratories are biased relative to each other. The 95% confidence interval estimates the bias to be between 1.041 and 2.069 percentage point units.

This case study illustrates the importance of graphical methods in data analysis. A formal statistical significance test on the data of Table 4.12 would have indicated a difference in analytical performance on the part of the two laboratories, but would not have given the important message conveyed by Figure 4.8, which is that the system of differences is not stable. In fact, this instability invalidates the formal test, based as it is on a premise of stable chance variation, only. The formal and graphical tools each have their own parts to play in the overall analysis. Graphical methods reveal trends in the data and departures from assumptions; formal methods ensure we are not misled by chance patterns in the graphs.

This example also extends our earlier discussion of control charts, in Chapter 2. There we focused on monitoring analytical performance in

Table 4.14 *Paired t-test and confidence interval for the laboratory data*

Paired T-Test and CI: Lab-B, Lab-A

	N	Mean	StDev	SE Mean
Lab-B	40	41.595	3.037	0.480
Lab-A	40	40.040	2.956	0.467
Difference	40	1.555	1.607	0.254

95% CI for mean difference: (1.041, 2.069)
T-Test of mean difference = 0 (*vs* not = 0): T−Value = 6.12 P−Value = 0.000

one laboratory, based on repeated analyses of a control material. Here, the control chart of Figure 4.8 was based on differences between the dissolution rate results produced by two different laboratories for many different batches of production material, rather than a single control material. A control chart based on the data of Table 4.13, either the individual laboratory data or the mean of both sets, would tell whether the production system is now in-control – Figure 4.7 shows that it was not stable earlier in the production sequence. QC laboratories are often required to monitor production results, or at least to provide the data with which production engineers and chemists will do so. Production control charts are powerful tools for such monitoring. Control charts may also be applied usefully to the data routinely produced in incoming inspection of raw materials.

Exercise

4.11 In Exercise 4.9 we carried out a paired *t*-test to determine whether or not the long-run mean percentage recoveries for two sugars, fructose and glucose, were the same. However, we did not investigate the assumptions underlying our statistical procedures. Draw an individuals chart of the run-differences for the 50 pairs of values – does the system of differences appear stable? Plot the differences against the means – does the variability of the differences appear to depend on the magnitude of the results? Draw a Normal plot of the differences – does the Normality assumption appear reasonable?

4.4 SAMPLE SIZE FOR COMPARATIVE STUDIES

The problem of determining appropriate sample sizes for situations in which a single sample is selected from a process or population was discussed in Chapter 3. The approach described there would be suitable, for example, to determine the number of measurements required for making accept/reject decisions in acceptance sampling of in-coming raw material or for testing for bias in intra-laboratory method validation studies. There are no new conceptual issues involved in determining sample sizes for comparing sample means in two-group studies: with slight modifications, the approach is the same as before. First, the use of sample size tables for studies involving two independent groups of measurements will be discussed. Following this, paired studies will be addressed. Finally, sample size for studies of precision will be considered.

4.4.1 Comparing Means of Two Independent Groups

The method development study of Section 4.2, which involved a comparison of the recoveries of an active ingredient in a HPLC assay using two different mobile phases, will be used to provide a focus for the discussion. When a statistical test is carried out to compare two sample means then, because of measurement error, the analyst could:

- incorrectly decide that the methods are biased relative to each other (*i.e.*, give different recoveries) when, in fact, they are not;
- incorrectly decide that the methods give the same recoveries, on average, when, in fact, they do not.

In practice, of course, only one of these errors can be made, since the methods either do or do not have the same long-run recovery rates. However, the analyst does not know which is the case and so, in deciding on an appropriate sample size, must protect himself or herself against the possibility of making either error.

Much of what follows simply repeats, with appropriate modifications, what was said in Chapter 3 about specifying the parameters of the problem. The sample size table (Table A7 in the Appendix) is similar to that previously used, but takes into account the fact that two sample means, each subject to chance variation, are involved in the study.

Specifying the Parameters of the Problem. Since the presence of random variation lies at the heart of the problem, some estimate of its magnitude is required before the appropriate number of measurements can be determined. The size of the standard deviation, σ, that describes the expected chance variation must be specified. The most commonly used test assumes that both methods have the same precision and that will be assumed here. Given this, three other aspects of the problem require specification:

- We must specify the significance level (α) that will be used and whether a one or two-tailed test will be carried out. The significance level defines the probability of incorrectly deciding a bias exists when, in fact, the methods are unbiased relative to each other. In method development and validation studies two-tail tests are usually appropriate;
- While, in principle, any relative bias between the two methods might be of interest, it is clear that a very large sample size will be required to detect a very small bias. Therefore, some bias (δ) which it is considered important to detect must be decided upon;

- Having specified δ, the risk (β) of failing to detect a bias of this size must be chosen.

The comments in Chapter 3 on choice of significance level (α) and on the importance of holding information on measurement and product variability remain relevant here. Since it will not be easy to specify δ and β, the suggestion that a range of calculations be done under different assumptions is just as appropriate for comparative studies as it is for single-sample studies.

Using the Sample Size Table. Suppose that in planning the study the analysts were concerned to detect a difference of 1 unit. Suppose, also, that they expected the standard deviation σ to be 1 unit for both mobile phases and they decided on $\alpha = \beta = 0.05$, *i.e.*, a 5% risk of erroneously deciding there was a bias where none existed and a 5% risk of failing to detect a bias of 1 unit.*

To use the sample size table (Table A7), the ratio:

$$D = \frac{\delta}{\sigma} = \frac{1}{1} = 1$$

is calculated. This determines the appropriate row of Table A7. The column entry is jointly determined by the significance level $\alpha = 0.05$ (two-tailed test) and by $\beta = 0.05$ and the table gives a sample size of:

$$n = 27.$$

Note that 27 is the number required for each method, so the full study size is 54.

The effects on the required sample size of the various parameter choices can now be examined. If β is allowed to increase to 0.10 there is only a marginal reduction in study size: each method requires 23 measurements. The standard deviation is a critical parameter. Thus, varying σ and leaving the other parameter values fixed (*i.e.*, $\alpha = 0.05$, $\beta = 0.05$ and $\delta = 1$) gives the following results:

- $\sigma = 1.1$ implies $D = 0.91$ and requires $n = 30 - 34$
- $\sigma = 1.2$ implies $D = 0.83$ and requires $n = 37 - 42$
- $\sigma = 1.5$ implies $D = 0.67$ and requires $n > 55$

*Note that if these specifications are implemented and the bias is greater than 1 unit, then the risk of failing to detect that bias will be less than 5%, so the specification implies a risk of at most 5% of failing to detect a bias of 1 unit or more.

The sample sizes are not exact here due to the limited extent of Table A7. Minitab allows exact sample size calculation and gives values of 33, 39 and 60, respectively, for the three sets of assumptions. Note the importance of the standard deviation, σ, in the calculations above: a 50% change from $\sigma = 1$ to $\sigma = 1.5$ more than doubles the sample size required to detect a difference of one percentage point in % recovery. This again underlines the importance of maintaining good records regarding the variability of measurements and products in different situations: doing so ensures the ready availability of fundamentally important data for planning future studies.

Since the study involved only 12 observations in each group, either the analysts planning the study did not make the same assumptions as were made above or, for some reason, decided to reduce the sample size or, as often happens, just plucked the number from the air! Sample size tables can be used to estimate the difference the study might have been expected to detect using a sample size of 12 for each method. Keeping $\alpha = \beta = 0.05$ and $\sigma = 1$, a sample size of 12 for a two-sided test corresponds to $D = 1.6$ and since $\sigma = 1$ this corresponds to:

$$\delta = D\sigma = 1.6(1) = 1.6$$

The study would have a probability of 0.05 of failing to detect a difference of 1.6 percentage point units in the long-run average recoveries using the two mobile phases, when using a t-test with a significance level of $\alpha = 0.05$, if the assumed standard deviation of 1 unit is correct. An alternative way of expressing this is to say that the power of the test for detecting a difference $\delta = 1.6$ would be 0.95. The power of the test is the probability of rejecting the null hypothesis when this is false. The power is $1 - \beta$.

Some Useful Graphs. Several graphical ways to examine the implications of the number of observations were discussed in Chapter 3. Such graphs can only be drawn with appropriate mathematical or statistical software, but the popular statistical packages are now beginning to include sample size and related calculations; Figures 4.12 and 4.13 were drawn in Minitab.

Figure 4.12 shows the power of the test to detect a long-run difference of $\delta = 1$ between the two method means plotted against sample sizes from 2 to 50. The plot is based on the assumptions that there really is a long-run difference of $\delta = 1$, that the standard deviations for both methods are $\sigma = 1$, and that a two-tailed t-test, using a significance level of $\alpha = 0.05$ will be carried out. Inspection of the graph shows that the value $n = 27$ (for each method) corresponds to a power of around 0.95; recall that this was specified in arriving at the sample size of 27.

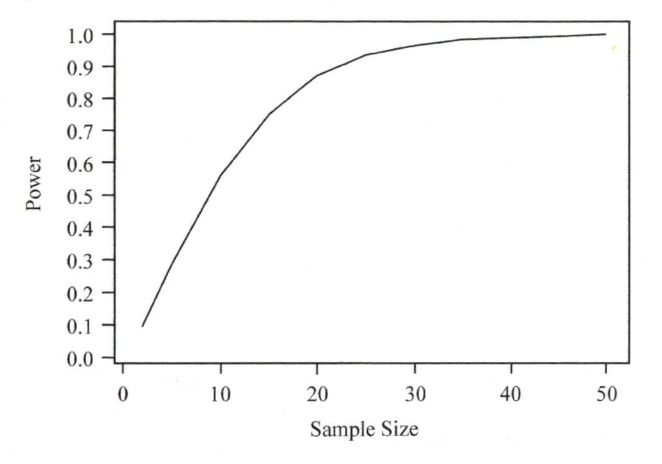

Figure 4.12 *Power of the test for a range of sample sizes ($\delta = 1$, $\alpha = 0.05$, $\sigma = 1$)*

Figure 4.12 also shows that the probability of detecting a difference of $\delta = 1$ would be only 0.65 for a sample size of $n = 12$ for each method.

Figure 4.13 shows the power plotted against the size of the difference between the two long-run means for three different sample sizes, again assuming the standard deviations for both methods are $\sigma = 1$, and a two-tailed t-test, using a significance level of $\alpha = 0.05$ is carried out. Thus, for a difference of 1 percentage point unit, a sample size of 10 gives a power of 0.56, a sample size of 20 a power of 0.87, while a sample size of 30 gives a power of 0.97. The graph shows that if differences smaller than 2 units were not considered important, there would be little point in using a sample size bigger than 10.

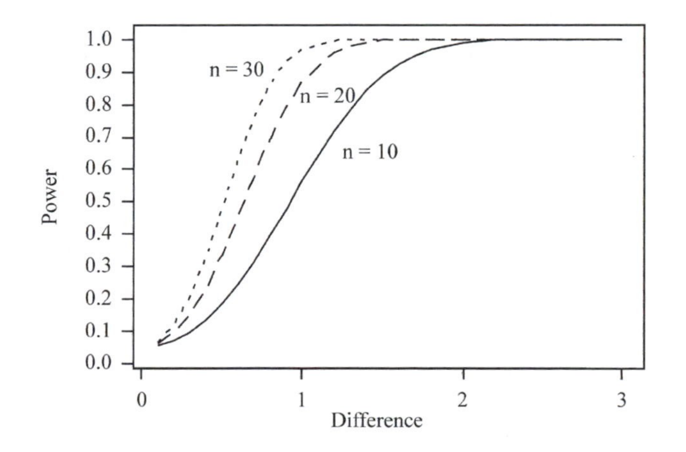

Figure 4.13 *Power of the test versus difference (δ) for three sample sizes ($\alpha = 0.05$, $\sigma = 1$)*

Exercises

4.12 A laboratory supervisor wishes to compare the results recorded on a control chart by two analysts to see if they are biased relative to each other (see Exercise 4.1, p 139). If the assay has an intermediate precision standard deviation of 0.25 (units are % potency), what is the minimum number of observations that you would recommend her to use if she is particularly interested in long-run differences of 0.5 units, or more.

By how much would the required sample size increase if she decided that a difference of 0.25 units was worth detecting?

4.13 Refer to Exercise 4.2, p 140 (analysis of coconut bars). Suppose the analysts had expected an intermediate precision standard deviation (within both laboratories) of somewhere between 0.15 and 0.20 units (% fat), based on preliminary work. What sample sizes would you recommend for studies required to detect inter-laboratory relative biases of 0.2 and 0.4 units? Use $\alpha = \beta = 0.05$.

4.4.2 Paired Studies of Relative Bias

The paired analysis of the laboratory comparison data of Section 4.3 was based on the six most recent batches of product. Suppose that, as a consequence of this analysis, it had been decided that a special study would be carried out which would involve both laboratories measuring a fixed number of batches of historical material within a specified time frame and the results being compared. It is normal that quantities of material from all batches of product are retained for stability testing and also for detailed investigation should problems arise subsequently with the product. Accordingly, a large number of batches is available for testing. The question is: how many should be included in the study?

The paired comparison analysis involves calculating the differences d_i ($i = 1, 2, \ldots n$) between the batch purity results at the two laboratories and then testing if the average difference $\bar{d}$ is statistically significantly different from $\mu_0 = 0$. Each pair of results is reduced to a single difference d_i. Accordingly, sample size calculations are carried out in the same way as for the single-sample problems of Chapter 3, using Table A4. Note, however, that the standard deviation used for paired designs is that of differences between test results.

The standard deviation of the differences obtained in the comparison of test results from the last six batches was 0.24; this will be taken as a starting point in our sample size analysis.* Suppose the intention was to

*Note that this standard deviation is based on results produced at different times, *i.e.*, in six different analytical runs in each of the two laboratories. The assumption here will be that the new study will involve measuring *n* batches, one in each of *n* different runs in the two laboratories. The design of studies of this type is discussed later, in Section 4.5.

carry out a two-tailed test using a significance level of $\alpha = 0.05$ and a probability $\beta = 0.05$ of failing to detect a relative bias of $\delta = 0.5\%$ purity. This gives:

$$D = \frac{\delta}{\sigma} = \frac{0.5}{0.24} = 2.08$$

which corresponds to $n = 6$, the nearest value in Table A4; this, coincidentally, was the sample size for the retrospective study of Section 4.3.

The standard deviation value used in the calculation ($\sigma = 0.24$) is, however, based on only six results (*i.e.*, differences). If a 95% confidence interval is calculated for the true standard deviation (see Chapter 3) based on this sample value, the upper 95% bound is 0.59. Using this in the sample size calculation gives:

$$D = \frac{\delta}{\sigma} = \frac{0.5}{0.59} = 0.85$$

which leads to a sample size of $n = 21$. The large difference between the two sample size values emphasizes, once again, the importance of retaining information on the magnitude of chance variation (be it sampling, product or measurement) so as to provide precise estimates of σ for use in planning future studies.

Here, a direct estimate of the standard deviation of differences was available. When this is not the case, an estimate may be obtained provided individual measures of precision for the two systems being compared (here laboratories) are available. If σ_1 is the intermediate precision standard deviation for replicate measurements made by Laboratory 1 and σ_2 the corresponding value for Laboratory 2, then these may be combined to give the standard deviation of differences:

$$\text{SD}(d_i) = \sigma_d = \sqrt{\sigma_1^2 + \sigma_2^2}$$

Of course, in practice, only estimates of these quantities will be available; these could be obtained from laboratory control charts, as discussed in Chapter 2.

Graphs similar to those for independent group studies can also be drawn for paired studies. Figure 4.14 is a variant on these: it shows the sample sizes required to detect a bias of 0.5 % purity when $\alpha = \beta = 0.05$, plotted against the assumed standard deviation of differences. Thus, it shows $n = 6$ for $\sigma = 0.24$ and $n = 21$ for $\sigma = 0.59$, the values obtained in the examples above. This graph makes clear that the sample size calculation is highly sensitive to the value of σ. In circumstances where

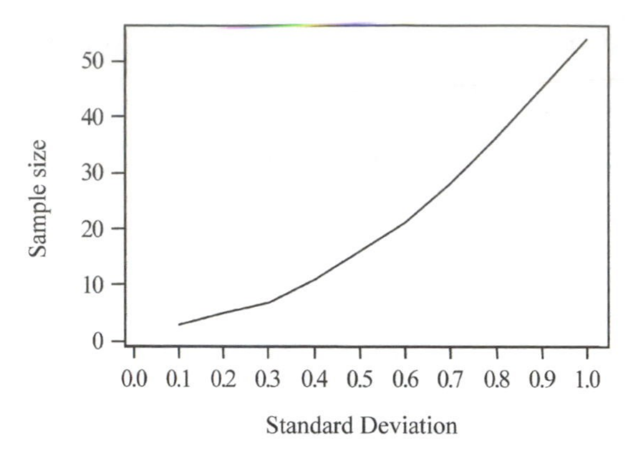

Figure 4.14 *Sample sizes as a function of the standard deviation* ($\alpha = \beta = 0.05, \delta = 0.5\%$)

information on the value of σ is unavailable or poor, pragmatic decisions regarding sample size will have to be made. The statistical calculations are, however, a useful aid in making these decisions.

Exercises

4.14 Suppose you were designing a study involving a paired comparison between the results obtained by two laboratories and that based on an analysis of the relevant analytical systems in the two laboratories you estimated that the standard deviation of differences between pairs of observations, one from each laboratory, would be in the region $\sigma = 1.5$ units. Assume it was required to detect a relative bias of 1 unit with a power of (a) 0.90 and (b) 0.95, using a two-tailed test with a significance level of 0.05. What sample sizes would you recommend?

4.15 Exercise 4.9 was concerned with the mean difference in percentage recovery for fructose and glucose in the routine analysis of sugars in orange juice. Suppose that it had been estimated in advance that the standard deviation of differences would be about 2.5 units. How many pairs of results would have been required to detect differences of either 2 or 3% in mean recovery with a power level of 0.95, if a two-sided t-test were to be carried out using significance levels of either 0.05 or 0.01?

4.4.3 Sample Sizes for Comparing Standard Deviations

In Section 4.2 the F-test for comparing standard deviations was introduced. The approach to sample size determination for these tests is similar to that used for comparing means; the only difference is that it is the ratio of the standard deviations, rather than their difference, that forms the basis for the analysis. The null hypothesis for the tests is that the ratio of the two long-run standard deviations is 1; in choosing a sample size we want to ensure that our test will be powerful enough to detect the fact that the ratio, R, differs from 1 by some specified amount.

The appropriate sample size table (A8 in the Appendix) requires specification of the significance level (α), whether the test is to be one or two-tailed, the ratio:

$$R = \frac{\sigma_1}{\sigma_2}$$

that it is considered important to detect, and the risk β which will be taken of failing to detect a ratio of size R, *i.e.*, the risk of deciding that the standard deviations are equal when, in fact, their ratio is R.* The probability of failing to detect a ratio bigger than R will be equal to or smaller than β, so the risk is *at most* β that a ratio of R *or greater* will be undetected.

Suppose the standard deviations for the precision achieved using two variants of an analytical method are to be compared and a high probability is required of rejecting the null hypothesis of equal standard deviations if either variant gives results with a standard deviation which is 20% greater than that of the other variant. If risks of $\alpha = \beta = 0.05$ are specified, then Table A8 shows that sample sizes of $n = 393$ will be required for each variant of the method. This is a huge sample size – unrealistic for most purposes. Clearly, it is not possible to guarantee detection of even moderate changes in precision without extraordinary analytical effort.

Figure 4.15 shows the sample size plotted as a function of R for a two-sided test with $\alpha = \beta = 0.05$. To detect a 50% increase in the standard deviation ($R = 1.5$) sample sizes of 82 are required; for a doubling ($R = 2$) or a tripling ($R = 3$) of the standard deviation, the corresponding sample sizes are 30 and 13, respectively. This graph reinforces the comment made above about the need for large sample sizes when comparing standard deviations, and the comments in Section 4.2 concerning the lack of power of the F-test. For moderate sample sizes the ability of the F-test to detect changes in standard deviation is poor. Accordingly, where a test, based on small sample sizes, is carried out and the result is not statistically significant, this cannot be taken as convincing support of an assumption of constant precision. In the

*As when carrying out F-tests, it is convenient here to deal with ratios greater than 1, by examining the ratio of the larger to the smaller standard deviation. The sample size table takes this into account.

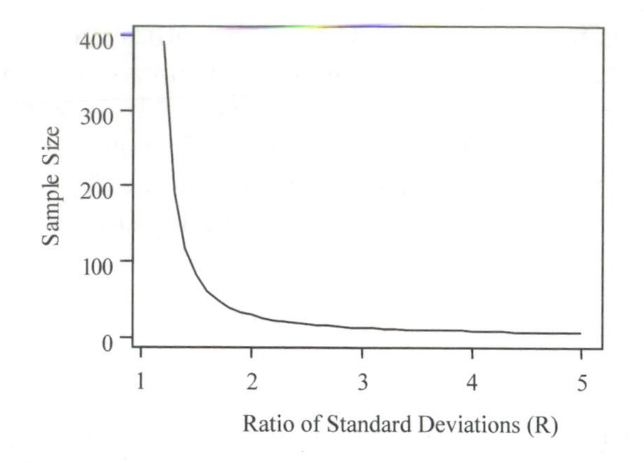

Figure 4.15 *Sample sizes for various standard deviation ratios*

absence of strong statistical evidence informed professional judgement becomes essential.

In cases where it is known in advance which standard deviation, if either, will be larger, then a one-tailed test might be carried out. For example, where the work of a trainee analyst is to be compared to that of an experienced analyst, it would be reasonable to assume that the experienced analyst will achieve better precision until the trainee is fully trained. The requirement will be to test if the trainee's standard deviation is the same or *bigger*. A two-tail test addresses the question as to whether the trainee's standard deviation is the same or *different*.

Exercise

4.16 From Table A8, select about 8 values for R and corresponding sample size n which will allow you to draw a rough sketch similar to Figure 4.15, but setting $\alpha = \beta = 0.10$ for a two-tailed test. Compare your sketch to Figure 4.15.

4.5 SOME COMMENTS ON STUDY DESIGN

The focus of Sections 4.2 and 4.3 was on the analysis of data and on validating the statistical assumptions underlying that analysis. Section 4.4 addressed the question of how much data would be required for a satisfactory study. Here, the concern will be that the right data are collected. Two main issues are considered. First, the randomization of the study sequence is recommended as a protection against unknown sources of bias. Then, the question of within-run *versus* between-run replicates, which was an issue in the setting up of control charts, will be re-examined in the context of simple comparative studies. A simulation study will illustrate how the apparently obvious approach to a simple comparison between the results produced by two laboratories can produce very high error rates and the underlying design issues will be discussed.

4.5.1 Experimental Run Order

In any study it is always possible that observed effects may be due, not to the condition under study, but to other factors which, unknown to those carrying out the work, influence the response of the system. For purely observational studies based on retrospective analysis of laboratory records, there is no built-in protection against this happening. A thorough analysis of all opportunities for biases to occur is, therefore, essential. When inferences are made from such studies, a confirmatory experimental study will often be appropriate in order to validate the conclusions. Experimental studies allow more control over the conditions under which data are collected. In particular, the opportunity to randomize the study sequence provides protection against sources of bias which are ordered in time. Before discussing how randomization is carried out, the implications of two systematic orderings of the study sequence will be examined.

Suppose two methods A and B are to be compared and that six replicate analyses using each method will be carried out. Figure 4.16 shows two possible systematic arrangements for the order in which the twelve test results will be generated. Time order (1) is a very natural ordering for the study – a tidy-minded analyst will naturally favour completing one set of analyses, all carried out in the same manner, before starting on the second set. This ordering, however, leaves the results of the study open to unknown sources of bias. Suppose that half-way through the study there is a once-off downwards shift in the results. This might be caused, for example, by impaired performance of a mechanical vibrator such that extraction vessels experience much reduced shaking. The shift will mean that all the B results will be reduced with respect to the A results, even if, fundamentally, there is no difference between

A A A A A A B B B B B B (1)

A B A B A B A B A B A B (2)

Time Order

Figure 4.16 *Two possible experimental sequences*

A and B. Where an A, B difference does exist, the shift could either counterbalance or exaggerate this difference. If the shift occurs at a point other than half-way through the experimental sequence it will have a lesser effect, since it will affect all analyses for one method and only some for the other method. Nevertheless, it will bias the test results and, potentially, jeopardize the outcome of the study.

Suppose that, instead of a sudden shift, there is a downwards drift in results. This might be due to failure of a temperature controller, leading to a drift in the temperature of a chromatographic column or water bath. In the case of the mechanical vibrator cited above, there might be a gradual decrease in the amount of shaking experienced by extraction vessels. The effect on test results will be as before – although test results for individual test portions may be affected to a different extent, overall, the B results will all be reduced relative to the A results, leading to a bias in the analysis.

Time order (2) is also a natural ordering for such a study, representing an attempt on the part of the analyst to be 'fair' to both methods: this ordering will protect against a once-off shift. Unless an A, B pair is split by the shift, then A and B will be equally represented before and after any shift: this means that the effect of the shift on the averaged results will be eliminated when the difference between the averages for A and B is calculated. Splitting a single A, B pair will be unlikely to have a major impact on the statistical analysis. Time order (2) does not, however, protect the study results from the effects of a drift. An upwards drift in test results, for example, will mean that the B results for all pairs will be incremented more than the A results. Consequently, the average for B will be raised relative to A. Again, this could either exaggerate or reduce any real difference between the methods. Time order (2) is also open to biases caused by a cyclical systematic effect.

Randomization. In the discussion above it is unknown sources of bias that are of concern. If known about, they can either be designed out or, at least in principle, allowed for in the analysis. Randomization of the order in which the analyses are carried out is the best protection against an unknown source of bias: it is very unlikely that the randomized sequence will coincide with the source of bias. In the two-sample example above, the two methods will tend to be affected equally by a shift or drift, under random allocation, and the effects of the bias will tend to cancel. From a purely technically statistical point of view, the process of randomization also underpins the assumption of independent observations, which is embodied in the statistical models underlying significance tests and confidence intervals; it is, therefore, a good thing!

Table 4.15 *Generating a randomized sequence for the study*

Random nos.	*Labels*	*Sorted nos.*	*Study sequence*
0.49100718	A	0.17139003	B
0.50394854	A	0.18515499	B
0.48992599	A	0.24861655	A
0.77446668	A	0.43278067	B
0.24861655	A	0.48992599	A
0.87444864	A	0.49100718	A
0.17139003	B	0.50394854	A
0.43278067	B	0.55030938	B
0.55030938	B	0.61849910	B
0.79971223	B	0.77446668	A
0.18515499	B	0.79971223	B
0.61849910	B	0.87444864	A

To carry out the randomization, the random number generator in either a statistical or spreadsheet package could be used. For example, column 2 of Table 4.15 contains labels for the two methods. Column 1 was generated using the random number generator in Excel.* this generates (uniform) numbers between 0 and 1, with all numbers in that interval being equally likely. Using the SORT command, columns 1 and 2 are sorted, using column 1 as a key: this results in columns 3 and 4. Since the magnitudes of the numbers in column 1 are randomly ordered, sorting column 2 using column 1 as a key, generates a randomized ordering for the A and B labels. The study should then be carried out using the randomized order of the rows of column 4.

Randomization is recommended virtually always for selection or allocation decisions in the laboratory, not just for time ordered situations. Thus, in Chapter 3 the need for random selection as part of acceptance sampling procedures was discussed. In experimental studies, unlike many sampling problems, the process of randomization is usually relatively simple to implement, as shown above. It should be incorporated automatically into study protocols unless there are very good reasons for not doing so.

4.5.2 Appropriate Measures of Precision

In the discussion of control charts in Chapter 2, a distinction was drawn between within-run variability (repeatability) and between-run variability (intermediate precision) for the purposes of establishing control

*Once the column of random numbers was generated it was copied and pasted back on itself using the 'paste special-values' option: otherwise, the SORT routine would generate and sort a new set of random numbers.

limits for an X-bar chart. It was shown there that using the wrong measure of variability could lead to seriously misleading control limits resulting in major disruptions to the laboratory's work routine. This distinction must also be borne in mind in planning and executing comparative studies, because, again, the wrong measure of chance variation can lead to incorrect conclusions being drawn from such studies.

Suppose that the QC Director of a pharmaceutical company asks for a special inter-laboratory study to assess the magnitude, if any, of the relative bias between two of the group's laboratories. Suppose, also, that it has been decided that each laboratory will assay the 20 most recent batches of the product that is supplied by one plant to the other and that the results will be compared using a paired *t*-test. How should the 20 measurements be made? Two possible study protocols suggest themselves. All $n = 20$ test materials might be measured within one analytical run in each laboratory. Alternatively, each material might appear in a different analytical run. The first protocol would be the typical approach taken, since it would be preferable in terms of completing the study quickly and would appear to imply a more efficient use of the analysts' time. However, the statistical analysis described later shows that this approach is intrinsically flawed and that the second protocol, despite being logistically more difficult and perhaps more costly, is preferable.

The discussion of the implications of the two study protocols will take place in two stages. First, the process of carrying out the study using the two protocols will be simulated using computer-generated random data. This will demonstrate the problem with the first protocol, which is that it produces extremely high error rates when statistical tests are carried out on the resulting data. Following the simulation study, an analysis of the underlying statistical models will give insight into the nature of the problem.

A Model of the Measurement Process. In Chapter 3 a simple conceptual model of the measurement process, proposed by Thompson,[4] was introduced briefly, in the context of placing error bounds on routine analytical results. Thompson suggested that each analytical result may be viewed as the sum of five components. These are:

$$\text{Result} = \text{true value} + \text{method bias} + \text{laboratory bias} + \text{run bias} + \text{repeatability error}$$

This model can be used as a basis for simulating the laboratory comparison study. To simplify matters, we will assume that both laboratories use the same unbiased method, so that this component can be dropped from the model. We assume, also, that the laboratories have insignificant bias levels (or, equivalently, that their biases are insignificantly different). The purpose of a practical study would be to discover if an inter-laboratory bias exists. Here, we assume it does not, and then use simulation to discover the consequences of using the two different protocols. With these assumptions the model reduces to:

$$\text{Result} = \text{true value} + \text{run bias} + \text{repeatability error}$$

The usual assumptions concerning the biases/errors are that they are independently Normally distributed about a mean of zero; if either mean were other than zero it would be pushed up what Thompson described as the 'ladder of errors'. Thus, if the repeatability long-run mean were other than zero, it would be identified with the run bias. Since the model has a separate term for run bias, the assumption of zero repeatability mean makes sense. Similarly, if the run mean were other than zero, it would be identified with the laboratory bias. Our assumption of no laboratory biases implies zero means for the run biases, also. The Normality assumption, on the other hand, is empirically based – stable analytical systems very often exhibit Normal error distributions.

A Context for the Simulation. In order to use the measurement model to simulate the use of the two protocols, we will require values for the three elements that together make up a test result as described above. We will use some real test results to provide these. Table 4.16 shows the means and standard deviations of three within-run replicate potency measurements on 20 batches of a pharmaceutical product. These are reproduced from Table 1.3 of Chapter 1 – the first 20 means were selected, ignoring batch 18 which was considered an outlier. The three measurements were made on single test portions, sub-divided for the

Table 4.16 *Summaries for twenty sets of three replicate potency results (mg)* (Reprinted from the *Analyst*, 2002, Vol. 127, pp 207. Copyright, 2002, Royal Society of Chemistry.)

Batch	*Mean*	*SD*	*Batch*	*Mean*	*SD*
1	243.11	1.56	11	238.74	1.82
2	242.39	1.11	12	239.85	0.64
3	245.65	1.45	13	236.93	2.24
4	242.90	1.76	14	243.83	0.41
5	245.04	0.32	15	242.52	0.57
6	244.99	0.10	16	246.18	2.01
7	239.64	0.90	17	235.51	1.18
8	237.75	0.97	18	238.52	1.99
9	238.48	1.68	19	243.42	0.67
10	236.11	0.47	20	240.50	1.13

purposes of analysis. The variation between them, therefore, reflects analytical chance variation only; the variation between the batch means reflects manufacturing variability, though, inevitably, it contains measurement error, also.

We will suppose that these batches have been selected for analysis in the study. The 20 batches were manufactured at different times and measured at time of manufacture – the replicate data, therefore, come from 20 different analytical runs. The means of the three replicates (batch means) will be taken as the 'true values' in the simulation. An estimate of the within-run standard deviation will be obtained from the data of Table 4.12 and will be used to simulate the repeatability errors. A combined estimate, based on all the replicate data, is given by:

$$s_w = \sqrt{(s_1^2 + s_2^2 + s_3^2 + \cdot \cdot \cdot + s_{20}^2)/20} = \sqrt{1.70} = 1.30$$

To simplify matters further, we will assume that the true within-run standard deviation σ_w is the same for both laboratories and we will use the estimated value $s_w = 1.30$ as σ_w in simulating the model. Thompson[4] presents a table of relative magnitudes of analytical errors/biases. These he considers a reasonable overall picture for the food and drugs sector for laboratories that are performing well. The factor for run bias is 0.8, meaning that the run-to-run standard deviation would be expected to be about 80% of the within-run or repeatability standard deviation. If we adopt this, then the run-to-run standard deviation for our simulation will be $\sigma_R = 0.8\sigma_W = 0.8(1.30) = 1.04$.

Simulating the Comparative Study. Tables 4.17 and 4.18 show how the comparative study can be simulated in the worksheet of a statistics

Table 4.17 *Outline of the simulation for protocol 1*
(Reprinted from the *Analyst*, 2002, Vol. 127, pp 207. Copyright, 2002, Royal Society of Chemistry.)

C1 Batch	*C2 True value*	*C3 Repeat-ability error*	*C4 Run bias*	*C5 Lab 1 results*	*C6 Repea-tability error*	*C7 Run bias*	*C8 Lab 2 results*	*C9 Differences*
1	243.11	1.08	0.33	244.52	−1.94	−0.36	240.81	3.71
2	242.39	−1.34	0.33	241.38	−0.40	−0.36	241.63	−0.25
*	*	*	*	*	*	*	*	*
*	*	*	*	*	*	*	*	*
*	*	*	*	*	*	*	*	*
19	243.42	−2.05	0.33	241.70	−2.77	−0.36	240.29	1.41
20	240.50	1.90	0.33	242.73	1.07	−0.36	241.21	1.52

Table 4.18 *Outline of the simulation for protocol 2*
(Reprinted from the *Analyst*, 2002, Vol. 127, pp 207. Copyright, 2002, Royal Society of Chemistry.)

C1 Batch	*C2 True value*	*C3 Repeatability error*	*C4 Run bias*	*C5 Lab 1 results*	*C6 Repeatability error*	*C7 Run bias*	*C8 Lab 2 results*	*C9 Differences*
1	243.11	−0.39	0.28	243.00	−2.14	−0.31	240.56	2.44
2	242.39	0.29	−0.42	240.26	−0.52	1.69	243.56	−3.30
*	*	*	*	*	*	*	*	*
*	*	*	*	*	*	*	*	*
*	*	*	*	*	*	*	*	*
19	243.42	0.73	−0.34	243.81	2.57	−1.19	244.80	−0.99
20	240.50	0.40	−0.16	240.74	1.00	−1.55	239.95	0.79

package or spreadsheet. Minitab was used here. Again to simplify the presentation, it will be assumed that each batch is measured once only in the two laboratories. Two protocols will be simulated: protocol 1 assumes all batches are measured in one analytical run, while protocol 2 assumes each batch is measured in a different analytical run, in each laboratory.

According to protocol 1, each result from Laboratory 1 will consist of a true value, varying from batch to batch, a single run bias, common to all batches, and a repeatability or within-run error, which varies from batch to batch. Table 4.17 shows the results of a single simulation of the measurement of twenty batches using protocol 1. Column 2 (C2) contains the assumed true values for the 20 batches: the measured mean potencies of Table 4.16 are taken to be the true values. Column 3 contains the within-run repeatability errors; these are 20 values generated as if from a Normal distribution with zero mean and standard deviation $\sigma_W = 1.30$. Column 4 contains the run bias associated with the single run in which all 20 measurements are made; a single value was generated as if from a Normal distribution with standard deviation $\sigma_R = 1.04$. Column 5 contains the 20 simulated measurements for Laboratory 1, *i.e.*, the sum of columns 2, 3 and 4. Exactly the same set of steps is applied to generate column 8 from column 2 (same true values), column 6 (repeatability errors) and column 7 (single run bias for laboratory 2). Column 8 then gives the corresponding simulated batch potencies as measured in Laboratory 2. Column 9 shows the differences between the two sets of measurements (column 5 minus column 8). In the simulation study, protocol 1 was simulated in this manner 100,000 times.

Table 4.18 shows the corresponding simulation using protocol 2. The steps involved are identical to those described above. The only differences appear in columns 4 and 7 – instead of a single run bias for each laboratory, there is a separate run bias for each analytical run (batch). Protocol 2 was also simulated 100,000 times.

The Simulation Results. The standard approach to the analysis of data such as those of Tables 4.17 and 4.18 would be to carry out a paired t-test on the differences in columns 9. As we have seen, the assumptions underlying the test are that both sets of $n = 20$ differences are samples from a Normal distribution with some mean μ_d and some standard deviation σ_d. The long-run mean difference, μ_d, is the relative bias between the two laboratories. Accordingly, a t-test of the null hypothesis that $\mu_d = 0$ against the alternative hypothesis $\mu_d \neq 0$ provides a test for relative bias. The t-statistic is calculated as:

$$t = \frac{\bar{d} - 0}{s_d/\sqrt{n}} = \frac{\bar{d} - 0}{\sqrt{s_d^2/n}} \tag{4.8}$$

where $n = 20$ is the number of batches measured, $\bar{d}$ is the mean of the n differences, and the denominator is the estimated standard error of $\bar{d}$. The zero is written explicitly to emphasize that the hypothesis being tested is $\mu_d = 0$. This statistic has a Student's t-distribution with $n - 1 = 19$ degrees of freedom, when the null hypothesis is true. The null hypothesis is rejected if the t-statistic exceeds in magnitude the critical values corresponding to a pre-selected significance level α – a value of $\alpha = 0.05$ is most commonly used.

The null hypothesis is true for all the simulation experiments, as no laboratory biases have been built into either protocol. Therefore, if a significance level of $\alpha = 0.05$ is used in the tests, we expect about 5% of the simulations, or 5000 in each case, to result in statistically significant t-values.

Table 4.19 shows the results of 100,000 simulations for each protocol. For each of the simulated tables a t-test was carried out on the twenty laboratory differences in Column 9. The table shows that the 20-run

Table 4.19 *Results of the simulation study*
(Reprinted from the Analyst, 2002, Vol. 127, pp 207. Copyright, 2002, Royal Society of Chemistry.)

	Protocol 1	*Protocol 2*
	Single-run design	*n-run design*
Signif. results	57 749	5021
Error rates	57.75%	5.02%

protocol gave virtually exactly what is predicted by theory – the hypothesis was erroneously rejected in 5.02% of cases. The single-run simulation, on the other hand, shows an error rate of nearly 58%. If a real study were run using protocol 1, it is likely that the conclusion would be drawn that there is a relative bias between the laboratories, *even though this is not the case.*

Why is the Single-run Protocol a Bad Design? The test statistic (Equation 4.8) is the ratio of the sample mean difference, $\bar{d}$, to its estimated standard error. For protocol 1 the standard error of $\bar{d}$ is $\sqrt{2(\sigma_R^2+\frac{\sigma_W^2}{n})}$. However, the standard deviation of the observed differences, s_d, contains no information about σ_R since all the observations come from the same run within each laboratory, and, hence, cannot measure run-to-run variation. In fact, s_d^2/n estimates $2\sigma_W^2/n$ only. Accordingly, when the study is run using protocol 1 and the t-value is calculated in the usual way, the denominator will be an underestimate of the correct value. Hence, the t-ratio will be bigger than expected and more statistically significant results will be obtained. This is what happened in our simulation. Because it was only a simulation, we know that there is no bias between the laboratories and, hence, that the results of the statistical tests are wrong in very many cases. In a real study we would not know this. The simulation shows that we would have a high probability of wrongly concluding the presence of bias.

Our simulation study was carried out for a single sample size, $n = 20$ batches. However, it can be shown that the probability of rejecting the true null hypothesis of no inter-laboratory bias will increase as the number of batches measured increases.* The more data that are collected, the more likely it is that an incorrect conclusion will be reached! At a time when automated sensors, linked to sophisticated analytical devices, together with virtually unlimited computer storage space and processing power, yield mega- or gigabytes of data, it is salutary to be confronted with the fact that more data do not, *necessarily*, mean better information. The data must be fit for the intended purpose. In our example, the single-run data are not fit-for-purpose in that, because of the nature of the design, the set of laboratory differences provides no information on the run bias, which is key to obtaining correct results from the t-test.

Single-run Studies? A natural question, which follows from the preceding discussion, is whether it ever makes sense that a study should be conducted entirely within a single analytical run, when run-to-run

*See Mullins[5] for a full discussion of the statistical models and further discussion of the simulation study.

variation is a part of the analytical system. The answer is yes – where it is appropriate, the effects of between-run variability will be eliminated and the sensitivity of the study will be increased.

For example, suppose a particular analysis involves an extraction stage, and a comparison is required between the effects of using two different solvents. The study might involve a given number of test portions being carried independently of each other through the complete analytical protocol and differing only in the solvent used. Here, solvent replaces laboratory as the system element to be compared. In such a case there is no reason why all stages of the analysis might not be carried out by the same analyst, within one analytical run. Since the test results would not then be subject to any between-run variability, the random error in the study would be smaller and it would be easier to detect a difference between the effects of using the two solvents, should one exist.

In general, when comparisons can be embedded, validly, within a single analytical run it makes sense to do so. Each case, however, needs careful consideration on its own merits. In considering whether such an approach is appropriate, the question to be addressed is whether each half of the resulting dataset might contain a bias component which will not be averaged out. Thus, under protocol 1 in the laboratory comparison study, a separate single-run bias influenced each half of the dataset and there was no possibility for these biases to be averaged out. No such bias is involved in the solvent comparison example and, hence, the entire study may be carried out within a single analytical run.

When prior information is available on the magnitudes of the run-biases, it might be incorporated into the analysis to compensate for its absence from the single-run data. Such information could be obtained from laboratory control charts, as discussed in Chapter 2. The standard t-test would no longer be applicable, however, and in such cases it would be appropriate to seek professional statistical advice.

Further Examples. There is often scope for essentially the same problem that occurred in the inter-laboratory comparison to arise within a single laboratory. For example, when new analysts are trained, it is a common procedure that before they are 'signed-off' as competent to perform an analysis, a comparative study will be carried out. This will involve a comparison between the results produced by the trainee and by an experienced analyst, for a fixed number of test materials, say n. If both analysts measure all n materials in one analytical run, *i.e.*, each carries out a single (separate) analytical run, then no averaging of the two separate run biases takes place. This is directly comparable to the inter-laboratory study, where analyst replaces laboratory in the model.

The same issues may be involved in a comparison between two pieces of equipment. This might arise in a regulated environment when a method is being re-validated as a consequence of replacing an instrument by a similar one from a different manufacturer. If n materials are each measured on a single set-up of each of the two systems, with a view to comparing the average results, then, if there is a 'set-up bias' which varies at random every time the instrument is set up or calibrated, the data structure is similar to that in the laboratory comparison study, with set-up or calibration bias corresponding to run bias.

It is clear, therefore, that the simulation example is not exceptional – there are many routine laboratory situations where the application of an apparently simple statistical test will be inappropriate, because the nature of the data collection is such as to make the test results unfit for the intended purpose. There are two morals to the story. Firstly, even simple statistical procedures should not be applied uncritically. More importantly, time spent thinking about the design of any study is well spent: it may be impossible to rescue the results of a badly designed study.

4.5.3 Representativeness

It is often the case that special studies are carried out by the most experienced analyst in the laboratory. While this will usually be appropriate for within-laboratory studies, such as the solvent comparison example above, in some cases it raises the question as to how representative the results of the study will be. In the laboratory comparison studies of Section 4.3 and that discussed above, for example, what is required is not a comparison between the *best* results that the laboratories can produce, but one between the standards of performance *routinely* achieved in the laboratories. This is most likely to be obtained if the data are generated blindly by several of the analysts who normally perform the analysis under study. Blindly means the analysts are unaware that the materials under analysis are in any way special. Ideally (though this will often be difficult to achieve in practice), the test materials should be introduced into the stream of routine test materials arriving at the laboratory, labelled in such a way that the analysts are unaware that they are part of a special study. This should be done over a sufficiently long period so that a representative selection of analysts will be involved in the study.

4.6 CONCLUDING REMARKS

In this chapter the simplest kind of laboratory experiments/studies were discussed. These involve changing a single aspect of the analytical system – such as the analyst who carries out the analysis, the method of filtration, or even the laboratory in which analysis takes place. The statistical methods usually used to analyze the resulting data are *t*-tests and confidence intervals. These depend for their validity on certain assumptions. Where a reasonable number of observations is available, these assumptions may be investigated using plots of residuals or, for paired data, of the differences on which the tests are based. Checking model assumptions is the statistical equivalent of analytical method validation – you should be as slow to use a statistical method blindly, as you would be to use an unvalidated analytical method.

A key aspect of the chapter was the discussion of a structured approach to sample size determination for comparative studies. Too often study size is plucked from the air, or, perhaps, decided by reference to earlier similar studies. In the absence of knowledge of the statistical ideas embedded in the approach to sample size determination discussed earlier, basing study sizes on those of earlier studies is perfectly rational. It does, however, make two major assumptions, both of which may be unjustified. It assumes that the systematic effect of interest and the magnitude of the chance variation in the prior study were both similar to the relevant quantities in the current study. It also assumes that the investigators who designed the prior study were better informed on appropriate sample size determination.

The final section of the chapter was concerned with aspects of study design and implementation. At first sight, recommendations that studies should be randomized, that several rather than one analyst carry out the work, that analyses are carried out in several rather than one analytical run, may appear over-elaborate. Also, the execution of a study becomes more time consuming, and very likely more expensive, as it becomes more complicated. Because of the resource implications, laboratory managers may decide not to follow all the procedures recommended above, however desirable. It should be borne in mind, though, that the extra costs of a more elaborate study may be a lot less than the costs of the consequences of poor study results; this needs to be considered in making such decisions.

4.7 REVIEW EXERCISES

4.17 Sorbic acid in foods is routinely measured by a Public Analyst's laboratory using steam distillation and HPLC. An individual's control chart is maintained to monitor the stability of the analytical system; this is based on spiking test portions of food with sorbic acid. Twenty-seven test results (mg kg^{-1}) are reported in Table 4.20.

Table 4.20 *Sorbic acid recoveries*

Run	*Sorbic acid*	*Run*	*Sorbic acid*
1	88.4	15	84.3
2	91.9	16	87.4
3	93.8	17	88.4
4	93.0	18	90.5
5	94.4	19	89.9
6	89.7	20	90.4
7	95.3	21	90.6
8	92.6	22	91.1
9	91.2	23	91.4
10	92.6	24	90.8
11	95.2	25	85.4
12	92.1	26	88.1
13	90.6	27	86.7
14	77.4		

- Draw a run chart of the data. Note the major outlier. On investigation, it was discovered that the steam distillation unit was leaking (this would give rise to losses of sorbic acid) and the tubing was replaced. Note that there appears to be a shift downwards in results for the data after this incident occurred.

- Draw dotplots of the before-after data.

- Carry out an F-test to compare the standard deviations of test results generated before and after the problem.

- Carry out a t-test to compare the means.

- If you decide that there has been a shift, calculate a 95% confidence interval to estimate its magnitude.

- Calculate residuals for each group of results, combine them into a single dataset, and draw a Normal plot to assess the Normality assumption underlying the tests and confidence interval.

4.18 Exercise 1.11 refers to data reported by Moffat *et al.* on the percentage of paracetamol in batches of Sterwin 500 mg tablets. The tablets were assayed by both the BP1993 UV method and by an alternative NIR reflectance spectrophotometric method. Duplicate measurements on 25 production batches are reproduced in Table 1.6, p 31.

Estimate the repeatability standard deviations for the two measurement systems using the method described in Section 1.6. How many degrees of freedom are associated with each of the two estimates? Carry out an F-test to compare the two estimates. What do you conclude?

4.19 In Exercise 4.11 we investigated the statistical assumptions underlying a paired t-test carried out to determine whether or not the long-run mean percentage recoveries for two sugars, fructose and glucose, were the same. This involved drawing an individuals chart of the run-differences for the 50 pairs of values, plotting the differences against the means, and drawing a Normal plot of the differences.

Sucrose recoveries were also monitored within the same analytical runs; the data are shown in Table 4.21. Carry out the same graphical analyses for a comparison between glucose and sucrose. Carry out the t-test – is there a recovery difference?

4.20 Duplicate measurements of three sugars (glucose, sucrose and fructose) are routinely measured on every tenth test material by a Public Analyst's laboratory. Glucose results are given in Table 1.7, p 32. Note that run 20 was found to be outlying. Sucrose data are given in Table 2.6, p 54; runs 5 and 33 were outliers. Fructose data are given in Table 2.13, p 82; all the data were considered in-control.

Exclude the outliers, calculate the within-run standard deviations and combine them, as discussed in Section 1.6. Carry out F-tests, to determine if the repeatability standard deviations are the same for the three sugars.

Table 4.21 *Sucrose recoveries*

Run	*Sucrose*	*Run*	*Sucrose*
1	99.00	26	98.35
2	103.50	27	96.20
3	99.50	28	97.55
4	99.00	29	100.43
5	92.00	30	100.53
6	102.50	31	96.17
7	97.00	32	94.50
8	96.00	33	98.20
9	106.55	34	97.00
10	103.45	35	99.95
11	91.40	36	97.05
12	91.35	37	101.55
13	106.70	38	100.20
14	94.45	39	99.55
15	95.50	40	99.80
16	102.00	41	100.40
17	97.05	42	94.55
18	96.40	43	98.35
19	99.45	44	100.20
20	101.85	45	96.90
21	102.85	46	99.60
22	97.67	47	98.40
23	100.28	48	103.05
24	100.45	49	100.70
25	97.53	50	103.95

4.8 REFERENCES

1. R. Pendl, M. Bauer, R. Caviezel and P. Schulthess, *J. AOAC Int.*, 1998, **81**, 907.
2. D.S. Moore, in *Statistics for the Twenty-first Century*, F. Gordon and S. Gordon (eds), MAA Notes, **26**, Mathematical Association of America, 1992.
3. Microsoft Corporation, *Microsoft Excel*, Seattle, WA, 2000.
4. M. Thompson, *Analyst*, 2000, **125**, 2020.
5. E. Mullins, *Analyst*, 2002, **127**, 207.

CHAPTER 5

Studying Complex Systems

5.1 INTRODUCTION

It is still a widely held view that the scientific approach to designing experiments is to investigate one system parameter at a time. That this is badly wrong is demonstrated below. This chapter introduces statistical strategies for the design of experiments for the systematic investigation of complex systems. It is not intended as a comprehensive account of experimental design, but it will give the reader a useful approach to studies requiring investigation of several system parameters. The 'factorial designs' covered here will be useful for method development, trouble-shooting, and robustness/ruggedness testing during method validation.

The ideas are introduced in the context of the simplest possible extension of the discussion of Chapter 4: two system parameters are to be studied and two settings for each will be investigated. The approach will then be extended to studying the effects of varying three or more system parameters. The 'fractional factorial designs', which are used for ruggedness testing, are also discussed in this chapter. These allow many system parameters to be investigated simultaneously, without requiring excessive experimental runs. An example in which twelve parameters were studied using only sixteen runs is described.

Two key aspects of multi-factor studies are the need to acquire the maximum information from a given amount of work and the importance of discovering whether varying one system parameter influences the effect of varying a second parameter. These are discussed in Section 5.2. The importance of graphical analysis as a method for validating statistical assumptions was a feature of Chapter 4. It plays a key role in this chapter, also. The approach to sample size determination, introduced in Chapter 3 and applied to simple comparative experiments in Chapter 4, will be extended in a simple way to more complex multi-factor studies in this chapter.

5.2 STATISTICAL OR TRADITIONAL DESIGNS?

The differences between the responses of a measurement system under two different settings of each of two system parameters are to be investigated. In a GC method development study, for example, the two parameters might be Injection Temperature (A) and Split Ratio (B) and the response might be peak area. The statistical terminology refers to the parameters as 'factors' and the settings as 'levels'; where there are just two levels, they are usually described as 'high' and 'low', even where no ordering is implied, *e.g.*, the two levels could be two different carrier gases.

Suppose that the budget for the project means that a total of twelve measurements may be made. The traditional approach to such a study might proceed as follows: having decided to investigate B first, A is held at its low level while four observations are made at each level of B. The next step would be to make the final four observations, holding B at its better level (low, say) but changing A to its high level. The full investigation is described schematically by Figure 5.1, where y represents the system response.

A natural performance measure of the effect of changing A or B is the difference between the average results at the two levels of the factor:

$$\text{effect A}: \frac{y_9 + y_{10} + y_{11} + y_{12}}{4} - \frac{y_1 + y_2 + y_3 + y_4}{4}$$

$$\text{effect B}: \frac{y_5 + y_6 + y_7 + y_8}{4} - \frac{y_1 + y_2 + y_3 + y_4}{4}$$

Note that in each case the average of 4 results is compared to the average of 4 results.

Consider Figure 5.2 which shows an alternative experimental design in which all possible combinations of the levels of A and B are studied.

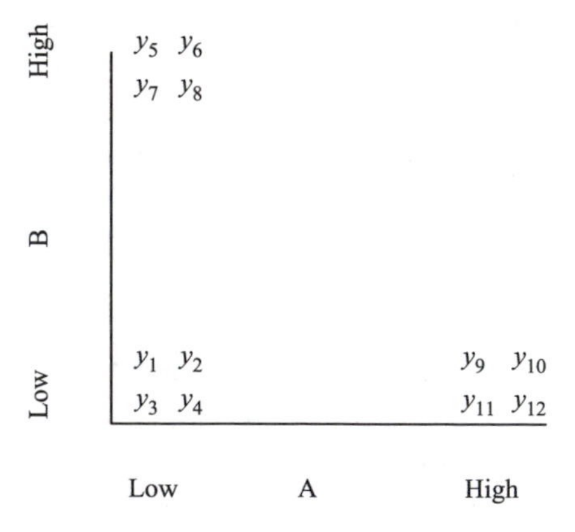

Figure 5.1 *A traditional experimental design for two factors*

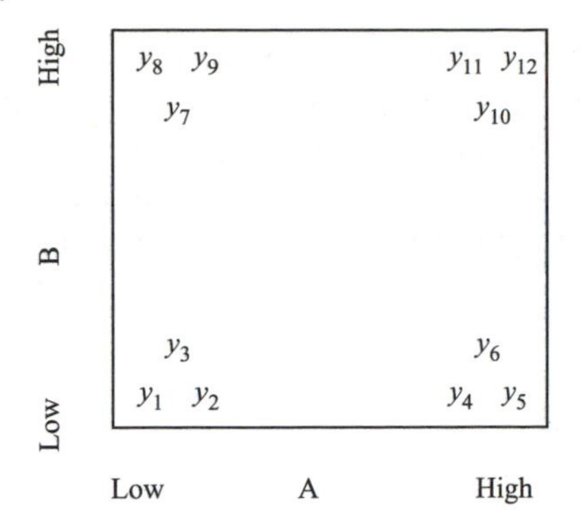

Figure 5.2 *A factorial design for two factors*

This is called a 'factorial design' and can be represented by the corners of a square; with 12 measurements in total, 3 are made at each corner. When compared to the traditional design, the factorial design does not appear radically different, but the difference is important.

As above, a natural measure of the effect of changing A or B is the difference between the average result when the parameter is at its high level and that obtained when it is low:

$$\text{effect A}: \frac{y_4 + y_5 + y_6 + y_{10} + y_{11} + y_{12}}{6} - \frac{y_1 + y_2 + y_3 + y_7 + y_8 + y_9}{6}$$

$$\text{effect B}: \frac{y_7 + y_8 + y_9 + y_{10} + y_{11} + y_{12}}{6} - \frac{y_1 + y_2 + y_3 + y_4 + y_5 + y_6}{6}$$

Note that in each case an average of 6 results is compared to an average of 6 results.

The superiority of the second experimental design should now be apparent. Although the same number of observations was made in both cases, the second design measures each effect as the difference between two averages of 6 measurements, whereas the first measures effects as the difference between two averages of 4 measurements. Expressed differently, the second design allows all 12 observations to be used in measuring the effects of changing the levels of both A and B, while the first design uses only 8 of the observations for each comparison, thus ignoring one third of the data. To get the same quality of comparison with the traditional design, 18 observations would be needed, 50% more than for the factorial design. The factorial design is much more efficient – this efficiency increases strongly with the number of factors studied.

Interactions. The factorial design, as well as being more efficient, has another advantage over the traditional design, which the following example demonstrates. Suppose that, in a study following the traditional approach, when A is at its low level the average responses at low and high B are 120 and 100, respectively, and that large responses are desirable. The system response when B is low and A is high is investigated next and the resulting average response is 140, as shown in Figure 5.3.

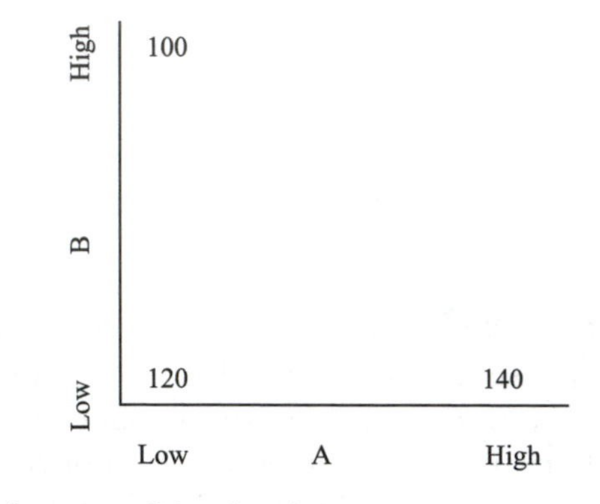

Figure 5.3 *Results from a traditional study*

If current operating conditions are low A, high B (*i.e.* with a system response of 100 units) a 40% improvement has been achieved and the experimenters will be very happy with the outcome of the study. However, a system configuration involving both A and B at their high levels has not been investigated and it might well be that such a setup would result in a system response of 180 units, as shown in Figure 5.4.

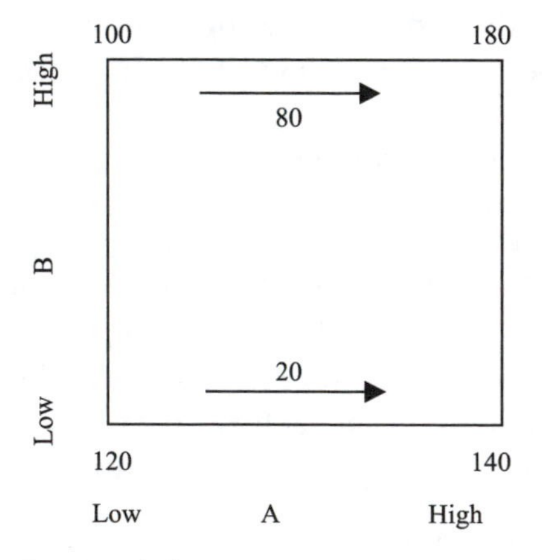

Figure 5.4 *The interaction revealed*

Figure 5.4 illustrates what statisticians call an 'interaction': the effect of changing A depends on the level of B. Thus, if A is changed from its low to its high level, when the level of B is low, the response increases by 20 units. If A is changed from its low to its high level, when the level of B is high, the response increases by 80 units. Such interactions are widespread in scientific and industrial applications and it makes sense that the experimental design be such that, if present, they will be detected.

Clearly, the 'study-one-factor-at-a-time' approach would have detected the best operating conditions if A had been investigated first, followed by B. However, it can hardly be considered a good strategy if finding the best operating conditions depends on haphazardly choosing the right factor to investigate first. Most analytical systems will have many factors worth investigating and, as the number of factors to be investigated increases, the chances of choosing a good sequence in which to carry out the investigation become more and more remote.

The factorial design requires all combinations of all levels of all factors to be run at least once; thus for 4 factors, each at 2 levels, this requires $2 \times 2 \times 2 \times 2 = 16$ runs. There is no reason why the factors must have the same number of levels – a three factor study could have factor A with 3 levels, B with 4 levels and C with 5 levels, leading to a design with $3 \times 4 \times 5 = 60$ design points. However, the class of designs with all factors at two levels has been found to be particularly useful and these designs will be explored here. Designs with more than two levels will be discussed in Chapter 7. The coverage of this chapter will move from a fairly extended account of the simple design outlined above to introduce the more sophisticated fractional factorial designs, which are commonly used in method development and ruggedness testing studies. Fractional factorial designs involve running only a (carefully selected) subset of the complete set of design points that constitute the full factorial design.

5.3 THE 2^2 DESIGN

The simplest case of a multi-factor study is where two factors are each investigated at two levels. In this section a structured approach to the analysis of the resulting data is introduced. This may appear over-elaborate for the analysis of such a simple design, but the approach generalizes in an obvious way to more complex studies and so is a convenient introduction to the analysis of factorial designs. The design is described as a 2^2 design: the superscript refers to the number of factors and there are two levels for each factor; thus, a 2^k design has k factors each at two levels.

5.3.1 An Example

At one stage in the development of a GC method for the analysis of trace compounds in distilled spirits a 2^2 factorial study was carried out. The factors studied were Split-Ratio (1:30, 1:25) and Injection Temperature (200 and 220 °C). Each factor combination was replicated three times and the twelve runs were carried out in randomized order. A single bottle of the spirit was used to provide a test portion for each run. Several response variables were studied: peak area was one of them and one objective was to choose the combination of the factors which would maximize the peak area. Many different components of the spirit were analyzed; peak areas for propanol are reported in Table 5.1. Figure 5.5 shows the average results at each of the four design points; the first (low) level of each factor listed above is labelled as '–', the high level as '+'.

Table 5.1 *The GC study data (peak areas, arbitrary units)*

Injection temperature/°C	*Split ratio*	
	1:30	*1:25*
	39.6	56.7
220	37.6	54.6
	42.4	51.2
	50.6	54.2
200	49.9	55.9
	46.8	56.6

Main Effects. It is clear that greater peak areas are obtained for the higher split ratio, as would be expected. The effect of higher temperature

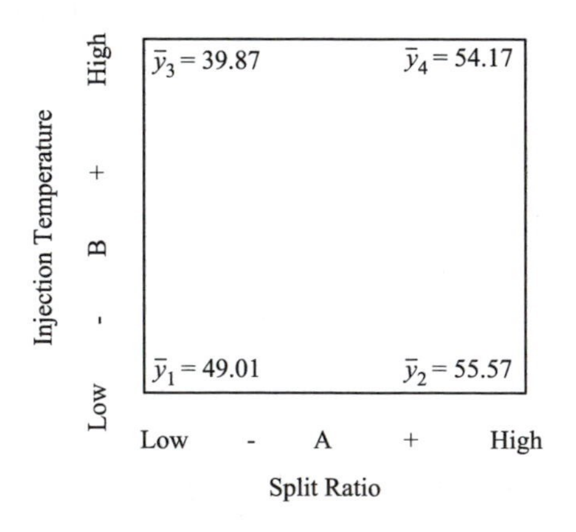

Figure 5.5 *A summary of the study results*

is not so clear-cut: we will discuss this further later, having first used the data to illustrate a structured approach to the analysis of such studies.

The average change in response as a factor is varied from its low to its high level is called its 'main effect'. The **Main Effect of Split Ratio (A)** is defined as:

$$\text{average result at high A} - \text{average result at low A}$$

$$\frac{\bar{y}_2 + \bar{y}_4}{2} - \frac{\bar{y}_1 + \bar{y}_3}{2} = \frac{54.17 + 55.57}{2} - \frac{39.87 + 49.10}{2}$$
$$1/2(\bar{y}_2 + \bar{y}_4 - \bar{y}_1 - \bar{y}_3) = (54.87 - 44.48)$$
$$1/2(-\bar{y}_1 + \bar{y}_2 - \bar{y}_3 + \bar{y}_4) = 10.38 \tag{5.1}$$

The peak area increased by 10.38 units, on average, when the Split Ratio was changed from its low level, 1:30, to its high level, 1:25.

Similarly, the **Main Effect of Injection Temperature (B)** is given by:

$$\text{average result at high B} - \text{average result at low B}$$

$$\frac{\bar{y}_3 + \bar{y}_4}{2} - \frac{\bar{y}_1 + \bar{y}_2}{2} = \frac{39.87 + 54.17}{2} - \frac{49.10 + 55.57}{2}$$
$$1/2(\bar{y}_3 + \bar{y}_4 - \bar{y}_1 - \bar{y}_2) = (47.02 - 52.34)$$
$$1/2(-\bar{y}_1 - \bar{y}_2 + \bar{y}_3 + \bar{y}_4) = -5.32 \tag{5.2}$$

The peak area decreased by 5.32 units, on average, when Injection Temperature was changed from its low level, 200 °C, to its high level, 220 °C.

Interactions. Calculating main effects as was done above essentially assumes that the two factors act independently of each other. Experience shows that very often the effect of changing one factor depends on the level of a second factor. In a case as simple as the present design, an interaction may be identified simply by inspecting the four means. It is convenient, however, to set up a formal definition of an interaction effect, just as main effects were defined above. Factors A and B are said to interact if the effect of changing A at low B is different from the effect of changing A at high B. The interaction is defined to be *half* this difference.

AB interaction:

$$1/2[\text{effect A at high B} - \text{effect A at low B}]$$
$$= 1/2[(\bar{y}_4 - \bar{y}_3)] - (\bar{y}_2 - \bar{y}_1)]$$
$$= 1/2(\bar{y}_1 - \bar{y}_2 - \bar{y}_3 + y_4) \tag{5.3}$$

The numerical results are:

$$\text{AB interaction} = {}^1\!/_2[\text{effect A at high B} - \text{effect A at low B}].$$

$$= {}^1\!/_2[(54.17 - 39.87) - (55.57 - 49.10)]$$

$$= {}^1\!/_2[(14.3) - (6.47)] = 3.92$$

Figure 5.5 illustrates the interaction: changing from low to high A leads to an increase of 14.3 units in peak area at high B, but an increase of only 6.47 units at low B. The effect of switching Split Ratio is clearly sensitive to the level of the Injection Temperature. These numerical results are, of course, subject to chance variation. The effects of this chance variation will be ignored for the moment, but will be considered later.

The interaction has been defined in terms of the difference between the effects of changing A at high and low B. It could just as well have been defined as the difference between the effects of changing B at high and low A: because of the symmetry of the design, the two definitions are equivalent and lead to the same numerical results (check it!).

Interpretation of Effects. Main effects tell what happens, on average, when a factor is changed. Where there is a strong interaction effect, as in this example, the main effects may not be very informative. Thus, Figure 5.5 shows that the effect of changing from low to high A, at low B, was to increase peak area by 6.47 units, while peak area increased by 14.3 units, when B was high. The average of the two changes is 10.38, which was the main effect of A. However, changing A does not give a change of 10.38 units in peak area: it gives a change of 14.3 units when the injection temperature is high (220 °C) and a change of 6.47 units when injection temperature is low (200 °C). The average of two different kinds of behaviour does not mean very much. Accordingly, in order to understand the response of the system, the nature of the interaction effect should be studied through examination of the four means on the corners of the square.

If there is no interaction it means that the effect of changing one factor, say A, is about the same (apart from chance variation) at both levels of the other factor, B. In such a case the main effect is the proper measure of what happens when a factor is changed.

Are the Effects Real? Until now, no account has been taken of chance variation. The question must be addressed, however, as to whether the calculated effects are real, or whether they are simply the result of the inevitable chance variability in the study results. This will be done using statistical significance tests. The tests will assess whether the calculated effects differ significantly from zero, *i.e.*, the null hypothesis will be that changing from one level of a factor to another has no systematic or long-run effect on the system response.

The model for the analysis assumes that the long-run average response may vary from design point to design point, but that the long-run standard deviation of replicate results, σ, is the same for all design points. This allows the sample standard deviations at the four design points to be combined to give an overall pooled value, as was done in Chapter 4 when independent group *t*-tests were discussed.

Table 5.2 shows the experimental data in a different format: each row corresponds to one design point. Thus, row one gives the results obtained when both factors were at their low levels. The last column of the table contains the row standard deviations, *i.e.*, the standard deviations of the three replicate results for each design point. Each of these has $n - 1 = 2$ degrees of freedom, where $n = 3$ is the number of replicates. The standard deviations are combined into a single estimate by calculating a simple average of their squares, the four sample variances:

$$s^2 = \frac{s_1^2 + s_2^2 + s_3^2 + s_4^2}{4}$$

$$= \frac{4.080 + 1.513 + 5.808 + 7.728}{4} = 4.78$$

The resulting value of 4.78 has $4(n - 1) = 8$ degrees of freedom.

Since each effect is the difference between two means, each of which is an average over half the dataset ($N/2$ values), the standard error of an effect is:

$$SE(effect) = \sqrt{\frac{\sigma^2}{N/2} + \frac{\sigma^2}{N/2}} = \sqrt{\frac{4\sigma^2}{N}}$$

where N is the total number of observations in the study. This formula is exactly equivalent to the standard error formula used for comparing two independent group means in Chapter 4. It will apply to all effects in all

Table 5.2 *Summary statistics for the GC method development study*

Split ratio	*Inject temp*	*Peak areas*			*Mean*	*s*
–	–	50.60	49.90	46.80	49.10	2.02
+	–	54.20	55.90	56.60	55.57	1.23
–	+	39.60	37.60	42.40	39.87	2.41
+	+	56.70	54.60	51.20	54.17	2.78

the designs discussed in this chapter, irrespective of the complexity of the study. The estimated standard error of any effect is calculated by replacing σ by s to give:

$$S\hat{E}(effect) = \sqrt{\frac{4s^2}{N}}$$

where N is the total number of observations in the study, 12 in our example.

A useful rule of thumb is that an effect looks interesting if it is at least two and a half times its estimated standard error. More precise statements can be made by comparing the ratio of the calculated effect to its estimated standard error (the t-statistic) with the appropriate t-distribution. The degrees of freedom of the t-distribution will be those of the s^2 used in computing the standard error estimate; here this is 8. The 'null hypothesis' that an effect is null (*i.e.* its long-run value is zero) is rejected only if the t-statistic lies beyond the critical values of the t-distribution curve. Here, an effect is considered statistically significant only if the corresponding t-statistic is either less than –2.31 or exceeds 2.31, which are the critical values corresponding to a two-tailed significance level of 0.05 for a t-distribution with 8 degrees of freedom.

Is the Split Ratio by Injection Temperature interaction real? Since

$$\text{AB} = 3.92 \text{ and } S\hat{E}(AB) = \sqrt{\frac{4(4.78)}{12}} = 1.26$$

the calculated effect is more than three times the estimated standard error (the t-statistic is $3.92/1.26 = 3.11$), so it is very likely a real, *i.e.* reproducible, effect. Deciding that the factors interact means that they need to be considered jointly. Accordingly, the main effects are not tested. If, however, the interaction had been non-significant, tests would have been carried out on the main effects to decide if either or both were statistically significant.

The Interaction Revisited. Figure 5.6 summarizes the results of the study, with the changes in the responses shown along the sides of the square. Thus, changing from low to high Temperature, when Split Ratio was 1:30 resulted in an average reduction of 9.23 units in peak area.

To determine if these changes are statistically significant, appropriate t-tests may be carried out. The estimated standard error* of the difference

*Note that in using $s^2 = 4.78$ we are using information about the magnitude of chance variability based on all four sets of replicates, even though our comparisons involve means at only two design points. This increases the power of our tests.

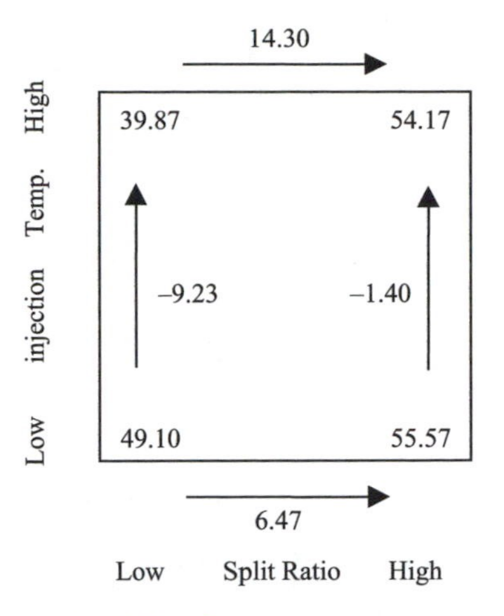

Figure 5.6 *Summary results for GC study*

between any two of the four means, $\bar{y}_i$ and $\bar{y}_j$, say, is:

$$SE(\bar{y}_i - \bar{y}_j) = \sqrt{\frac{2S^2}{3}} = \sqrt{\frac{2(4.78)}{3}} = 1.63$$

since each of the four means is based on 3 observations. The smallest difference, –1.4 is clearly not significant, since its ratio to the standard error will be less than 1. The next largest gives a t-value of 6.47/1.63 = 3.97, which is clearly statistically significant (the critical value is still 2.31 since s has 8 degrees of freedom) and so all the others must be statistically significant also.

The nature of the interaction is now clearer. When the Split Ratio (A) is 1:25, the effect of changing the Injection Temperature (B) from 200 to 220 °C is not statistically significant. This means that the evidence from the study is consistent with the temperature change having no effect on peak area, for this split ratio. On the other hand, when the Split Ratio is 1:30 there is a marked decrease in peak area when the Injection Temperature is raised. Since large peak areas are desirable, the study points to a Split Ratio of 1:25 as giving better results, which would be expected. The purpose of the investigation was to determine if the Split Ratio interacted with Injection Temperature and, if so, to identify the best operating conditions. The study results suggest that either Injection Temperature level could be used, but the analysts felt that lower temperature was likely to prolong the lifetime of the injection port. Figure 5.7, which is called an 'interaction plot', is an alternative representation of the information displayed in Figure 5.6.

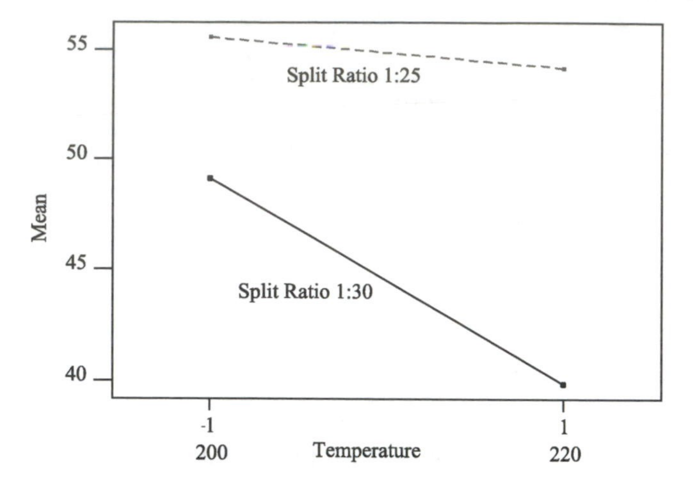

Figure 5.7 *An alternative display of the interaction effect*

The fact that the lines are not parallel indicates the presence of an interaction. Thus, changing Split Ratio at low Temperature shows a smaller jump in peak area than is seen at higher Temperature on the right-hand side of the Figure. In such plots, parallel lines indicate a lack of interaction between the two factors. If this occurred in our case it would mean that when a change is made from low to high Split Ratio, the increase in peak area would be the same at both levels of Injection Temperature. Here, of course, the lines are not parallel, since there is a highly significant interaction.

Sometimes it is desirable to obtain a confidence interval for the long-run mean response at the best combination of the factors studied. In this case, there would be little interest in such an estimate, but, nevertheless, one will be calculated to illustrate the details of the calculations. A 95% confidence interval for the long-run average peak area for propanol at the chosen set of parameter settings (Injection Temperature 200 °C and Split Ratio 1:25) is given by:

$$\bar{y}_1 \pm t_{.025}\sqrt{\frac{s^2}{3}}$$

since $\bar{y}_1$ is based on 3 observations. Note that, as above, the t-distribution has 8 degrees of freedom since s^2 has this number of degrees of freedom. The calculated interval is

$$55.57 \pm 2.31\sqrt{\frac{4.78}{3}} \quad \text{or} \quad 55.57 \pm 2.92$$

5.3.2 Model Validation

As outlined earlier, the model underlying our analysis assumes that at each design point the observations vary independently about some mean value, which depends on the design point, and come from a Normal distribution with a long-run standard deviation, σ, that is the same for all design points. The assumption of constant standard deviation implies that while changing the settings of the factors in the study may change the average response, the variability of test results about the mean values is unaffected by the parameter changes. This is usually a reasonable assumption for 2^k studies, as the factor level changes are normally quite modest. The assumptions can be investigated by residual analysis as was done in Chapter 4.

If, from each of the three replicates at each corner of the design square, the corresponding mean is subtracted, a set of twelve residuals will be calculated. These residuals should behave approximately like a sample from a Normal distribution with mean zero and constant standard deviation. If they do not do so, the model assumptions are questionable.

Figure 5.8 shows a plot of the residuals *versus* the mean results at the four corners of the design square, shown in Figure 5.6. There is no indication of systematic variation of the residuals with the means. In particular, there is no tendency for the variability to increase with the magnitude of the measurements, which is commonly observed in measurement systems. This is not surprising here, as the range of the measured values is quite narrow. Figure 5.9 shows a Normal probability plot of the residuals. The residuals lie close to a straight line and the

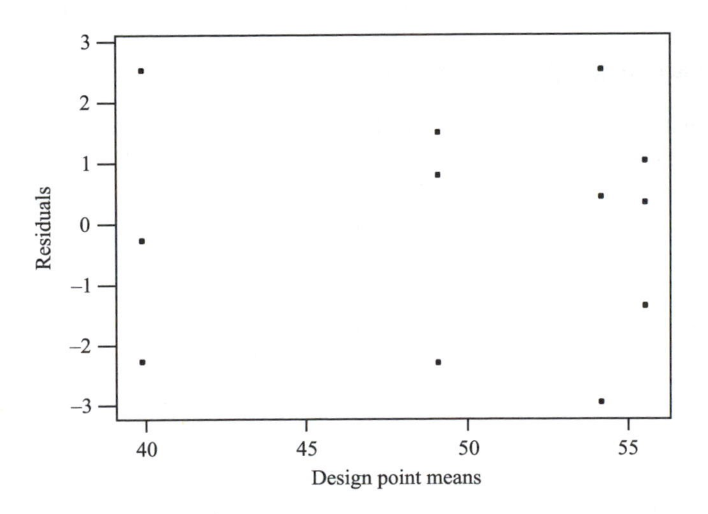

Figure 5.8 *Residuals versus design point means*

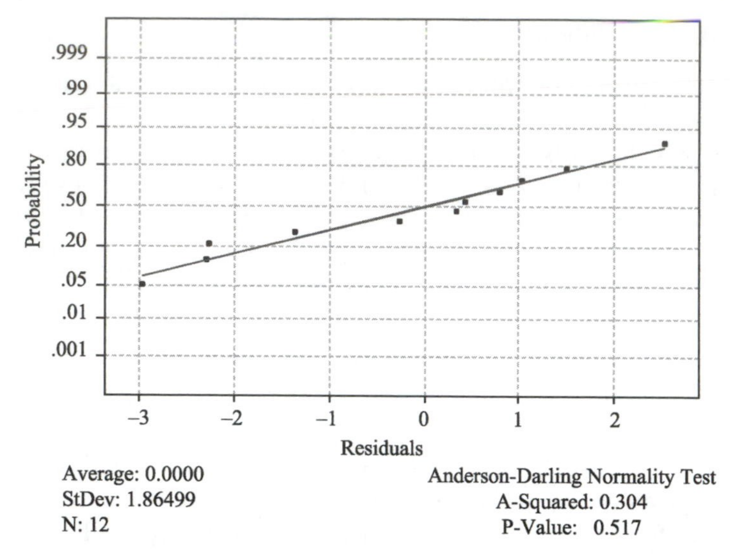

Figure 5.9 *A Normality probability plot of the residuals*

Normality test *p*-value is quite large – this plot does not call into question the Normality assumption underlying our tests and confidence intervals.

The simple plots discussed above suggest that the assumptions underlying our statistical tests and confidence intervals are likely to hold in this case. However, with so few observations we could not expect to detect anything other than large violations of the assumptions. When the statistical evidence is weak it places responsibility on the analyst to justify such assumptions from prior knowledge of the system.

5.3.3 Organizing the Calculations

The information displayed in Figures 5.5–5.7 can be presented in tabular form, as shown in Table 5.3, where the low level of each factor is labelled '–' and the high level '+'. The block of signs in the table is described as a 'design matrix' in the statistical literature.

Table 5.3 *The study design matrix, with results*

Design point	*A*	*B*	*Result*
1	–	–	$\bar{y}_1$
2	+	–	$\bar{y}_2$
3	–	+	$\bar{y}_3$
4	+	+	$\bar{y}_4$

The table has two principal uses. Firstly, each row of the design matrix describes one experimental run that is to be carried out. Thus, row 3 indicates that a run with A low and B high is required. In a case such as this one, which involves so few runs, assistance in organizing the experimental details is not required. If, however, six factors each at two levels were being studied, such an array would be useful in keeping track of, and recording the results of the $2^6 = 64$ runs. Note the pattern in the order of the design points: the levels of A alternate from row to row, while those of B alternate in pairs. This is a standard format for displaying design matrices. In practice, however, the run order should be randomized in carrying out the study. The role of randomization was discussed in Chapter 4.

The array is useful, also, when it comes to analyzing the results. To calculate any main effect, we subtract the average of the results corresponding to the negative signs from those corresponding to the positive signs, in the corresponding columns. A simple expansion of the array facilitates measuring the interaction effect. An extra column, labelled AB is inserted, with entries being the product of corresponding entries in columns A and B, to give Table 5.4.

Table 5.4 *The expanded design matrix*

Design Point	*A*	*B*	*AB*	*Result*
1	−	−	+	$\bar{y}_1$
2	+	−	−	$\bar{y}_2$
3	−	+	−	$\bar{y}_3$
4	+	+	+	$\bar{y}_4$

The difference between the averages of the means corresponding to the plus and minus signs in the AB column gives the same result as given by the definition of an interaction effect: $AB = 1/2(\bar{y}_1 - \bar{y}_2 - \bar{y}_3 + \bar{y}_4)$, see Equation (5.3), page 191.

Organizing the data into a design matrix table gives a convenient and simple way of both displaying the results and calculating the effects. Of course, the analyses described in the preceding pages are rather elaborate, given that only four means need to be compared. However, the benefits of this approach will be seen later when more complicated experimental designs will be discussed.

Exercises

5.1 A study was conducted to investigate the influence of various system parameters on the performance of a programmed temperature vaporizer (PTV)

Table 5.5 *Peak areas for GC study*

Solvent flow rate / mL min^{-1}	*Injection volume/μL*	
	100	*200*
400	13.1	126.5
	15.3	118.5
	17.7	122.1
200	48.8	134.5
	42.1	135.4
	39.2	128.6

operated in the solvent split mode. The purpose was to develop a GC method of analyzing trace aroma compounds in wine, without the need for prior extraction. One phase of the investigation involved studying two factors each at two levels; A: Injection Volume (100 or 200 μL), B: Solvent Elimination Flow Rate (200 or 400 mL min^{-1}). The four design points were replicated three times, giving a total of 12 observations. The order in which the runs were carried out was randomized. One of the responses studied was the total peak area given by a synthetic mixture of aroma compounds in ethanol–water (12:88 v/v). The results (absolute peak areas) are shown in Table 5.5. What conclusions can be drawn from these data? Check model assumptions by drawing appropriate residual plots.

This illustrative example (see also Exercises 5.2 and the Example of Section 5.4), was constructed using information taken from Villen *et al.*[1] Readers with an interest in the substantive issues should consult the original paper.

5.2 In another phase of the method development, a second study was carried out. This study investigated the same levels of Injection Volume (*i.e.* 100 or 200 μL) but this time the second factor was PTV Initial Temperature (20 or 40 °C). There were four replicates; the sixteen runs were carried out in a randomized order. The response is again peak area; the data are shown in Table 5.6. Analyze the results of the study.

Table 5.6 *Peak areas for GC study*

Initial temperature/° C	*Injection volume/μL*	
	100	*200*
40	22.0	125.5
	30.1	131.8
	21.3	124.1
	27.2	117.8
20	31.4	125.5
	31.3	121.6
	35.0	128.7
	23.6	127.7

5.3 During development of a multi-residue supercritical fluid extraction method for the extraction of benzimidazole residues from animal liver, Danaher *et al.*[2] investigated the influence on recovery rates of varying several method parameters, each over two levels. The percentage recovery results of a replicated 2^2 study are shown in Table 5.7 for triclabendazole (TCB). The factors varied were CO_2 volume (40, 80 L) and pressure (248, 543 bar). What can we learn from the study?

Table 5.7 *Triclabendazole (TCB) recoveries*

Pressure	*Carbon dioxide*	
	Low	*High*
	41	71
High	60	81
	43	83
	66	67
	8	0
Low	6	15
	25	24
	8	10

5.4 THE 2^3 DESIGN

Experimental studies involving more than two factors, all at two levels, are a straightforward generalization of the 2^2 design: for k factors all 2^k combinations of the different levels must be run, preferably several times each, in randomized order. The analysis follows that discussed above, but will require measures of two-factor interactions for all pairs of factors, three-factor interactions for all sets of three factors and so on. The three-factor case will be discussed now; larger designs will be discussed later.

5.4.1 An Example

A study was conducted to investigate the influence of various system parameters on the performance of a programmed temperature vaporizer (PTV) operated in the solvent split mode. The purpose was to develop a GC method of analyzing trace aroma compounds in wine, without the need for prior extraction. One phase of the investigation involved studying three factors each at two levels; A: Injection Volume (100 or 200 μL), B: Solvent Elimination Flow Rate (200 or 400 mL min^{-1}), C: PTV Initial Temperature (20 or 40 °C). The eight design points were replicated twice, giving a total of 16 observations. The order in which

Table 5.8 *Design matrix and results for PTV development study*

Design point	*A*	*B*	*C*	*Results*	
1	–	–	–	37.5	24.6
2	+	–	–	149.4	146.5
3	–	+	–	24.9	28.8
4	+	+	–	117.3	114.0
5	–	–	+	27.5	32.2
6	+	–	+	129.5	136.6
7	–	+	+	23.7	22.4
8	+	+	+	111.0	112.6

the 16 runs were carried out was randomized. One of the responses studied was the total peak area given by a synthetic mixture of aroma compounds in ethanol–water (12:88 v/v). The results (absolute peak areas) are shown in standard order in Table 5.8.

When there are three factors, each at two levels, there are $2 \times 2 \times 2$ combinations, *i.e.*, eight design points. If n replicate measurements are made at each design point, the average results may be displayed at the corners of a cube, as shown in Figure 5.10. In the diagram '–' represents the 'low' level of a factor and '+' the 'high' level. Note that a triple of signs describes each corner of the cube. Thus '– – –' describes the lower left-hand side (LHS) design point at the front of the cube. This is row one of the design matrix in Table 5.8. Here, the factor levels are naturally ordered, but often the low and high labels are arbitrary.

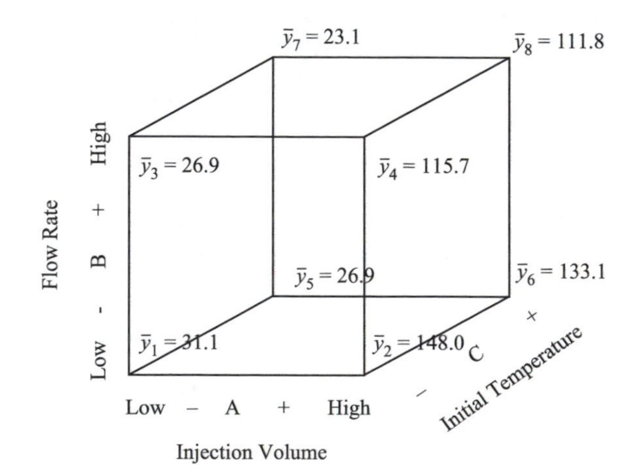

Figure 5.10 *The 2^3 design and study results*

For this design there are three main effects corresponding to the three factors, three two-factor interactions and a three-factor interaction. The main effects and two-factor interactions are defined exactly as they were in the 2^2 case; the three-factor interaction is a simple generalization of the two-factor interaction.

The Main Effects. The right-hand side (RHS) of the cube corresponds to those runs of the system under study in which factor A is at its high level, the LHS to those where it is at its low level. Accordingly, the difference between the averages of the results at these two sets of four design points is a measure of the average effect of changing from the low to the high level of factor A. The **main effect of Injection Volume (A)** is defined as:

$$\begin{aligned}&\text{average result at high A} - \text{average result at low A}\\&= \frac{\bar{y}_2 + \bar{y}_4 + \bar{y}_6 + \bar{y}_8}{4} - \frac{\bar{y}_1 + \bar{y}_3 + \bar{y}_5 + \bar{y}_7}{4} = 127.1 - 27.4\\&= 1/4(-\bar{y}_1 + \bar{y}_2 - \bar{y}_3 + \bar{y}_4 - \bar{y}_5 + \bar{y}_6 - \bar{y}_7 + \bar{y}_8) = 99.4 \qquad (5.4)\end{aligned}$$

The level of Flow Rate (B) is high at the top of the cube and low at the bottom. Accordingly, the difference between the averages of the results at the top and the bottom of the cube gives the main effect of B. By examining Figure 5.10, verify that this definition of the main effect of B leads to:

$$\textbf{Main effect B:}\quad 1/4(-\bar{y}_1 - \bar{y}_2 + \bar{y}_3 + \bar{y}_4 - \bar{y}_5 - \bar{y}_6 + \bar{y}_7 + \bar{y}_8) = -16.1 \qquad (5.5)$$

Similarly, Initial Temperature (°C) is high at the back of the cube and low at the front and its main effect is given by:

$$\textbf{Main effect C:}\quad 1/4(-\bar{y}_1 - \bar{y}_2 - \bar{y}_3 - \bar{y}_4 + \bar{y}_5 + \bar{y}_6 + \bar{y}_7 + \bar{y}_8) = -5.9 \qquad (5.6)$$

Note that the patterns of signs of $\bar{y}$ values in Expressions (5.4)–(5.6) correspond to the patterns in the columns of the design matrix shown in Table 5.8. Thus, for factor A, the − and + signs alternate, for B they alternate in pairs, while for C there are four − signs followed by four + signs.

Two-Factor Interactions. In a three-factor study three two-factor interactions, corresponding to all possible pairs of factors, may be estimated. To define any of these, the cube is collapsed over the third factor and then the interaction is defined exactly as was done for the 2^2 design. This process will be illustrated for the AB interaction; the reader is invited to carry through the same steps for the AC and BC interactions. Figure 5.11 shows the results of collapsing over factor C.

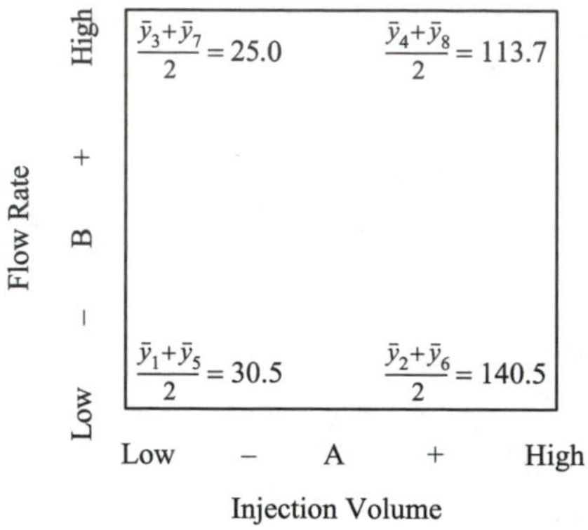

$\frac{\bar{y}_3+\bar{y}_7}{2} = 25.0$ $\frac{\bar{y}_4+\bar{y}_8}{2} = 113.7$

$\frac{\bar{y}_1+\bar{y}_5}{2} = 30.5$ $\frac{\bar{y}_2+\bar{y}_6}{2} = 140.5$

Flow Rate: Low – B + High

Low – A + High

Injection Volume

Figure 5.11 *The AB interaction*

The Injection Volume-Flow Rate or **AB interaction** is defined as:

$$\tfrac{1}{2}(\text{effect A at high B} - \text{effect A at low B}).$$

The two parts of this expression are given by:

$$\text{effect A at high B}: \frac{\bar{y}_4+\bar{y}_8}{2} - \frac{\bar{y}_3+\bar{y}_7}{2} = 113.7 - 25.0 = 88.7$$

$$\text{effect A at low B}: \frac{\bar{y}_2+\bar{y}_6}{2} - \frac{\bar{y}_1+\bar{y}_5}{2} = 140.5 - 30.5 = 110.0$$

and the interaction is the calculated as half the difference between the effects:

$$\tfrac{1}{2}\left\{\left(\frac{\bar{y}_4+\bar{y}_8}{2} - \frac{\bar{y}_3+\bar{y}_7}{2}\right) - \left(\frac{\bar{y}_2+\bar{y}_6}{2} - \frac{\bar{y}_1+\bar{y}_5}{2}\right)\right\}$$

$$\tfrac{1}{2}\{88.7 - 110.0\} = -10.6$$

which gives:

$$\textbf{AB interaction}: \tfrac{1}{4}\{\bar{y}_1 - \bar{y}_2 - \bar{y}_3 + \bar{y}_4 + \bar{y}_5 - \bar{y}_6 - \bar{y}_7 + \bar{y}_8\} = -10.6 \qquad (5.7)$$

The other two-factor interactions can be obtained in a similar manner; they are:

$$\textbf{AC interaction} : {}^1\!/_4\{\bar{y}_1 - \bar{y}_2 + \bar{y}_3 - \bar{y}_4 - \bar{y}_5 + \bar{y}_6 - \bar{y}_7 + \bar{y}_8\} = -3.4 \qquad (5.8)$$

$$\textbf{BC interaction} := {}^1\!/_4\{\bar{y}_1 + \bar{y}_2 - \bar{y}_3 - \bar{y}_4 - \bar{y}_5 - \bar{y}_6 + \bar{y}_7 + \bar{y}_8\} = 2.1 \qquad (5.9)$$

Note that in all cases the roles of the factors are symmetrical: for example, the AB interaction could have been defined as half the difference between the effects of B at high and low A, and the resulting definition would have been identical to (5.7).

The Three-Factor Interaction. Consider now the possibility that the effect of changing from low to high A depends not only on the level of B, *i.e.*, that there is an AB interaction, but also that the size of this interaction depends on the level of factor C. This would imply the existence of a complex interdependence between factors A, B and C. A three-factor interaction is defined similarly to a two-factor interaction. A two-factor interaction is said to exist if the main effects of one factor (A) are different at the two levels of a second factor (B). A three-factor interaction exists if the two-factor interaction AB measured at the high level of C is different from the AB interaction measured at the low level of C. The AB interaction at the high level of C is based on the results corresponding to the back of the cube of Figure 5.10, shown below as Figure 5.12.

The two-factor interaction at the back of the design cube is:

AB interaction at high C : ½ (effect A at high B – effect A at low B)

$${}^1\!/_2\{(\bar{y}_8 - \bar{y}_7) - (\bar{y}_6 - \bar{y}_5)\} = {}^1\!/_2(88.75 - 103.20) = -7.225$$

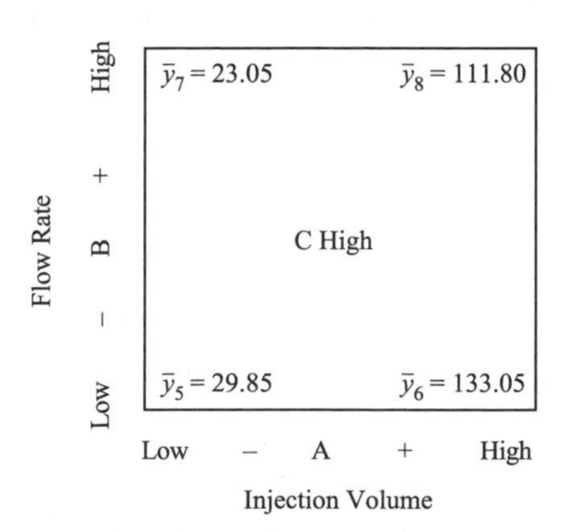

Figure 5.12 *The back of the design cube: C at its high level*

The AB interaction at the low level of C is based on the results corresponding to the front of the cube of Figure 5.10, shown below as Figure 5.13.

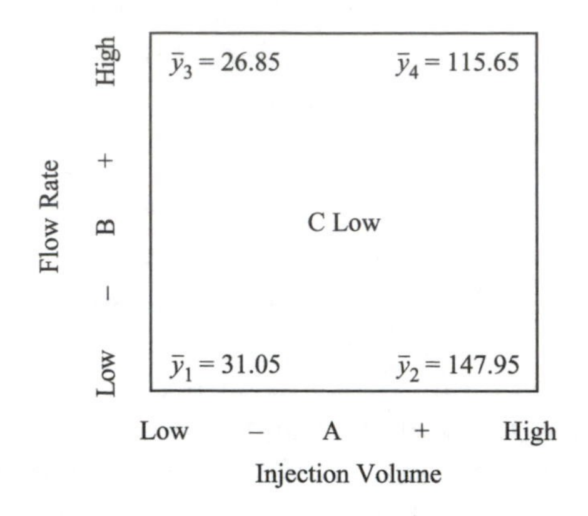

Figure 5.13 *The front of the design cube: C at its low level*

The two-factor interaction at the front of the design cube is:

$$\textbf{AB interaction at low C} : \tfrac{1}{2}(\text{effect A at high B} - \text{effect A at low B})$$
$$\tfrac{1}{2}\{(\bar{y}_4 - \bar{y}_3) - (\bar{y}_2 - \bar{y}_1)\} = \tfrac{1}{2}(88.80 - 116.90) = -14.05$$

The three-factor ABC interaction is then defined as half the difference between the pair of two-factor interactions:

$$\textbf{ABC interaction} : \tfrac{1}{2}\{\text{AB interaction at high C} - \text{AB interaction at low C}\}$$
$$\tfrac{1}{2}\{-7.23 - (-14.05)\} = 3.41$$

Symbolically this is:

$$\tfrac{1}{2}\{\tfrac{1}{2}[(\bar{y}_8 - \bar{y}_7) - (\bar{y}_6 - \bar{y}_5)] - \tfrac{1}{2}[(\bar{y}_4 - \bar{y}_3) - (\bar{y}_2 - \bar{y}_1)]\}$$

which gives:

$$\textbf{ABC interaction} : \tfrac{1}{4}\{-\bar{y}_1 + \bar{y}_2 + \bar{y}_3 - \bar{y}_4 + \bar{y}_5 - \bar{y}_6 - \bar{y}_7 + \bar{y}_8) = 3.41 \qquad (5.10)$$

Note that the definition of the three-factor interaction could have been developed through a comparison of BC interactions at high and low A or of AC interactions at high and low B, in both cases leading to the same expression (5.10).

All seven possible effects have now been defined. Note that each of the definitions involves the averages at all eight design points, *i.e.*, all the data are used in measuring each of the seven effects. This is a highly efficient use of the data collected in the experiment. As will be seen below, Expressions (5.4)–(5.10) indicate the level of complexity of the results of the study. Thus, if interaction terms are large, the experimental factors must be considered *together*, if the nature of the system is to be understood. If all interaction effects are negligible, then the main effects indicate how the system responds as each factor is changed, irrespective of the levels of the other factors.

5.4.2 Data Analysis

The effects have already been calculated from first principles. They might have been calculated, alternatively, by using the ± sign patterns of the expanded design matrix, which is shown in Table 5.9, together with the means and variances of the replicates at each design point. The ± patterns for the main effects correspond to the design matrix (the first three columns of Table 5.9). The ± pattern for any interaction effect is found by multiplying the columns of signs corresponding to the terms in that interaction. Thus, the pattern of signs for the ABC interaction is obtained by multiplying together the first three columns of the table. The differences between the means of the results corresponding to ± signs give the effect estimates. These are:

$$A = 99.4 \quad B = -16.1 \quad C = -5.9$$

$$AB = -10.6 \quad AC = -3.4 \quad BC = 2.1$$

$$ABC = 3.4$$

Since the experimental conditions are fixed for each row (design point), the variances of the replicates measure purely random variation. They

Table 5.9 *The expanded design matrix and summary statistics for the study*

A	*B*	*C*	*AB*	*AC*	*BC*	*ABC*	*Mean*	s^2
−	−	−	+	+	+	−	31.10	83.21
+	−	−	−	−	+	+	148.00	4.21
−	+	−	−	+	−	+	26.90	7.61
+	+	−	+	−	−	−	115.70	5.45
−	−	+	+	−	−	+	29.90	11.05
+	−	+	−	+	−	−	133.10	25.21
−	+	+	−	−	+	−	23.10	0.85
+	+	+	+	+	+	+	111.80	1.28

can be combined into a single estimate, s^2, by taking a simple average:

$$s^2 = \frac{\sum s_i^2}{8} = 17.35$$

The resulting value has 8 degrees of freedom, $2 - 1 = 1$ from each design point. The standard error of the effects is estimated by:

$$S\hat{E}(effect) = \sqrt{\frac{4s^2}{N}} = 2.08$$

where N is the total number of observations, 16 here.

The statistical significance of the effects can be tested using t-tests. In doing this, the most complex interaction should be tested first. If the three-factor interaction is statistically significant, it means that to understand the behaviour of the system the three factors must be studied *simultaneously*. In such a case, the two-factor interactions and the main effects need not be tested. If the three-factor interaction is not statistically significant, the three two-factor interactions should then be tested. Any main effect not involved in a significant interaction should be tested next.

Ultimately, the experimenter will want to study the data themselves, rather than deal with the effects, which are only statistical constructs. However, the process outlined above will identify the best summaries to extract the information contained in the raw data: if there are no interactions the overall averages at high and low levels of the significant main effects tell all. Where a two-factor interaction exists collapsing the data onto the appropriate square will reveal the more complex behaviour of the system, and so on.

For the PTV data the t-tests have 8 degrees of freedom; hence, the ratio of an effect to its estimated standard error (which is 2.08 for all effects) will need to be greater in magnitude than 2.31 to be statistically significant. The three-factor interaction does not meet this criterion. Of the three two-factor interactions only AB produces a t-ratio whose magnitude is greater than 2.31. Factor C (Initial temperature) is not involved in a significant interaction, but it is clear that the main effect (−5.9) is statistically significant. As the PTV Initial Temperature is increased from 20 to 40 °C the peak areas decrease by about 6 units (apart from random variation). This is true irrespective of the settings of the Injection Volume or Solvent Elimination Flow Rate. Obviously, this and any other statement based on the data analysis can only be relied upon within the ranges of the system parameters that were studied in the experiment.

To interpret the two-factor interaction between Injection Volume and Solvent Elimination Flow Rate, the results may be collapsed onto the

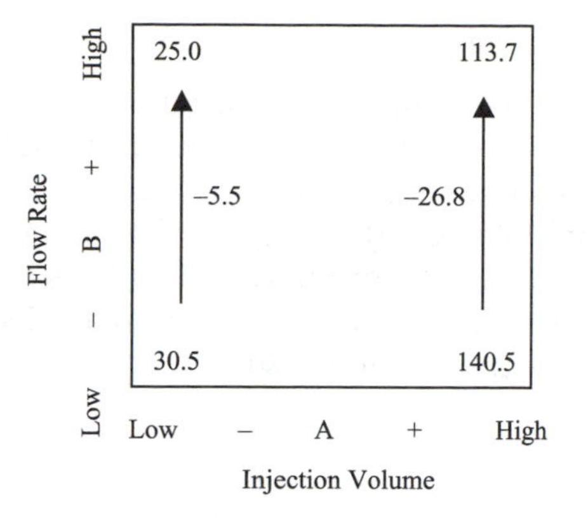

Figure 5.14 *Average results at the four AB design points*

square defined by the four combinations of the two factors; this is done in Figure 5.14.

Figure 5.14 shows that the total peak area decreases by an estimated 5.5 units if the solvent elimination flow rate is increased from 200 to 400 mL per min, when an injection volume of 100 μL is used; it decreases by an estimated 26.8 units for the same change in flow rate when the injection volume is 200 μL. Thus, while the injection volume effect is dominant (the main effect is 99.4), the solvent flow rate does influence the peak area when injection volume is at its high level. Before settling on this description of the outcome of the study, however, we must recognize that these estimated changes are subject to chance variation. The standard error of differences along the arms of the square is $\sqrt{\frac{2\sigma^2}{4}}$, since each mean is based on 4 observations;* its estimate is 2.95. The change from low to high solvent flow rate, at low injection volume, is not statistically significant, since $t = -5.5/2.95 = -1.90$, which lies in the body of the t-distribution (the critical values are ± 2.31). At high injection volume the change from low to high solvent flow rate is highly statistically significant ($(113.7 - 140.5)/2.95 = -9.08$) and, therefore, low solvent flow rate would be expected to give better peak areas.

In order to achieve the greatest peak area, the system should be operated at high Injection Volume (200 μL), low Solvent Elimination Flow Rate (200 mL per min) and low PTV Initial Temperature (20 °C).

*Recall from Chapter 4 that the standard error of the difference between two means is $\sqrt{\frac{2\sigma^2}{n}}$, where each mean is based on n results.

5.4.3 Model Validation

Figure 5.15 is a plot of the residuals against the design point means; the residuals are the differences between the two test results and their mean, at each of the eight design points. The graph displays rather odd patterns. The problem here is that at each of the eight design points there are only two test results. Accordingly, the two residuals will be mirror images of each other – each one having a paired value of opposite sign. This creates the odd patterns in the graph.

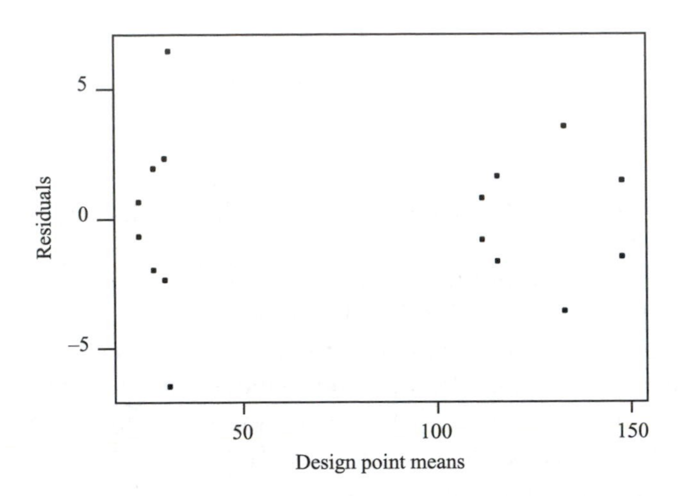

Figure 5.15 *Residuals versus design point means for the PTV study*

One pair of residuals appears to show greater variability than the others. The standard deviation or range, based on only two observations, is a very uncertain quantity and it would be unwise to read too much into the apparently larger pair of residuals in Figure 5.15. If it is thought that conditions at this design point could lead to greater test result variability then it would be appropriate to collect more data in order to verify that this is so. Here, a statistical test (not shown) does not call into question the assumption of constant standard deviation for all design points. The assumption of Normality, that underlies our analysis, is supported by the Normal probability plot of Figure 5.16.

Note that the Normal plot is subject to the same mirror image effect as Figure 5.15 – if you examine Figure 5.16 carefully, you will see that every point has a corresponding value reflected through the origin. Here, it does not cause a problem, but it is something to be aware of when dealing with duplicate data.

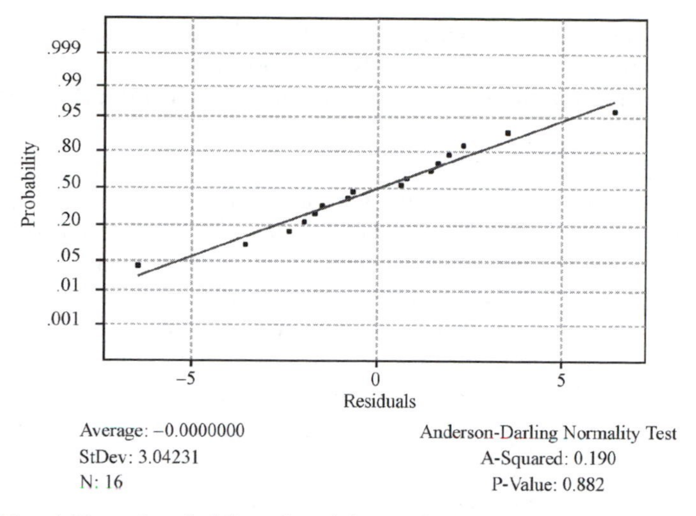

Figure 5.16 *A Normal probability plot of the residuals*

Exercises

5.4 The data shown in Table 5.10 come from a study of the factors that influence the performance of a formic acid biosensor, which is used for air monitoring. For more information and a reference to the original study see Exercise 5.8; more factors are given there and the labelling here is consistent with that in Exercise 5.8. For the factors listed below, the first level is coded –1 and the second level as +1 in the data table:

A: Concentration of Medola's blue (mg g^{-1}) 0.3, 1.5
C: Concentration of nicotinamide adenine dinucleotide (mg g^{-1}) 1.2, 4.8
E: Amount of solution placed on the electrode (μL) 8, 12

The response is the current (nA) at steady state when the biosensor was exposed to a formic acid vapour concentration of 1.4 mg m^{-3}. Large responses are

Table 5.10 *Biosensor development study results (nA)*
(Reprinted from the *Analyst*, 2001, Vol. 126, pp 2008. Copyright, 2001, Royal Society of Chemistry.)

A	*C*	*E*	*Results*			
–1	–1	–1	213	217	238	263
1	–1	–1	296	266	250	314
–1	1	–1	285	302	264	244
1	1	–1	257	375	291	310
–1	–1	1	258	285	313	296
1	–1	1	278	362	277	320
–1	1	1	259	302	264	214
1	1	1	298	289	302	307

desirable. There are four replicates at each of the 2^3 design points. Analyze the data, draw interaction plots, where appropriate, carry out a residual analysis and set out your conclusions as to what are the important features of the study results.

5.5 During development of a multi-residue supercritical fluid extraction method for the extraction of benzimidazole residues from animal liver, Danaher *et al*[2]. investigated the influence on recovery rates of varying several method parameters, each over two levels. The percentage recovery results of a replicated 2^3 study are shown in Table 5.11 for albendazole sulfoxide (ABZ-SO). The factors varied were temperature (60, 100 °C), CO_2 volume (40, 80 L) and pressure (248, 543 bar). What can we learn from the study?

Table 5.11 *Albendazole sulfoxide (ABZ-SO) recoveries (%)*

Press	*Temp*	*CO_2*	*ABZ–SO recoveries*	
247.5	60	40	10	20
542.5	60	40	36	32
247.5	100	40	7	0
542.5	100	40	46	50
247.5	60	80	11	16
542.5	60	80	56	54
247.5	100	80	13	8
542.5	100	80	58	67

5.5 SAMPLE SIZE FOR FACTORIAL DESIGNS

An approach to determining sample sizes for simple comparative studies was introduced in Chapter 4. This can be applied to 2^k designs also, and, as we will see, it leads to the same sample sizes as a more sophisticated calculation embodied in standard statistical software. The reason why this works is that when testing effects in 2^k designs, we compare two halves of the dataset – this involves essentially the same two-sample t-test that was discussed in Chapter 4. For example, in the 2^3 PTV study discussed in the last section, a test of the statistical significance of any effect involves a comparison between the means of the responses corresponding, respectively, to the 4 plus signs and the 4 minus signs of the appropriate column of Table 5.9. The t-statistic was:

$$t = \frac{\bar{x}_+ - \bar{x}_-}{\sqrt{\frac{4s^2}{N}}} = \frac{\bar{x}_+ - \bar{x}_-}{\sqrt{\frac{s^2}{N/2} + \frac{s^2}{N/2}}}$$

in which $N = 16$ is the total number of observations in the study. This test statistic has exactly the same form as the two-sample t-statistic used in Chapter 4.

Our objective here is to determine the sample size that would be required to detect an effect of a given size in a 2^k study. To provide a focus for our discussion, we assume the analysts expected a standard deviation for replicate results of $\sigma = 4$ (the estimated s^2 was 17.35 in the PTV study), that they wanted to use two-tailed tests with a significance level of $\alpha = 0.05$, and required a power of $1-\beta = 0.95$ of detecting effects of size δ. Column 1 of Table 5.12 shows the sample sizes given by the sample-size table for two-sample t-tests (Table A7). The other columns show the implications of using these sample sizes as guides for 2^2 and 2^3 studies. We will consider three different values of δ, the effect size. The power values shown in Table 5.12 are the probabilities of detecting a shift of size δ, if the total sample size, N_T, as given by Table A7, is used to implement the three designs represented by the three columns of Table 5.12.

$\delta = 10$

When $\delta = 10$, the value of D required for using the sample size table is $D = \frac{\delta}{\sigma} = \frac{10}{4} = 2.5$, which gives a sample size of $n = 6$ for each group, *i.e* ., a total study size of $N_T = 12$ would be required in a two-sample t-test designed to detect a difference of $\delta = 10$ between the two long-run means. Since sample size can only jump in integer values, the solution which gives a power of at least 0.95 of detecting an effect of size δ, usually involves rounding up to the nearest integer value. Consequently, the actual power will be slightly higher than the specified value of 0.95. In this case the actual power was 0.9730.

Table 5.12 *Sample size and power calculations; $\sigma = 4$ in all cases*

Sample sizes using two-sample t-test table	*2^2 Designs*	*2^3 Designs*
$\delta = 10$ $n = 6$ $N_T = 12$ *Power = 0.9730*	Replicates = 3 $N_T = 12$ *Power = 0.9649*	Replicates = 2 $N_T = 16$ *Power = 0.9914*
$\delta = 6$ $n = 13$ $N_T = 26$ *Power = 0.9561*	Replicates = 6 $N_T = 24$ *Power = 0.9373* *Replicates = 7* $N_T = 28$ *Power = 0.9675*	Replicates = 3 $N_T = 24$ *Power = 0.9314* *Replicates = 4* $N_T = 32$ *Power = 0.9825*
$\delta = 4$ $n = 27$ $N_T = 54$ *Power = 0.9501*	Replicates = 13 $N_T = 52$ *Power = 0.9421* *Replicates = 14* $N_T = 56$ *Power = 0.9565*	Replicates = 6 $N_T = 48$ *Power = 0.9222* *Replicates = 7* $N_T = 56$ *Power = 0.9560*

If, instead of a two-sample t-test designed to detect a difference of $\delta = 10$ between two long-run means, what we require is a 2^2 study in which effects of size $\delta = 10$ will be detected with power $= 0.95$, then we simply allocate the total sample size $N_T = 12$ to the four design points. This gives $n = N_T/4 = 12/4 = 3$ replicates at each of the four design points. The power of the test for detecting an effect of size $\delta = 10$ will then be 0.9649, which is virtually the same as for the two-sample t-test. Thus, by using the total study size, N_T, from the sample-size table (A7), we are led to the correct sample size for the 2^2 study.

A 2^3 study will require $n = N_T/8 = 12/8 = 1.5$ replicates at each design point. We round this up to two, which results in a total sample size of 16. With two replicates, the power will be 0.9914 for detecting an effect of size $\delta = 10$.

$\delta = 6$

For $\delta = 6$, the two-sample table gives $N_T = 26$. This suggests either 6 or 7 replicates for a 2^2 study, giving either $N_T = 24$ or 28. Intuitively, we might expect to have to round up ($26/4 = 6.5$) to the nearest multiple of 4 – the number of design points. The second column shows a power of 0.9373 if we round down to 6 replicates or 0.9675 if we round up to 7. For a 2^3 study the Table A7 sample size value of $N_T = 26$ suggests $26/8 = 3.25$ replicates. We obtain a power of 0.9314 if this is rounded down to 3 replicates and 0.9825 if it is rounded up to 4 replicates.

Our intuition is correct – to get a power of at least 0.95, rounding up will be required always. However, if we are not concerned that the power will be slightly less than 0.95, rounding down gives satisfactory sample sizes, also, in both cases.

$\delta = 4$

The same pattern emerges when an effect of size $\delta = 4$ is required to be detected. Table A7 suggests $N_T = 54$. For a 2^2 study we are led to $54/4 = 13.5$, which we round to 14 replicates and $54/8 = 6.75$, which we round to 7 for a 2^3 study.

Note that when a sample size is requested with the specifications $\alpha = \beta = 0.05$, $\sigma = 4$ and $\delta = 10, 6, 4$ used above, the rounded-up values suggested by Table A7 are the same in all cases as those produced by the sample size generator for factorial designs, available in Minitab.

The approach to sample size determination discussed above is easy to use and applies in exactly the same way to 2^4, 2^5 and higher order designs. It is unlikely, though, that more than a single replicate would be run at each design point for higher order designs, as the total sample size increases very rapidly with the number of factors. Such designs are considered in the next section.

5.6 EXPERIMENTS WITH MANY FACTORS

The design and analysis of studies involving many two-level factors is a natural generalization of the approach taken for two and three-factor experiments. The design simply requires studying all possible combinations of the various factor levels. This can be set up very conveniently by extending the matrix of signs as was done in moving from the 2^2 to the 2^3 design. The problem, of course, is that the number of combinations grows very rapidly. With five factors there are 32 combinations and, with even two replicates at each design point, this requires 64 runs. In many situations this would be regarded as an unrealistically large study. As the number of factors increases, various strategies are used to reduce the number of runs required. The simplest of these is to run the study with only one observation at each design point. This means that there is no replication in the study and, hence, the study does not provide a direct estimate of the standard error of the effects. If an estimate of the standard deviation of replicate observations is available from prior work, then the standard error can be calculated and used in the analysis. If not, then a Normal probability plot of the effects is often useful in identifying the important effects. This approach is illustrated here in the context of a five-factor study.

A 2^5 Study. Table 5.13 shows data simulated to illustrate the use of Normal plots of effects. The five factors are labelled A–E. The design matrix of Table 5.13 can be expanded by column multiplication to generate 31 columns in total. For a five-factor design there are 5 main effects, 10 two-factor interactions (2-fis), 10 3-fis, 5 4-fis and a single five-factor interaction. The four-factor interaction is defined by analogy with the three-factor case: a four-factor interaction exists if a three-factor interaction (say ABC) is different at the two levels of the fourth factor (D); the four-factor interaction is defined as half this difference. The five-factor interaction is defined similarly. Note the way the design was generated: the signs for A alternate, those for B alternate in pairs, for C in sets of four, for D in sets of eight, while E has sixteen – and sixteen + signs. To generate designs for larger numbers of factors, this pattern is continued in an obvious way.

The effects shown in Table 5.14 were estimated by subtracting the averages of results corresponding to the minus signs from the averages corresponding to the plus signs, as before. If all the effects are 'null', *i.e.*, if varying the five study parameters has no systematic

Table 5.13 *A design and results for a five-factor study*

A	*B*	*C*	*D*	*E*	*Results*
−	−	−	−	−	54.1
+	−	−	−	−	58.6
−	+	−	−	−	47.1
+	+	−	−	−	57.0
−	−	+	−	−	41.2
+	−	+	−	−	45.3
−	+	+	−	−	50.3
+	+	+	−	−	60.2
−	−	−	+	−	53.2
+	−	−	+	−	61.6
−	+	−	+	−	45.5
+	+	−	+	−	52.7
−	−	+	+	−	39.5
+	−	+	+	−	51.1
−	+	+	+	−	46.0
+	+	+	+	−	55.5
−	−	−	−	+	50.7
+	−	−	−	+	59.5
−	+	−	−	+	49.1
+	+	−	−	+	55.5
−	−	+	−	+	45.0
+	−	+	−	+	48.5
−	+	+	−	+	48.9
+	+	+	−	+	56.6
−	−	−	+	+	55.5
+	−	−	+	+	57.5
−	+	−	+	+	51.4
+	+	−	+	+	56.5
−	−	+	+	+	42.0
+	−	+	+	+	48.7
−	+	+	+	+	45.4
+	+	+	+	+	53.9

influence on the response of the system, then the effect estimates will be expected to vary at random around zero. All effects have the same standard error:

$$SE(effect) = \sqrt{\frac{4\sigma^2}{N}}$$

where N is the total number of observations in the study and σ is the standard deviation of replicate observations at any design point. Accordingly, when the estimated effects are plotted on a Normal probability plot they should behave like a random sample from a single Normal distribution, *i.e.*, they should lie along a straight line, apart from

Table 5.14 *The 31 effects as estimated from the study results*

A	7.113	ABC	0.300
B	1.225	ABD	–0.713
C	–5.463	ABE	–0.075
D	–0.725	ACD	1.125
E	0.362	ACE	–0.063
AB	0.913	ADE	–0.775
AC	0.575	BCD	–0.563
AD	0.263	BCE	–1.800
AE	–1.025	BDE	1.138
BC	5.713	CDE	–0.875
BD	–1.500	ABCD	–0.575
BE	0.013	ABCE	0.362
CD	–1.013	ABDE	1.100
CE	–0.375	ACDE	0.388
DE	0.362	BCDE	0.075
		ABCDE	–0.412

chance variation. Large deviations from a line suggest non-null effects, *i.e.*, ones that indicate systematic changes in the system responses.

Figures 5.17 and 5.18 are two Normal probability plots of the effects from the study. Figure 5.17 includes all the effects of Table 5.14 and shows three effects that stand out clearly from the others: A, C and BC. These were deleted and the plot was redrawn as Figure 5.18, which shows

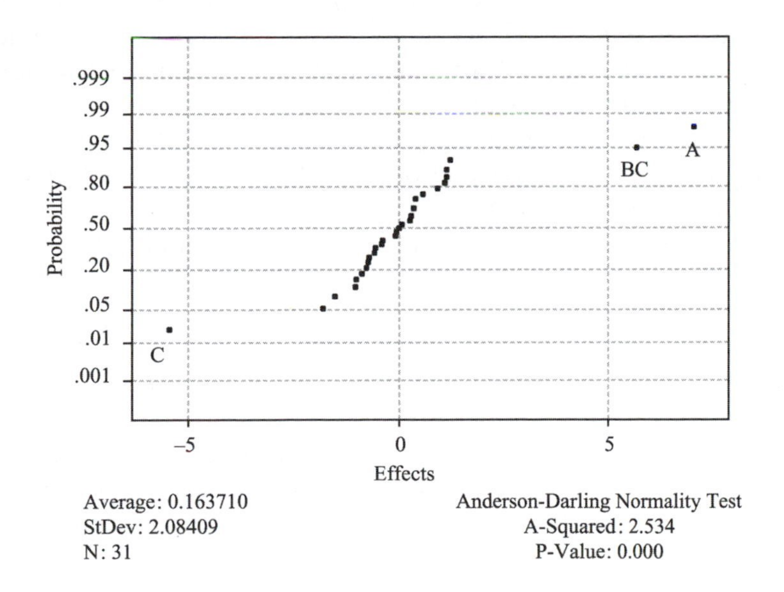

Figure 5.17 *A Normal probability plot of the effects*

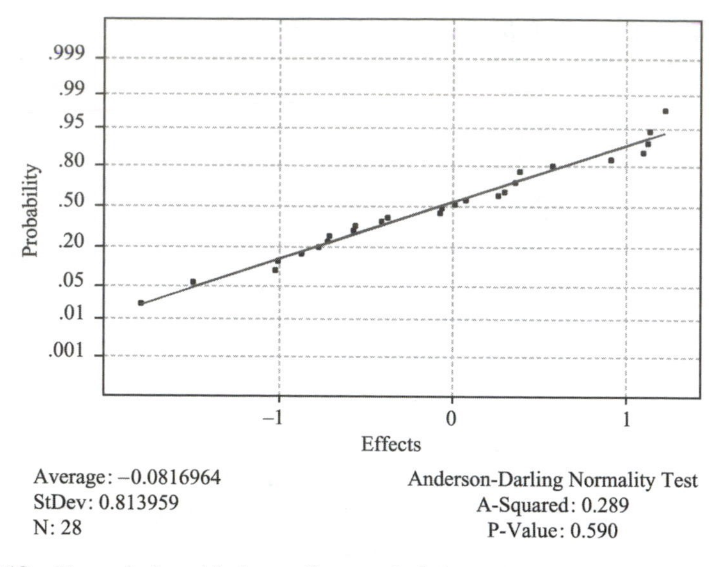

Figure 5.18 *Normal plot with three effects excluded*

no exceptional variation. Thus, although the study allowed for the estimation of 31 effects, 28 do not appear to have any influence on the response. Three effects A, C and BC do appear influential and their influences need further exploration.

Analysis of Effects. Had this been a real study, instead of a simulation, our results would mean that the analyst would be free to set parameters D and E at whichever level was more convenient or more economical, since the two main effects are not statistically significant and neither parameter is involved in a significant interaction. Factor A is not involved in a significant interaction, but its main effect is statistically significant. Hence, comparison of the mean results at its high and low levels should indicate the better setting. Factors B and C interact and must be studied together. Inspecting columns B and C of Table 5.13, ignoring the other three columns, shows that there are eight replicate results corresponding to each of the four corners of the BC design square. Collapsing onto the square allows interpretation of the nature of the BC interaction, exactly as before.*

The Normal probability plot has allowed us to identify the important effects in this study. In general, Normal plots will identify large effects, but they may fail to identify smaller ones. In cases like the current one,

*This analysis is not carried through here, since the data are simulated and not real study results. However, the reader might like to do the analysis as an exercise.

where two factors and all their interactions were considered null, a second approach is possible. Having identified D and E as null, we could have re-analyzed the data as a 2^3 design in the other three factors – this design would have four replicates at each design point. Sometimes such an approach will identify smaller effects as being statistically significant. It also allows residual analysis to be carried out.

5.7 FRACTIONAL FACTORIAL DESIGNS

Analysts will often want to study many variables simultaneously. The classic example is robustness or ruggedness testing, where it is desired to demonstrate that small variations in many system parameters do not lead to substantial changes in the analytical results. While full factorial designs may be extended to include any number of factors, it follows that the number of design points grows very rapidly, resulting in unrealistically large study sizes. For example, a full factorial design for seven factors, each at two levels, involves $2^7 = 128$ design points. In this and the next section alternative designs for ruggedness tests will be introduced. These same designs are suitable in any screening situation, *e.g.*, at the early stages of method development, where many parameter settings have to be selected, and in trouble-shooting, where one or more of a large number of candidate factors may be responsible for the problem being investigated.

The designs are known as 'fractional factorial designs', as they involve using a fraction of the full set of factorial design points. The principle behind the construction of these designs will be introduced first, using an extremely simple example (which is probably too small for practical use). Following this, more practically useful designs will be discussed.

5.7.1 A Simple Example

Suppose an experiment has been carried out, involving three factors A, B, C. This produces four observations, one corresponding to each of the four design points shown in Table 5.15. We will discuss later how this design matrix was constructed, but first we will examine the implications of using such a design.

As for full factorial designs, calculating the three main effects involves a comparison between the average results at the high and low levels of the factors. This gives Expressions (5.11) – (5.13):

$$\mathrm{A}: \frac{y_2 + y_4}{2} - \frac{y_1 + y_3}{2} \tag{5.11}$$

Table 5.15 *The design matrix for the study*

Design point	*A*	*B*	*C*	*Result*
1	–	–	+	y_1
2	+	–	–	y_2
3	–	+	–	y_3
4	+	+	+	y_4

$$\mathrm{B}: \frac{y_3 + y_4}{2} - \frac{y_1 + y_2}{2} \tag{5.12}$$

$$\mathrm{C}: \frac{y_1 + y_4}{2} - \frac{y_2 + y_3}{2} \tag{5.13}$$

To calculate interactions the array is expanded as shown in Table 5.16, where, as before, the interaction columns are the product of the corresponding main effects columns.

The two-factor AB interaction may now be calculated as:

$$\mathrm{AB}: \frac{y_1 + y_4}{2} - \frac{y_2 + y_3}{2} \tag{5.14}$$

To verify that this does, in fact, measure the interaction, note that (5.14) may be written as:

$$\tfrac{1}{2}[(y_4 - y_3) - (y_2 - y_1)]$$

where $(y_4 - y_3)$ is the effect of A at high B, and $(y_2 - y_1)$ is the effect of A at low B (see the first two columns of Table 5.16). This corresponds to the definition of an interaction. *i.e.*, half the difference between the effects of A at high and low B. The other two interactions are calculated in a similar way:

$$\mathrm{AC}: \frac{y_3 + y_4}{2} - \frac{y_1 + y_2}{2} \tag{5.15}$$

$$\mathrm{BC}: \frac{y_2 + y_4}{2} - \frac{y_1 + y_3}{2} \tag{5.16}$$

Table 5.16 *The expanded design matrix*

Design point	*A*	*B*	*C*	*AB*	*AC*	*BC*	*ABC*	*Result*
1	–	–	+	+	–	–	+	y_1
2	+	–	–	–	–	+	+	y_2
3	–	+	–	–	+	–	+	y_3
4	+	+	+	+	+	+	+	y_4

Comparing these Expressions with (5.11)–(5.13), or equivalently comparison of the corresponding columns of signs in Table 5.16, shows, for example, that the expression:

$$\frac{y_2 + y_4}{2} - \frac{y_1 + y_3}{2} \qquad (5.11) = (5.16)$$

measures both A and BC, *i.e.*, measuring the main effect of A automatically involves measuring the BC interaction, also. The two effects are said to be 'confounded' or 'aliased', *i.e* they are irretrievably mixed up with each other and cannot be estimated separately. In fact, it can be shown that expression (5.11) measures the sum of the two effects, *i.e.*, A + BC; the other expressions correspondingly measure B + AC and C + AB. Note that the ABC column contains only + signs. This means we cannot estimate the ABC interaction; we will return to this aspect of the design in the next section.

In a situation where there is good reason to believe that none of the two-factor interactions exist, *i.e.*, that the three factors act independently of each other, then the design shown in Table 5.15 allows the three main effects to be estimated, while carrying out only four runs. This is very efficient. The principle involved here can be extended to larger designs and, where higher-order interactions can be assumed not to exist, main effects and lower-order interactions can then be measured without investigating all the combinations of factor levels required for full factorial designs. This can often result in a very substantial reduction in the amount of experimental work required.

A comment on terminology is appropriate here. In full factorial designs the columns of signs do not coincide with other columns and we are able to estimate the effects unambiguously. In fractional factorial designs, where two columns coincide (*e.g.*, C and AB) we will describe the result of subtracting the average of the test results for negative signs from the average for positive signs as a 'contrast'. While all effects are contrasts (they involve a comparison between the two halves of the dataset), in fractional factorial designs the contrasts estimate a mixture of effects.

Table 5.17 shows a full 2^3 design, which allows the separate measurement of all three main effects, all three two-factor interactions and the three-factor interaction. Comparison of Tables 5.15 and 5.17 shows that the design points of Table 5.15 correspond to half of the design points in Table 5.17: the four design points of Table 5.15 corresponding to runs 5, 2, 3, 8, respectively, of Table 5.17. The design shown in Table 5.15 is said to be a half-fraction of the full factorial design of Table 5.17. It is usually labelled 2^{3-1}, meaning that there are three factors, each at two levels, but only $2^2 = 4$ runs.

Table 5.17 *The full 2^3 design matrix*

Design point	*A*	*B*	*C*	*Result*
1	−	−	−	y_1
2	+	−	−	y_2
3	−	+	−	y_3
4	+	+	−	y_4
5	−	−	+	y_5
6	+	−	+	y_6
7	−	+	+	y_7
8	+	+	+	y_8

Constructing the Fractional Factorial Design. Note that the sign pattern in column C of Table 5.15 is the product of those of columns A and B. This is how Table 5.15 was constructed. Starting from a 2^2 design for factors A and B, the column of signs for C was assigned to be the same as that of the AB interaction. Each row of Table 5.15 then specified the levels of A, B and C to be used in a single experimental run of the system being studied. Thus, the first run involved A and B being set at their low levels, while C was set at its high level.

If a bold capital letter is taken to represent a column of signs, then the assignment described above may be written as:

$$\mathbf{C} = \mathbf{AB} \tag{5.17}$$

where **AB** represents column **A** multiplied by column **B**. This assignment is called the 'design generator' for obvious reasons.

The Confounding Patterns. The confounding patterns can be found, as was done earlier, by expanding the design matrix and then noting which columns of signs are identical to each other. This is easy in the current example, but would become tedious in larger studies; for example, a design involving seven factors would have a design matrix which could be expanded to 127 columns. Fortunately, there are simple rules, based on the design generator(s), which can be used to work out the confounding patterns for any fractional factorial design.

The first step is to obtain the 'defining relation' from the design generator(s). If both sides of (5.17) are multiplied by C, then the defining relation of the design is obtained:

$$\mathbf{C} \times \mathbf{C} = \mathbf{C} \times \mathbf{AB}$$

which gives

$$\mathbf{I} = \mathbf{ABC} \tag{5.18}$$

where **I** is a column of + signs, the result of multiplying **C** by itself. Now, if we wish to know with which factor(s) the AC interaction is confounded, we multiply both sides of (5.18) by **AC**:

$$\mathbf{AC} \times \mathbf{I} = \mathbf{AC} \times \mathbf{ABC}$$

$$\mathbf{AC} = \mathbf{AABCC}$$

which gives **AC** = **B**. Multiplying **AC** by a column of plus signs leaves it unchanged, so the left-hand side is **AC**. Multiplying any column of signs by itself produces a column of plus signs, thus **A** × **A** and **C** × **C** both give **I** and, hence, the right-hand side is just **B**. The other confounding patterns can be found similarly.

Exercise

5.6 Use the method just described to verify the confounding patterns A = BC and C = AB for the design of Table 5.15.

The Second Half-fraction. The remaining four design points in Table 5.17, corresponding to rows 1, 4, 6, 7, make up the second half-fraction. These may be obtained from Table 5.15 by multiplying the signs of column **A** by those of column **B** to get **AB**, and then reversing the signs of the **AB** column (*i.e.*, by multiplying them all by – 1) and assigning this new column to factor C. This may be written as: **C** = – **AB**. The defining relation is then **I** = – **ABC**. If a study is run using this half-fraction, then subtracting the averages of the test results corresponding to the – signs from those corresponding to the + signs in the three columns of this new design matrix will produce estimates of A – BC, B – AC and C – AB. Again, if the two-factor interactions are negligible, the design allows the three main effects to be estimated using only four runs.

This very simple example shows the underlying principle of fractional factorial designs: many factors may be investigated in few runs, but at the cost of having main effects or lower order interactions confounded (*i.e.*, mixed up) with higher order interactions. If all interactions are negligible then the designs provide a very efficient means of measuring the main effects. If the interactions are non-negligible then further experimental runs may allow them to be disentangled from the main effects. This is discussed further below.

Exercise

5.7 Write down the table of signs that results from the design generator **C** = –**AB** and verify that its rows correspond to the remaining four runs of Table 5.17.

5.7.2 The 2^{5-1} Design

While the 2^{3-1} design is useful as a basis for introducing the ideas underlying fractional factorial designs, it is too small for most practical purposes. A half-fraction of a 2^5 study, on the other hand, can be a very useful design. Table 5.18 shows a sixteen-run design matrix, which is, in fact, a half-fraction of the 32 run 2^5 design* of Table 5.13. The corresponding results are also shown.

Table 5.18 *The 2^{5-1} design and study results*

A	*B*	*C*	*D*	*E*	*Response*
−	−	−	−	+	50.7
+	−	−	−	−	58.6
−	+	−	−	−	47.1
+	+	−	−	+	55.5
−	−	+	−	−	41.2
+	−	+	−	+	48.5
−	+	+	−	+	48.9
+	+	+	−	−	60.2
−	−	−	+	−	53.2
+	−	−	+	+	57.5
−	+	−	+	+	51.4
+	+	−	+	−	52.7
−	−	+	+	+	42.0
+	−	+	+	−	51.1
−	+	+	+	−	46.0
+	+	+	+	+	53.9

Inspection of the columns of signs shows that columns A, B, C and D correspond to a full factorial design in four factors – there are $2^4 = 16$ design points. The fifth column is just the product of the first four. Thus, this half-fraction was 'generated' by setting **E** = **ABCD**, which gives the defining relation **I** = **ABCDE**. The confounding patterns are easily obtained from this defining relation and these are shown, together with the estimates, in Table 5.19. For example, multiplying both sides of the defining relation by **A** gives:

$$\mathbf{A} \times \mathbf{I} = \mathbf{A} \times \mathbf{ABCDE}$$

which gives **A** = **BCDE**. Thus, the contrast corresponding to the first column estimates A + BCDE.

*The sixteen runs of the half-fraction correspond to runs 17, 2, 3, 20, 5, 22, 23, 8, 9, 26, 27, 12, 29, 14, 15 and 32 of the full factorial design.

Table 5.19 *Summary of results for the 2^{5-1} study*

Effects	*Estimates*
A + BCDE	7.1875
B + ACDE	1.6125
C + ABDE	– 4.3625
D + ABCE	– 0.3625
E + ABCD	– 0.2125
AB + CDE	0.0375
AC + BDE	1.7125
AD + BCE	– 1.5375
AE + BCD	– 1.5875
BC + ADE	4.9375
BD + ACE	– 1.5625
BE + ACD	1.1375
CD + ABE	– 1.0875
CE + ABD	– 1.0875
DE + ABC	0.6625

If, as discussed in Section 5.6, varying the five parameters has no systematic influence on the responses, then the contrasts should behave like a random sample from a single Normal distribution and give an approximately straight line on a Normal probability plot. Figure 5.19 is a Normal probability plot of the 15 contrasts. It shows two contrasts as markedly different from the others, *i.e.*, they do not appear to fall on the

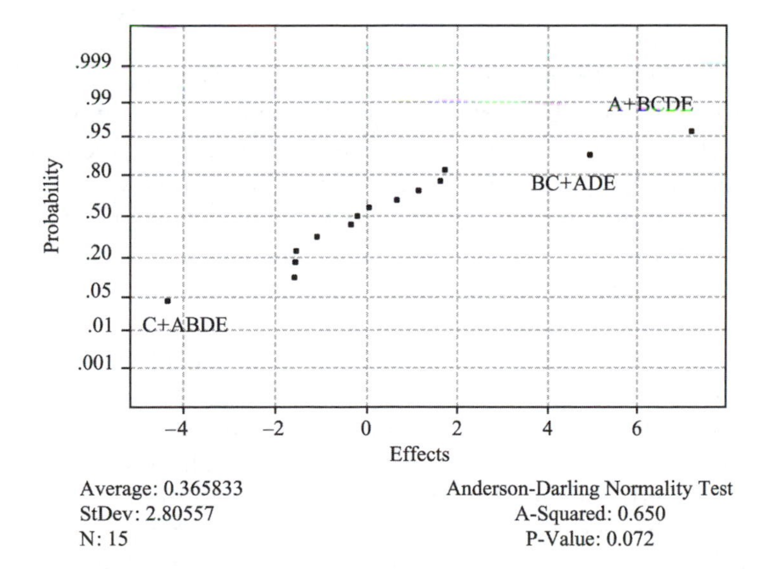

Figure 5.19 *Normal plot of the 2^{5-1} contrasts*

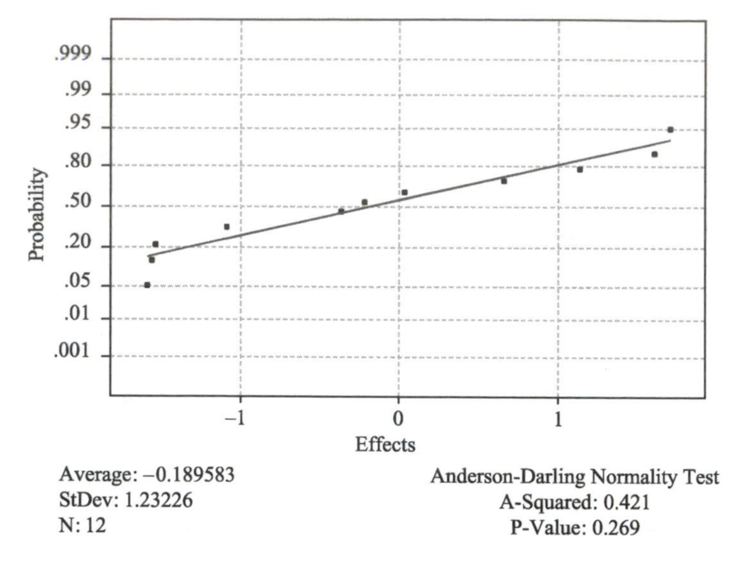

Figure 5.20 *Normal plot after exclusion of three contrasts*

same straight line. A third contrast (C + ABDE) is possibly different also, though it does not deviate markedly from the line defined by the main group of twelve contrasts. However, since factor C is contained in the BC interaction, this ambiguity need not concern us. When the three large ones are excluded, Figure 5.20 shows the remaining contrasts forming an approximately straight line – they are consistent with the corresponding effects being null.

When the full 32 run dataset was analyzed in Section 5.6, we found that factors C, A and BC were statistically significant. Here, these same three effects stand out from the rest, but this time the three effects are confounded with higher order interactions. If we can assume that three and four-factor interactions are unlikely to exist, then we have obtained the same information as before, while running only half of the full set of design points – a considerable saving.

The Second Half-fraction. Table 5.20 shows the other half of the full factorial design of Section 5.6. This was generated by setting **E** = **–ABCD**, *i.e.*, the first four columns were multiplied, as before, to produce the fifth column, but this time the signs of all the elements of the newly created fifth column, **E**, were changed. The confounding is the same as in Table 5.19 except that the plus signs are replaced by minus signs. For example, the first column now estimates A – BCDE instead of A + BCDE. The Normal probability plot of the calculated contrasts is shown in

Table 5.20 *The second half-fraction of the 2^{5-1} study*

A	*B*	*C*	*D*	*E*	*Results*
−	−	−	−	−	54.1
+	−	−	−	+	59.5
−	+	−	−	+	49.1
+	+	−	−	−	57.0
−	−	+	−	+	45.0
+	−	+	−	−	45.3
−	+	+	−	−	50.3
+	+	+	−	+	56.6
−	−	−	+	+	55.5
+	−	−	+	−	61.6
−	+	−	+	−	45.5
+	+	−	+	+	56.5
−	−	+	+	−	39.5
+	−	+	+	+	48.7
−	+	+	+	+	45.4
+	+	+	+	−	55.5

Figure 5.21. The pattern is similar to that in Figure 5.19 – three contrasts stand out. These are associated with the C, A and BC effects, if the three and four-factor interactions can be assumed null. This shows that the same information would have been obtained from both half-fractions.

Combining Half-fractions. An experimenter might decide to run the second half-fraction having discovered interesting effects in the first half.

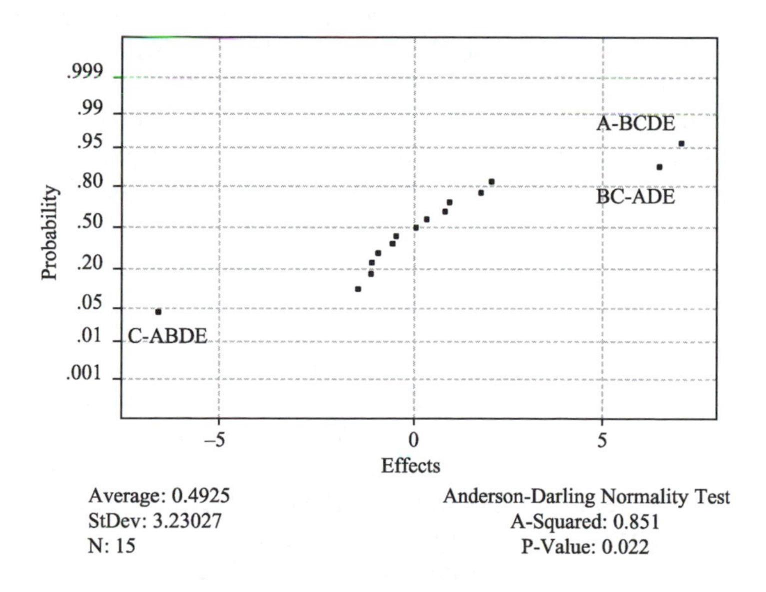

Figure 5.21 *Normal probability plot for the second half-fraction of the 2^{5-1} study*

This might be done as a confirmatory study, or simply to check that it is, in fact, BC rather than ADE which is the significant effect. If this is done, then the effect estimates can be calculated directly from the full 2^5 dataset, as before, or estimates can be calculated from the second half-fraction separately and the results then combined with those from the first half-fraction. An example of such a calculation is discussed in the next section – see also the exercise which follows that section and which refers back to this 2^{5-1} study. The results will be the same irrespective of which method of calculation is used.

5.7.3 Blocking

Under some circumstances, even where it is intended that the full design will be deployed, it is advantageous to run the study in two half-fractions. Suppose that only half of the experimental runs can be made initially and that the second half will have to wait for a lull in the laboratory workload. If a randomly selected set of the design points is run first and if, over the time gap, there is a shift in the performance of the system under study, then it will not be clear what implications this shift in results will have for the calculated effects. If, on the other hand, our first half-fraction had been run first and the second half-fraction run after the time gap, the shift in system response would be confounded with the ABCDE interaction, and all the other effects would remain uncontaminated by the shift.

To see why this is the case, consider what happens when the ABCDE column of signs in the combined 2^5 design is formed from the two half-fractions. Since the defining relation is **I** = **ABCDE** for the first half-fraction and **I** = **–ABCDE** for the second, the ABCDE column in the full design has sixteen + signs followed by sixteen – signs. Hence, the ABCDE contrast measures the shift in results with time, the 'time' effect, as well as the ABCDE interaction. If, as would usually be the case, the five-factor interaction is negligible, then this column measures the time shift in results, only.

What is more important, though, is that no other factor will be affected by the time-shift. This is the case because for every other column of the combined 32-run design, there will be 8 + signs and 8 – signs before the time shift, *i.e.*, in the first half-fraction, and similarly in the second half-fraction. This means that the + and – levels of each factor are equally represented before and after the time shift. Hence, the change in system response will be averaged out in the calculations. This holds for all columns of the expanded design matrix, other than the five-factor interaction. In a busy laboratory, where producing routine results will very often take precedence over special projects, the possibility of being able to split an experimental study in this way can be valuable.

Deliberately running the experiment in two halves, as just described, is known as 'blocking' the design. Whenever a study can be influenced by shifts in results due to factors not intrinsic to the objectives of the study, such as use of different batches of reagents or product, different machines, analysts, laboratories, times, *etc*., the possibility of improving the quality of the results by blocking the study should be considered. Blocking essentially means grouping the experimental runs in such a way that the study conditions are more homogeneous within the 'blocks'. In the present case the blocks are the two time periods in which the two half-fractions were run. Blocking is discussed again in Chapter 7.

Exercise

5.8 Sandstrom *et al.*[3] investigated the factors that influence the performance of a formic acid biosensor, which is used for air monitoring. Many system parameters and several response variables were studied and a more sophisticated statistical analysis than is considered here was carried out. The data used in this exercise were extracted from their paper; for full details see the paper. For the factors listed in Table 5.21, the first level is coded –1 and the second level as +1:

A: Concentration of Medola's blue (mg g^{-1}) 0.3, 1.5
B: Concentration of the enzyme formate dehydrogenase (mg g^{-1}) 0.36, 0.792
C: Concentration of nicotinamide adenine dinucleotide (mg g^{-1}) 1.2, 4.8
D: Concentration of the phosphate buffer (M) 0.048, 0.144
E: Amount of solution placed on the electrode (μL) 8, 12
F: Electrochemical potential (V) 0, 0.2

The response C1 reported here is the current (nA) at steady state when the biosensor was exposed to a formic acid vapour concentration of 1.4 mg m^{-3}. Large responses are desirable.

Verify that the design generator was **F** = **ABCDE** and by examining the effects that are confounded with the contrasts A, AB, ABC, show that main effects are confounded with 5-factor interactions (5-fis), 2-fis are confounded with 4-fis and that 3-fis are confounded with other 3-fis.

Expand the array to get columns corresponding to all 15 2-fis and the 10 3-fis that include A (*i.e.*, ABC, ABD...). These 3-fis will be confounded with the other 10 3-fis. Note that with 32 runs, 31 contrasts (apart from the overall average) may be obtained. By subtracting the average response at the low level for each column of the expanded array from the average at the high level, calculate the contrasts. Note that all of this will be automatically done for you by appropriate statistical software, but if this is not available to you, you could easily set up the arrays in a spreadsheet. Draw a Normal probability plot of the contrasts and interpret the plot.

Table 5.21 *Biosensor development study (nA)*
(Reprinted from the *Analyst*, 2001, Vol. 126, pp 2008. Copyright, 2001, Royal Society of Chemistry.)

A	*B*	*C*	*D*	*E*	*F*	*C1*
−1	−1	−1	−1	−1	−1	213
1	−1	−1	−1	−1	1	296
−1	1	−1	−1	−1	1	217
1	1	−1	−1	−1	−1	266
−1	−1	1	−1	−1	1	285
1	−1	1	−1	−1	−1	257
−1	1	1	−1	−1	−1	302
1	1	1	−1	−1	1	375
−1	−1	−1	1	−1	1	238
1	−1	−1	1	−1	−1	250
−1	1	−1	1	−1	−1	263
1	1	−1	1	−1	1	314
−1	−1	1	1	−1	−1	264
1	−1	1	1	−1	1	291
−1	1	1	1	−1	1	244
1	1	1	1	−1	−1	310
−1	−1	−1	−1	1	1	258
1	−1	−1	−1	1	−1	278
−1	1	−1	−1	1	−1	285
1	1	−1	−1	1	1	362
−1	−1	1	−1	1	−1	259
1	−1	1	−1	1	1	298
−1	1	1	−1	1	1	302
1	1	1	−1	1	−1	289
−1	−1	−1	1	1	−1	313
1	−1	−1	1	1	1	277
−1	1	−1	1	1	1	296
1	1	−1	1	1	−1	320
−1	−1	1	1	1	1	264
1	−1	1	1	1	−1	302
−1	1	1	1	1	−1	214
1	1	1	1	1	1	307

The study included five extra measurements which were pure replicates, *i.e.*, five replicate measurements were made at a fixed set of conditions, so that their standard deviation $s = 22.49$ measures the chance variation in the system and may be used to estimate the standard error of the effects, using $S\hat{E} = \sqrt{4s^2/32}$. Calculate the standard error and use it to identify the influential factors. What do you conclude?

5.8 RUGGEDNESS TESTING

It is a common requirement of method validation protocols that an analytical method be shown to be robust or rugged. This means that small changes to system parameters will not result in substantial changes to the system responses. Accordingly, ruggedness testing studies usually require making small changes to a large number of system settings and demonstrating that the test results obtained, when standard material is analyzed, do not vary beyond what would be expected under chance analytical variation.

Fractional factorial designs are ideally suited to such investigations. To illustrate this we will first discuss the ideas in the context of a very commonly used design, which allows investigation of up to seven system parameters simultaneously. Following this, an example in which twelve parameters of a GC system were investigated will be described.

5.8.1 Designing Ruggedness Tests

Consider the design shown in Table 5.22. This design is just the standard design matrix for a 2^3 study, expanded to include columns for the three two-factor interactions and the three-factor interaction. If it is used to investigate three factors (A, B, C) then it allows the measurement of the three main effects and the four interactions. It may also be used to study up to seven factors, by assigning the extra four factors to the interaction columns, *e.g.*, factor D to the **AB** interaction column, E to **AC**, F to **BC** and G to **ABC**, as shown in the table. Thus, the **AB** column of signs is used to define the levels of factor D that will be used in the various experimental runs. Similarly, the **AC**, **BC**, and **ABC** columns define the levels for factors E, F and G. This means that each row of Table 5.22 specifies the levels of the seven factors for one experimental run. Thus, row 1 requires that A, B, C and G are run at their low levels, while D, E and F are run at their high levels.

A full factorial design in seven factors would require $2^7 = 128$ runs. Such a design would allow the calculation of 7 main effects, 21 two-factor interactions (2-fis), 35 3-fis, 35 4-fis, 21 5-fis, 7 6-fis and a single 7-fi.

Table 5.22 *Design matrix for a ruggedness study*

Design point	*A*	*B*	*C*	*D = AB*	*E = AC*	*F = BC*	*G = ABC*
1	−	−	−	+	+	+	−
2	+	−	−	−	−	+	+
3	−	+	−	−	+	−	+
4	+	+	−	+	−	−	−
5	−	−	+	+	−	−	+
6	+	−	+	−	+	−	−
7	−	+	+	−	−	+	−
8	+	+	+	+	+	+	+

Very high order interactions rarely exist: 2-fis are common, 3-fis less so, and interactions of higher order are very much less common. The full design, therefore, is capable of estimating very many more effects than are likely to be required in practice. The design of Table 5.22 is a 1/16 th fraction of the full factorial design; it is described as a 2^{7-4}, *i.e.*, 7 factors in $2^3 = 8$ runs. Note that 16 different arrays may be generated by the different combinations of assignment: **D** = ± **AB**, **E** = ±**AC**, **F** = ±**BC** and **G** = ± **ABC**.

Analysis. The difference between the averages of the test results corresponding to the positive and negative signs in any column is a 'contrast' (l) and measures a complicated combination of main effects and interactions. Thus the contrast, l_1, corresponding to the first column of signs, measures:

$$l_1 \rightarrow \mathrm{A + BD + CE + FG + BCG + BEF + CDF + DEG + ABCF + ABEG + ACDG + ADEF + ABCDE + ABDFG + ACEFG + BCDEFG}$$

If it can be assumed that interactions higher than those between pairs of parameters do not exist, then panel A of Table 5.23 shows the combinations of effects measured by the seven contrasts.

When a design like this is used for ruggedness testing, the hope is that *none* of the effects will be large, *i.e.*, the responses at each of the eight design points will be the same, apart from chance variation. An experienced analyst may be able to confirm that this is so simply by inspecting the eight results. If the variation between the results at different design points is similar to that expected between replicates at any one design point, then the system is rugged, *i.e.*, the test results are not materially affected by small changes to the system settings. A more formal assessment may be made, by carrying out statistical tests on the seven contrasts. The null hypothesis would be that the calculated values are only randomly different from zero. To carry out these tests an

Table 5.23 *The confounding patterns for two designs*

A	B
$l_1 \rightarrow$ A + BD + CE + FG	$l'_1 \rightarrow$ A − BD − CE − FG
$l_2 \rightarrow$ B + AD + CF + EG	$l'_2 \rightarrow$ B − AD − CF − EG
$l_3 \rightarrow$ C + AE + BF + DG	$l'_3 \rightarrow$ C − AE − BF − DG
$l_4 \rightarrow$ D + AB + CG + EF	$l'_4 \rightarrow$ D − AB − CG − EF
$l_5 \rightarrow$ E + AC + BG + DF	$l'_5 \rightarrow$ E − AC − BG − DF
$l_6 \rightarrow$ F + AG + BC + DE	$l'_6 \rightarrow$ F − AG − BC − DE
$l_7 \rightarrow$ G + AF + BE + CD	$l'_7 \rightarrow$ G − AF − BE − CD

estimate of the chance variation in the system is required. This will often be available from replicate data obtained during the method development phase of the analytical system. If, for any reason, the prior data are considered unsuitable for use in estimating the standard deviation under the current operating conditions, replicate runs may be included in the ruggedness study for this purpose. When the study involves larger numbers of design points (an example is given below) then the ruggedness may be assessed using a Normal probability plot.

Sequential Studies. If one of the contrasts in panel A of Table 5.23 is large, then it will not be clear whether this is due to the relevant main effect, or to one or more of the associated 2-fis. A second experiment can help to resolve the issue. One way to design the follow-up experiment is to change the signs of every design point in Table 5.22 – this is referred to as 'folding' the design. Panel B of Table 5.23 shows the resulting confounding patterns for the second design. The results from running the second study can then be combined in a simple way with those of the first to give unambiguous information on the main effects. Thus:

$$\tfrac{1}{2}(l_1 + l_1') = \mathrm{A}$$
$$\tfrac{1}{2}(l_1 - l_1') = \mathrm{BD} + \mathrm{CE} + \mathrm{FG}$$

For an excellent example of the use of a design of this type in a sequential study, see Box, Hunter and Hunter,[4] Chapter 13.3. The attraction of the sequential approach is that if the first design shows the system to be rugged, the second study is *not* required. If the first shows that the system is *not* rugged, then the second design may be deployed and insight into the causes of the lack of ruggedness may be obtained from the accumulated data.

Exercise

5.9 Expand Table 5.20, p 227, which gives the second half-fraction of the 2^{5-1} study, and calculate all the contrasts. Write out the confounding patterns for this half-fraction. Using the method just described, combine the information from both half-fractions and verify that it gives the same results as were obtained when the full 2^5 factorial design was analyzed in Section 5.6.

5.8.2 Example 1

Weyland *et al.*[5] describe a ruggedness study of a method involving liquid chromatography with electrochemical detection that is used to measure preservatives in cosmetics. Seven system parameters were studied and three preservatives were measured. The responses were the percentages of Bronopol, Bronidox and MDBGN (methyl dibromoglutaronitrile) in the sample tested. The system parameters studied were:

A/a Amount of sample (g) 2 or 1.5
B/b Heating time in water bath during extraction (min) 5 or 10
C/c Shaking time during extraction (min) 1 or 2
D/d Composition of extraction solvent (water:methanol, v/v) 2 + 8 or 3 + 7
E/e Column oven temperature 40 °C or room temperature
F/f Pulse time of electrochemical detector (ms) 100 or 50
G/g Working potential (V) – 0.4 or – 0.5.

In each case the first level is the one prescribed in the method studied. Table 5.24 shows the design matrix and outcomes for the ruggedness study.

Table 5.24 *The design matrix and outcomes for the ruggedness study* (Reprinted from the Journal of AOAC INTERNATIONAL, 1994, Vol. 77, pp 1132. Copyright, 1994, by AOAC INTERNATIONAL.)

A	*B*	*C*	*D*	*E*	*F*	*G*	*Bronopol*	*Bronidox*	*MDBGN*
A	B	C	D	E	F	G	0.0328	0.0322	0.0319
A	B	c	D	e	f	g	0.0328	0.0297	0.0311
A	b	C	d	E	f	g	0.0313	0.0288	0.0294
A	b	c	d	e	F	G	0.0327	0.0312	0.0315
a	B	C	d	e	F	g	0.0314	0.0290	0.0300
a	B	c	d	E	f	G	0.0320	0.0284	0.0296
a	b	C	D	e	f	G	0.0311	0.0299	0.0304
a	b	c	D	E	F	g	0.0301	0.0296	0.0299

The differences between the factor levels were purposely made large to identify parameters that would affect the results. If, from a practical point of view, shifts in responses of less than 5% are regarded as acceptable (given the large changes in the parameters), should we consider the system robust?

This is a 2^{7-4} design, *i.e.*, one in which 7 factors are each measured at two levels, but involving only 8 runs. If the average results are calculated at the two levels of each factor and the average for the low level (arbitrarily defined) is subtracted from the average at the high level the result will be the contrast corresponding to each column. This is done below for the first response, Bronopol, where the capital letters have been taken as the low level for each column.

Term	Contrast
A	−0.001250
B	−0.000950
C	0.000250
D	0.000150
E	0.000450
F	0.000050
G	−0.000750

If we regard shifts in responses of less than 5% as acceptable then the system can be regarded as robust, since the average response for Bronopol is 0.032 and the largest contrast is only 3.9% of this.

Exercise

5.10 Replace the labels in Table 5.24 by ±1 and determine the design generators, i.e., if the first three columns are taken as the basic 8-run 2^3 design, how were the columns D–G constructed? The assignment of the labels is arbitrary, so it does not matter which way you do it, provided you are consistent.

Carry out the same analysis as above for the other two preservatives, Bronidox and MDBGN, and determine if the system is robust when measuring these responses.

5.8.3 Example 2

A ruggedness study[6] was carried out on a GC system routinely used to measure the aromatic compounds that are responsible for the taste of whiskey. Twelve system parameters were varied, each over two levels. The parameters and their settings are listed in Table 5.25.

Table 5.25 *The ruggedness study settings*

Label	*Parameter*	*'Low'*	*'High'*
A	Injection temperature	200 °C	220 °C
B	Sample washes	3	6
C	Sample pumps	3	6
D	Pre-injection solvent	3	6
E	Split ratio	1:25	1:30
F	Flame composition	10%	8%
G	Makeup flow	45 mL min^{-1}	50 mL min^{-1}
H	Post-injection solvent	3	6
J	Column flow	1.7	1.8
K	Initial temperature	40 °C	45 °C
L	Temperature rate	4 °C min^{-1}	5°C min^{-1}
M	Detector temperature	200 °C	250 °C

A brief explanation of the parameter list is given below.

- A is the temperature of the injection port.
- B is the number of times the injection syringe is rinsed with the sample before injection.
- C is the number of times the syringe plunger moves up and down in the sample to expel air bubbles from the syringe.
- D is the number of times the syringe is rinsed with a solvent before a sample is drawn.
- E is the split ratio, which is the ratio of the flow of the carrier gas onto the column and that which is 'purged', *i.e.*, expelled from the injection port and discarded.
- F is the ratio of hydrogen to air in the Flame Ionization Detector (FID).
- G refers to the 'makeup' gas, which is used to speed up the passage of the samples through the detector, to maintain the separation achieved on the column.
- H is the number of times the syringe is rinsed with solvent after a sample is drawn.
- J refers to the flow of gas through the column.
- K is the initial temperature of the oven in which the column resides. Low temperatures are suitable for the separation of volatile compounds, higher temperatures are required for compounds with higher molecular weight. Whiskey contains both types of component.
- L is the rate at which the oven temperature is increased after an initial period at the initial temperature.
- M refers to the FID detector temperature.

A full factorial study of twelve parameters, each at two levels, would be a formidable undertaking: $2^{12} = 4098$. Fortunately such a study is not required, as a fractional factorial design will produce sufficient information to answer the practical questions concerning ruggedness.

The Fractional Factorial Design. A full factorial design in four factors requires $2^4 = 16$ runs. The four column standard design matrix (see first four columns of Table 5.26) may be expanded by column multiplication to fifteen columns, allowing all the possible main effects and interactions to be measured. Alternatively, a fractional factorial design, which can be used to investigate more than four factors, can be generated by assigning some or all of the interaction columns to extra factors. Here, not all fifteen possible columns are required for the study and different assignments of factors to interaction columns are possible.

Table 5.26 *A fractional design for the twelve factor ruggedness study*

A	B	C	D	$E = ABC$	$F = ABD$	$G = ACD$	$H = BCD$	$J = ABCD$	$K = AB$	$L = AC$	$M = AD$	*Response ratio for propanol*
−	−	−	−	−	−	−	−	+	+	+	+	0.778
+	−	−	−	+	+	+	−	−	−	−	−	0.789
−	+	−	−	+	+	−	+	−	−	+	+	0.816
+	+	−	−	−	−	+	+	+	+	−	−	0.813
−	−	+	−	+	−	+	+	−	+	−	+	0.800
+	−	+	−	−	+	−	+	+	−	+	−	0.784
−	+	+	−	−	+	+	−	+	−	−	+	0.759
+	+	+	−	+	−	−	−	−	+	+	−	0.777
−	−	−	+	−	+	+	+	−	+	+	−	0.773
+	−	−	+	+	−	−	+	+	−	−	+	0.797
−	+	−	+	+	−	+	−	+	−	+	−	0.766
+	+	−	+	−	+	−	−	−	+	−	+	0.786
−	−	+	+	+	+	−	−	+	+	−	−	0.765
+	−	+	+	−	−	+	−	−	−	+	+	0.775
−	+	+	+	−	−	−	+	−	−	−	−	0.771
+	+	+	+	+	+	+	+	+	+	+	+	0.785

Various statistical packages contain allocation rules that make the assignment in an efficient way; the design used in this case was generated by Minitab. The eight extra system parameters were assigned to the columns of Table 5.26 using the 'design generators' shown as the headings of the table. Thus, factor E was assigned to have the pattern of signs used to estimate the ABC interaction. Each row of the table determines one experimental run of the GC system. Thus, the first row requires the system to be run with factors A–H at their low levels, while J–M are at their high levels.

When this assignment of parameters to columns of signs is used, then, ignoring possible 3-fis and higher order interactions, the contrasts associated with the columns of Table 5.26 measure the combinations of effects shown in Table 5.27. Table 5.26 shows only twelve of the possible 15 columns that can be generated by multiplication. The last three rows of Table 5.27 correspond to the other three columns – these measure three different combinations of two-factor interactions, again on the assumption that higher-order interactions are negligible.

Many components of the whiskey were measured in the study. The last column of Table 5.26 shows the results for one compound – the response ratios for propanol. An experienced analyst, who was familiar with this particular analytical system, would be in a position to judge whether the differences between the results at the sixteen design points of Table 5.26 are within the range that would be expected from chance variation only. If they are, then the system may be considered rugged. In the absence of such experience, a statistical analysis can help us interpret the data.

Table 5.27 *The contrasts measured by the design of Table 5.26*

Estimated effect combinations	*Contrasts*
A + BK + CL + DM + HJ	0.00973
B + AK + EL + FM + GJ	0.00154
C + AL + EK + FJ + GM	– 0.01267
D + AM + EJ + FK + GL	– 0.01233
E + BL + CK + DJ + HM	0.00697
F + BM + CJ + DK + HL	– 0.00263
G + BJ + CM + DL + HK	– 0.00165
H + AJ + EM + FL + GK	0.01809
J + AH + BG + CF + DE	– 0.00499
K + AB + CE + DF + GH	0.00207
L + AC + BE + DG + FH	– 0.00340
M + AD + BF + CG + EH	0.00706
AE + BC + DH + FG + JM + KL	– 0.00962
AF + BD + CH + EG + JL + KM	– 0.00207
AG + BH + CD + EF + JK + LM	0.00616

Table 5.27 shows the calculated contrasts associated with each column of the design matrix – these are simply the differences between the averages of the eight results from the design points with a '–' label for the relevant columns and the averages of the results corresponding to the eight '+' labels. They are all small, suggesting that the parameter changes have had little effect on the system responses. As discussed in Sections 5.6 and 5.7, if the calculated contrasts measure only null effects, they should fall on an approximately straight line when plotted on a Normal probability plot.

Figure 5.22 suggests that the fifteen contrasts may be considered to be a sample from a Normal distribution with mean zero – there is no evidence of any of the parameter changes having had a significant effect on the system response. Accordingly, the GC system may be considered robust with respect to changes of the magnitudes indicated in Table 5.25 for the twelve system parameters included in this study.

What has just been described is a fairly typical outcome to a robustness study – the system was not sensitive to small changes in the parameters studied. This is what we would expect of a method which has undergone a carefully executed development phase. If the method shows itself as not being rugged, it means either that the developers have not made extensive studies of its performance, or that they have been unlucky in there being an interaction between system parameter settings, which did not become apparent at the development stage.

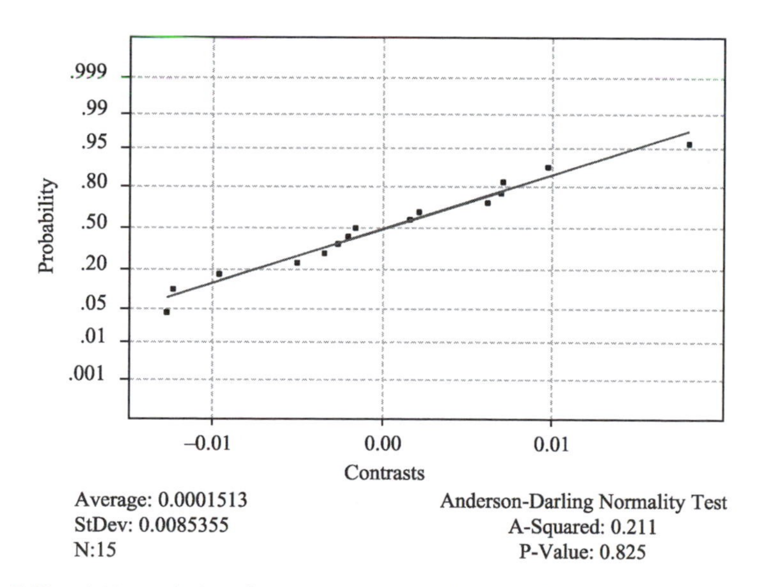

Figure 5.22 *A Normal plot of the contrasts from the ruggedness study*

Table 5.28 *Capsule potency ruggedness study*

Acetonitrile	*Water*	*GAA*	*Wavelength*	*Column*	*Flow rate*	*Results*
55	54	1.05	259	2	1.3	492.47
45	44	0.95	259	2	1.3	491.85
45	54	1.05	249	1	1.3	492.56
55	44	0.95	249	1	1.3	493.86
45	54	0.95	249	2	1.1	494.30
45	44	1.05	259	1	1.1	490.00
55	54	0.95	259	1	1.1	484.77
55	44	1.05	249	2	1.1	490.69

Exercise

5.11 Table 5.28 shows data generated in a robustness study of a HPLC chromatographic method for measuring the potency of 500 mg capsules. The responses are reported in units of mg per capsule. Six system parameters were varied, three of the six related to the composition of the mobile phase; the factors, the two levels for each factor and their coding are shown below.

Coding	**–1**	**+1**
Mobile phase composition:		
Acetonitrile/mL	45	55
Water/mL	44	54
Glacial acetic acid (GAA)/mL	0.95	1.05
Detection wavelength/nm	249	259
Column supplier	current	alternative
Flow rate/mL min^{-1}	1.1	1.3

The 2^{6-3} eight-run design was generated in Minitab. Calculate the contrasts as the differences between the two factor level averages for each column. Express the contrasts as a percentage of the nominal dose (500 mg per capsule). Do you think that the size of the largest contrast is likely to be of any practical importance?

Minitab gives the following alias patterns (up to order 2) for the contrasts. Confirm by multiplication of the columns of signs that the alias pattern for the first of these is correct, *i.e.*, that the pattern of signs corresponding to the column

Acetonitrile + Water*Wavelength + GAA*Column
Water + Acetonitrile*Wavelength + GAA*Flow
GAA + Acetonitrile*Column + Water*Flow
Wavelength + Acetonitrile*Water + Column*Flow
Column + Acetonitrile*GAA + Wavelength*Flow
Flow + Water*GAA + Wavelength*Column
Acetonitrile*Flow + Water*Column + GAA*Wavelength

associated with acetonitrile is the same as the product of the columns for water and wavelength and for the product of GAA and column.

5.9 CONCLUDING REMARKS

This chapter has focused on two-level factorial experimental designs as simple but powerful tools for investigating complex systems. Several aspects of the designs are worth re-emphasizing. They are highly efficient in their use of the experimental data and they are designed to detect interactions when present. These properties were discussed in Section 5.2, where factorial designs were contrasted with the traditional approach of optimizing the settings of one system parameter at a time. Full factorial designs, while capable of being used to investigate any number of experimental factors, become unrealistic in their requirements for very large numbers of observations, as the number of factors being studied increases. Fractional factorial designs are more useful in such situations. These designs involve carefully selected subsets of the full set of factorial design points. They have been deployed successfully in method development, trouble-shooting and, particularly, in ruggedness testing, which is a common requirement of method validation protocols.

The full factorial designs discussed in this chapter can be extended to include any number of factor levels. Thus, instead of a 2^3 design to investigate three factors at each of two levels, there is no reason why an $m \times n \times p$ design might not be used. This would involve all combinations of m levels of the first factor, n levels of the second factor and p levels of the third factor. The only difference is that the very simple and intuitive effects analysis of the 2^3 design must be replaced by what is known as an 'Analysis of Variance' (ANOVA). This is discussed in Chapter 7.

The two-level designs discussed in this chapter are especially useful in exploratory work, *i.e.*, for identification of important variables and their interactions and for ruggedness studies. Where the important variables influencing the response of a system are already known and what is desired is to characterize *in detail* their influence on the system response, a different approach is required. Typically this will involve collecting data at several *quantitative* levels of each parameter (*e.g.*, increasing levels of temperature or pressure) and building a predictive equation, which will allow a quantitative description of how the system responds to specified changes in the parameter settings. This is known as Response Surface Modelling (RSM). It is based on the technique of Regression Analysis, which is introduced in Chapter 6. A brief introductory account of the ideas underlying RSM will also be given in Chapter 6.

5.10 REVIEW EXERCISES

5.12 The data shown in Table 5.29 come from a robustness/ruggedness study of a four-hour dissolution test of a tablet product. The numbers represent the percentages of the active ingredient released at the sampling time points. The test involves dissolving a single tablet in each of the six vessels of a USP Type 1 dissolution apparatus and taking samples at the four time points shown in the table headings. The test portions are measured on a UV spectrophotometer and the six values are averaged to give the final test result. The factors studied and their coding were:

- pH of the buffer (–1 = 7.15, 1 = 7.25).
- Temperature of the bath (–1 = 36.5 °C, 1 = 37.5 °C).
- Sampling time – the method SOP requires that samples should be taken within 2% of the specified sampling time (–1 = 2%, 1 = 4%).
- Sample weighing – either the nominal or actual tablet weight may be used in calculating the final results. Weighing would mean that tablets could be out of their containers for several minutes before being put into the dissolution bath, with the possibility of moisture being absorbed (–1 = do not weigh, 1 = weigh).
- Measurement delay. For the earlier sampling times, the test portions were either measured immediately after sampling or were retained until all

Table 5.29 *Tablet dissolution ruggedness study*

pH	*Temp*	*Sample time*	*Sample weigh*	*Measure delay*	*RPM*	*30 mins*	*1 hour*	*2 hour*	*4 hour*
−1	−1	1	1	−1	1	25.8	51.5	74.3	96.4
1	−1	1	1	1	1	24.2	51.0	75.7	93.5
−1	1	−1	−1	−1	−1	26.8	54.6	78.9	93.6
1	−1	1	−1	1	−1	25.7	54.5	78.4	95.5
−1	−1	1	−1	−1	−1	23.8	49.4	75.5	92.7
−1	−1	−1	−1	1	1	26.4	51.4	74.9	94.3
1	−1	−1	−1	−1	1	26.8	52.3	74.7	99.0
1	1	−1	−1	1	−1	27.2	50.8	77.4	96.4
−1	1	−1	1	−1	1	26.8	52.1	77.5	97.9
1	1	−1	1	1	1	26.5	53.0	75.8	97.5
−1	1	1	−1	1	1	25.1	50.6	76.6	96.2
1	−1	−1	1	−1	−1	25.4	52.1	77.6	95.3
1	1	1	−1	−1	1	26.4	52.5	77.5	94.3
−1	−1	−1	1	1	−1	23.1	50.3	73.4	92.0
−1	1	1	1	1	−1	27.2	52.1	76.3	91.7
1	1	1	1	−1	−1	26.3	51.4	76.5	97.5

samples were taken (*i.e.*, 4 hours) so that all test portions could be measured together (–1 = no delay, 1 = measure at 4 hours).
- RPM is the paddle speed in revolutions per minute (–1 = 73 rpm, 1 = 77 rpm).

The 16 run design is known as a 2^{6-2} design, *i.e.*, 6 factors were studied in $2^4 = 16$ runs. The design was generated using Minitab which gave the following confounding or alias patterns.

pH
Temperature
Sampling. time
Weigh
Delay
RPM
pH*Temperature – Sampling.time*Delay
pH* Sampling.time – Temperature*Delay
pH*Weigh + Delay*RPM
pH*Delay – Temperature* Sampling.time + Weigh*RPM
pH*RPM + Weigh*Delay
Temperature*Weigh – Sampling.time*RPM
Temperature*RPM – Sampling.time*Weigh

This list indicates that the main effects are not confounded with two-factor interactions but that the two factor interactions are confounded with each other in the patterns shown. In the above, a minus sign indicates that the first interaction term has the same pattern of signs as the second, but with all the signs reversed. Note that the assumption is that three and higher-order interactions do not exist. Confirm some of these aliases by multiplying the correspond columns of signs.

Expand the array by creating columns of signs corresponding to the interaction terms – for example, the product of the columns of signs associated with pH and Temperature give the signs for the pH*Temp – Sampling.time*-Delay contrast. Calculate the contrasts by subtracting the average result at the minus signs from that at the plus signs for each contrast. Plot them on a Normal probability plot. If the criterion had been set that no contrast should be greater than 5% of the average dissolution at the corresponding time point, would you accept that the system can be considered robust? If not which system parameters would you flag as either needing further investigation or particular care in setting up the routine analysis?

5.13 In Table 5.21, p 230, data were reported from a study that investigated the factors that influence the performance of a formic acid biosensor, which is used for air monitoring. Several response variables were studied; a second response C2 is reported in Table 5.30 – it is the current (nA) at steady state when the biosensor was exposed to a formic acid vapour concentration of 2.9 mg m^{-3}.

Table 5.30 *Biosensor development study*
(Reprinted from the *Analyst*, 2001, Vol. 126, pp 2008. Copyright, 2001, Royal Society of Chemistry.)

A	*B*	*C*	*D*	*E*	*F*	*C2*
–1	–1	–1	–1	–1	–1	377
1	–1	–1	–1	–1	1	512
–1	1	–1	–1	–1	1	441
1	1	–1	–1	–1	–1	534
–1	–1	1	–1	–1	1	528
1	–1	1	–1	–1	–1	445
–1	1	1	–1	–1	–1	589
1	1	1	–1	–1	1	697
–1	–1	–1	1	–1	1	425
1	–1	–1	1	–1	–1	476
–1	1	–1	1	–1	–1	540
1	1	–1	1	–1	1	570
–1	–1	1	1	–1	–1	518
1	–1	1	1	–1	1	485
–1	1	1	1	–1	1	544
1	1	1	1	–1	–1	579
–1	–1	–1	–1	1	1	491
1	–1	–1	–1	1	–1	455
–1	1	–1	–1	1	–1	570
1	1	–1	–1	1	1	662
–1	–1	1	–1	1	–1	443
1	–1	1	–1	1	1	521
–1	1	1	–1	1	1	590
1	1	1	–1	1	–1	542
–1	–1	–1	1	1	–1	384
1	–1	–1	1	1	1	465
–1	1	–1	1	1	1	553
1	1	–1	1	1	–1	590
–1	–1	1	1	1	1	497
1	–1	1	1	1	–1	548
–1	1	1	1	1	–1	391
1	1	1	1	1	1	610

Large responses are desirable. For the factors listed below, the first level is coded –1 and the second level as +1 in the data table:

A: Concentration of Medola's blue (mg g^{-1}) 0.3, 1.5
B: Concentration of the enzyme formate dehydrogenase (mg g^{-1}) 0.36, 0.792
C: Concentration of nicotinamide adenine dinucleotide (mg g^{-1}) 1.2, 4.8
D: Concentration of the phosphate buffer (M) 0.048, 0.144
E: Amount of solution placed on the electrode (μL) 8, 12
F: Electrochemical potential (V) 0, 0.2

The design is the same as before so the main effects are confounded with 5-factor interactions (5-fis), 2-fis are confounded with 4-fis and the 3-fis are confounded with other 3-fis.

Expand the array to get columns corresponding to all 15 2-fis and the 10 3-fis that include A (*i.e.*, ABC, ABD...). These 3-fis will be confounded with the other 10 3-fis. By subtracting the average response at the low level for each column of the expanded array from the average at the high level, calculate the contrasts. Draw a Normal probability plot of the contrasts and interpret the plot.

The study included five extra measurements which were pure replicates, *i.e.*, five replicate measurements were made at a fixed set of conditions, so that their standard deviation $s = 42.0$ measures the chance variation in the system and may be used to estimate the standard error of the effects, using $S\hat{E} = \sqrt{4s^2/32}$. Calculate the standard error estimate and use it to identify the influential factors. What do you conclude?

5.14 Suppose now that only three factors had been studied, instead of six. Ignore C, D, E in Table 5.30 and re-analyze the data as a 2^3 design with four replicates at each design point. Compare the results you obtain to those found in the analysis of the full dataset. Because there are replicates you will be able to obtain residuals and draw the usual plots. Do the plots support the standard assumptions for the effects analysis?

5.15 Table 5.31 shows data generated in a robustness study of a HPLC method for measuring the potency of 80 mg capsules. The responses are reported in units of percentage of label claim. Six system parameters were varied; the factors, the two levels for each factor and their coding are shown below.

Coding	**–1**	**+1**
pH of sodium acetate buffer	6.5	7.5
Mobile phase ratio (sodium acetate buffer:acetonitrile)	25:75	35:65
Buffer concentration/gL^{-1}	1.56	1.72
Flow/mL min^{-1}	1.4	1.6
Temperature/°C	Ambient	30
Column type	current	alternative

Table 5.31 *HPLC ruggedness study*

pH	*Mobile phase ratio*	*Buffer conc*	*Flow*	*Temp*	*Column*	*Potency*
7.5	35:65	1.56	1.6	A	1	101.13
6.5	35:65	1.72	1.4	A	1	101.69
7.5	25:75	1.56	1.4	A	2	102.12
6.5	25:75	1.56	1.6	30	1	101.93
6.5	25:75	1.72	1.6	A	2	102.06
7.5	25:75	1.72	1.4	30	1	101.83
7.5	35:65	1.72	1.6	30	2	101.89
6.5	35:65	1.56	1.4	30	2	101.46

The 2^{6-3} eight-run design was generated in Minitab. Calculate the contrasts as the differences between the two factor level averages for each column. Since the contrasts are expressed as percentage changes of the nominal dose (80 mg per capsule), do you think that the size of the largest contrast is likely to be of any practical importance?

Minitab gives the following alias patterns (up to order 2) for the contrasts. Confirm by multiplication of the columns of signs that the alias pattern for the first of these is correct, *i.e.*, that the pattern of signs corresponding to the column associated with pH is the same as the product of the columns for mobile phase ratio and flow and for the product of buffer concentration and temperature.

Ph + Mobile*Flow + Buffer*Temp
Mobile + Ph*Flow + Temp*Column
Buffer + Ph*Temp + Flow*Column
Flow + Ph*Mobile + Buffer*Column
Temp + Ph*Buffer + Mobile*Column
Column + Mobile*Temp + Buffer*Flow
Ph*Column + Mobile*Buffer + Flow*Temp

5.11 REFERENCES

1. J. Villen, F.J. Senorans, M. Herraiz, G. Reglero and J. Tabera, *J. Chromatogr. Sci.*, 1992, **30**, 574.
2. M. Danaher, Personal communication.
3. K.J.M. Sandstrom, R. Carlson, A-L. Sunesson, J-O. Levin and A.P.F. Turner, *Analyst*, 2001, **126**, 2008.
4. G.E.P. Box, W.G. Hunter and J.S. Hunter, *Statistics for Experimenters*, Wiley, New York, 1978.
5. J.W. Weyland, A. Stern and J. Rooselaar, *J. AOAC Int.*, 1994, **77**, 1132.
6. J. Vale, Personal communication.

CHAPTER 6

Fitting Equations to Data

6.1 INTRODUCTION

There are many occasions on which it is of interest to investigate the possible relationship between two variables, where one variable, the predictor variable (often denoted X), is thought of as driving the second variable, the response, Y. For example, in Chapter 5 several examples referred to the influence of quantitative variables such as injection temperature, split ratio, or injection volume, on the peak areas produced by a GC system. There the concern was exploratory – to determine if moderate changes to particular variables influenced the response of the system, and whether any influences were dependent on the levels of other variables. If the intention were to investigate *in detail* the influence of, say, split ratio, then instead of just two levels, the study might involve five or six different split ratio values. The analysis would involve fitting a graphical or numerical relation between the response, peak area, and the predictor, split ratio. Either of these would provide a description of how peak area changes as split ratio is varied, and would allow prediction of the peak area that would be expected at any given split ratio level.

This chapter will be concerned with the investigation of, mainly, straight-line relationships. The first step in any such investigation should be to plot the data. In many cases this will answer the questions of interest: does there appear to be a relationship between the variables? If yes, is it linear? Do the variables increase together or does one decrease as the other increases? And so on. If it is required to go beyond such qualitative questions then regression analysis is the most commonly used method for fitting predictive models. This method will be explained in some detail and the standard regression output from general-purpose statistical software will be described and discussed. One of the most common uses of regression in analytical chemistry is for fitting calibration lines – this is considered in Section 6.3. Section 6.4 discusses the concept of the detection limit of an analytical system, which is important in trace analysis. Section 6.6 introduces weighted least squares which is a modification of the standard approach to fitting regression models and is used when response variability increases with the concentration or amount being measured. Section 6.7 provides a short introduction to multiple regression models and, in particular, to fitting curves and response surfaces.

Throughout the chapter there will be an emphasis on drawing pictures; graphical residual analysis, in particular, is discussed in Section 6.5 and is strongly recommended as the most valuable tool available to the analyst for assessing the validity of regression models.

6.2 REGRESSION ANALYSIS

The term 'regression analysis' refers to the fitting of equations to statistical data. Originally, 'regression' referred to the observation that when relationships between variables such as heights of fathers and sons were investigated, it was found that while tall fathers tended to have tall sons, the sons tended to be shorter than the fathers – there was a 'regression' towards the population mean height. The word regression has attached itself to the statistical methods used in investigating such relationships, but it conveys no useful meaning anymore, beyond being a label for a body of statistical techniques.

6.2.1 Introductory Example

The 1993 International Cooperation on Harmonization (ICH) guidelines on Stability Testing of New Drug Substances and Products[1] (which are closely based on the corresponding 1987 US Food and Drugs Administration guidelines) use regression analysis to define the shelf lives of drug products. The statistical ideas involved in regression analysis will be introduced in this context.

Figure 6.1(a) shows data from a stability testing programme in which samples of a drug product were stored at 25°C and 60% relative humidity. Determinations of the potency of the drug product were made at seven time points, *viz*., zero, 3, 6, 9, 12, 18 and 24 months after manufacture. The reported results are 'percentage of label claim'. Figure 6.1(b) shows a scatterplot of the data; there is a general tendency for the potency values, Y, to decrease with time, X. Figure 6.1(c) shows a regression line fitted to the scatterplot. The criterion used to fit the line was 'least squares'. In principle, this involves calculating the vertical distances from the observed data points, Y_i, to the corresponding

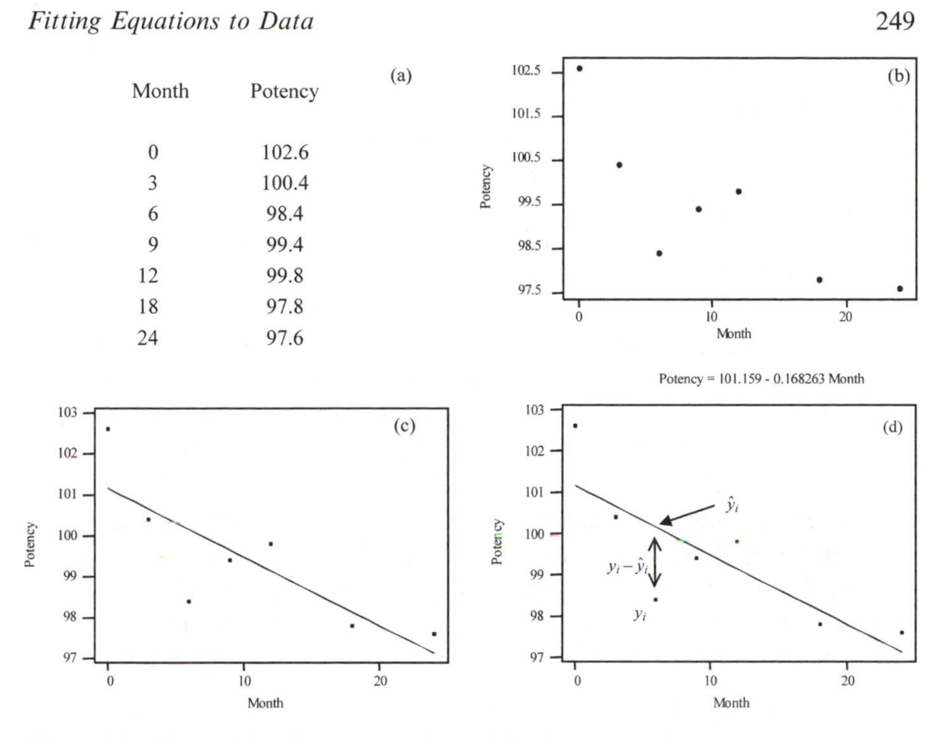

Figure 6.1 *The stability data, scatterplot and fitted regression line*

points, $\hat{Y}_i$, on any candidate line (as illustrated in Figure 6.1(d)) and finding the sum of the squares of these distances. The line that minimizes this sum is known as the least squares line. In practice, simple mathematical formulae give the required line; these are built into most statistical software packages. There are many other possible ways to fit a line through the data of Figure 6.1(b), but the vast majority of reported analyses are based on least squares and this is the default method used by virtually all statistical packages.

The regression line for the data of Figure 6.1 is:

$$\text{Predicted potency} = 101.16 - 0.168 \text{ time}$$

where potency is measured as a percentage of label claim and time in months elapsed from the date of manufacture. The interpretation of this equation is that the average potency is 101.16% of label claim at time of manufacture and that this decreases by 0.168 percentage points for every extra month that elapses. Note that the negative slope means that the product potency is degrading with time. The main reasons for collecting data like these are to help make judgements on the likely shelf life of the product and, thereafter, to monitor its stability over time. Consequently, the slope of the line is a key product performance indicator.

The line may be used for prediction in two ways. We might ask what the average potency will be at some particular time point. Thus, the line predicts an average potency of $101.16 - 0.168(24) = 97.13\%$ two years after manufacture. Alternatively, we might ask after how long the potency will have fallen to 95% of its label claim. By solving the equation $95 = 101.16 - 0.168X$, we find that the regression equation predicts that this occurs 36.7 months after manufacture.

Using the equation directly, as was done above, takes no account of the chance variation in the data; it is obvious from Figure 6.1 that there is considerable chance variation present. Had a second set of measurements been made at the same time points and a regression line been fitted, it is reasonable to conclude that the second line would be different from the first. The questions arise as to how different it might be and what the implications would be for the practical questions we answer in using the line. By studying the magnitude of the chance variation in the data to hand, we can address these issues. In order to do this we build a statistical model of the data and then consider the practical questions in terms of this model.

A Statistical Model for the Potency Data. We model the potency data by assuming that at each time point there is a 'true' mean batch potency value, which is related to the time elapsed from manufacture. The relation between mean batch potency and time is not immediately obvious from the data, because the potency results contain both sampling variability (the material is not perfectly homogeneous) and measurement error. The model may be written as:

$$Y_i = \beta_0 + \beta_1 X_i + \varepsilon_i \tag{6.1}$$

where Y_i is the measured potency at time point X_i and ε_i is the unknown chance variability component in Y_i. This chance component is usually assumed to be Normally distributed.

Figure 6.2 shows pictorially the statistical assumptions. At any time point the observed potency will vary randomly around the true mean value. The variation follows a Normal distribution with a standard deviation σ, which is the same at all time points. The 'true' line simply joins the means of these distributions. The parameter β_0 is the intercept, *i.e.*, the mean batch potency at time of manufacture, $X = 0$. The second parameter, β_1, is the slope of the line, *i.e.*, the amount by which mean batch potency changes (here decreases) per month.

The Regression Output. Table 6.1 shows some output produced by Minitab for a regression analysis of the stability data; this output is

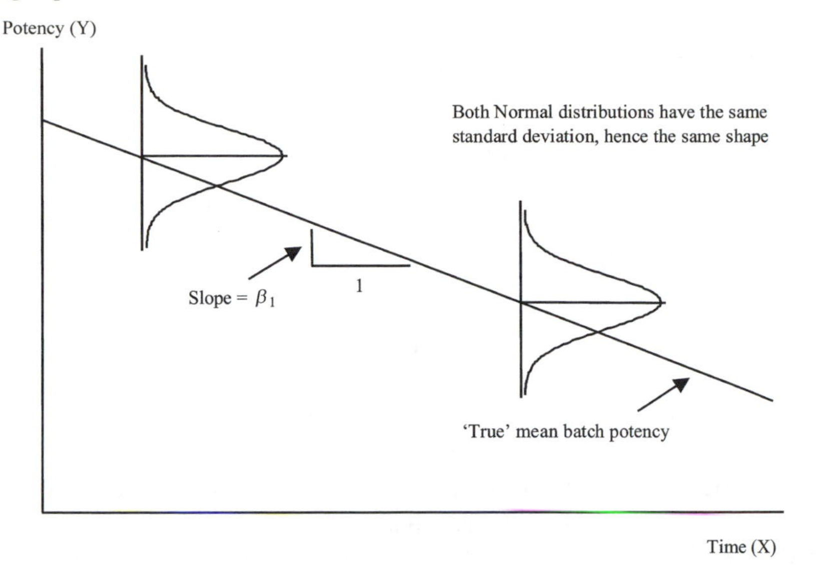

Figure 6.2 *A graphical representation of the statistical model*

typical of what would be produced by any of the very many statistical packages used in laboratories. The label 'constant' refers to the intercept. The fitted or estimated parameter values (also called the regression coefficients) are $\hat{\beta}_0 = 101.16$ and $\hat{\beta}_1 = -0.168$, where the 'hats' on the βs indicate that the values are estimates and not true model parameters. Thus, the fitted line

$$\hat{Y} = 101.16 - 0.168X_i \tag{6.2}$$

may be used to predict the batch potency at any particular time point, as discussed earlier.

The column header 'SE Coef' gives the estimated standard errors of the regression coefficients. Recall that the standard error of a sample mean $\bar{Y}$, based on n randomly selected values from a given population or process, is $\sigma/\sqrt{n}$ or, equivalently, $\sqrt{\sigma^2/n}$. This is used in making inferences about the corresponding population mean μ (see discussion of

Table 6.1 *Regression analysis of the stability testing data*

Predictor	Coef	SE Coef	T	P
Constant	101.159	0.685	147.58	0.000
Time	−0.16826	0.05302	−3.17	0.025
S = 1.099	R−Sq = 66.8%			

significance tests and confidence intervals in Chapter 3). Similarly, the standard error of the slope coefficient is

$$SE(\hat{\beta}_1) = \sqrt{\frac{\sigma^2}{\sum_i^n (X_i - \bar{X})^2}} \tag{6.3}$$

while that for the intercept coefficient is

$$SE(\hat{\beta}_0) = \sqrt{\sigma^2 \left(\frac{1}{n} + \frac{\bar{X}^2}{\sum_i^n (X_i - \bar{X})^2} \right)} \tag{6.4}$$

These depend on σ, the standard deviation of the chance component of variation in Y_i, on how many results, n, are obtained in the study and on how spread out the X values are. The more observations that are available and the more spread out the X values (*i.e.*, the bigger $\sum_i^n (X_i - \bar{X})^2$ becomes) the smaller the standard errors and the closer the regression coefficients $\hat{\beta}_0$ and $\hat{\beta}_1$ are likely to be to the true parameters β_0 and β_1, respectively. Standard errors provide a means for making inferences on model parameters and, hence, on the nature of the relationship between batch potency and elapsed time from manufacture for the current example.

The standard deviation, σ, that appears in both equations must be estimated from the data. Here, the estimate is given by:

$$s = \sqrt{\frac{\sum_{i=1}^n (Y_i - \hat{Y}_i)^2}{n-2}} = \sqrt{\frac{\sum_{i=1}^n (Y_i - \hat{\beta}_0 - \hat{\beta}_1 X_i)^2}{n-2}} \tag{6.5}$$

This is just the square root of the average of the squared deviations of the data points from the regression line. Recall that for a sample of n results from a single Normal distribution with standard deviation σ we estimated σ by:

$$s = \sqrt{\frac{\sum_{i=1}^n (Y_i - \bar{Y})^2}{n-1}} \tag{6.6}$$

that is, by the square root of the average of the squared deviations of the data points from the sample mean. For the regression data, we have a mean (the line) which varies with time, X. When averaging, we divide by the number of degrees of freedom. For a sample from a single distribution, the degrees of freedom are $n-1$: one degree of freedom is 'lost' in using $\bar{Y}$ to estimate μ. For the regression line, the degrees of freedom are $n-2$: two degrees of freedom are lost because we estimate the intercept, β_0, and the slope, β_1, from our data. It should be clear, however, that s is essentially the same estimator in both cases. Table 6.1 gives $s=1.099$; this is used in all the statistical tests and confidence intervals that are discussed below.

6.2.2 Using the Regression Line

We are now in a position to make allowances for the chance variation in the data when using the fitted regression line. Regression analysis is useful for description, prediction and control; these uses are illustrated below in the context of the shelf-life data.

Does Potency Degrade with Time? The fitted slope coefficient of -0.168 is negative, suggesting that potency degrades with time. However, since we have only a small dataset, $n=7$, and since Figure 6.1 shows considerable chance variation affecting the test results, the question naturally arises as to whether the negative slope is a systematic effect or whether it is simply the result of chance variation. We can carry out a statistical significance test to address this question, *i.e.*, to determine if the true slope is zero. The appropriate test is a t-test which, apart from the formula for the standard error and the number of degrees of freedom, is of exactly the same form as those encountered in Chapters 3, 4 and 5. The null and alternative hypotheses* are:

$$H_o : \beta_1 = 0$$
$$H_1 : \beta_1 \neq 0$$

The test statistic is

$$t = \frac{\hat{\beta}_1 - 0}{S\hat{E}(\hat{\beta}_1)} = \frac{\hat{\beta}_1 - 0}{\sqrt{s^2 \Big/ \sum_{i=1}^{n} (X_i - \bar{X})^2}} = \frac{-0.16826}{0.05302} = -3.17 \qquad (6.7)$$

The package inserts $s=1.099$ into the standard error formula Equation (6.3) and calculates the estimated standard error of $\hat{\beta}_1$ as 0.05302; this results in a t-statistic of $t=-3.17$. If a significance level of $\alpha=0.05$ is to be

*A one-sided alternative hypothesis $\beta_1 < 0$ would be appropriate here if it were known from the chemistry that the potency could not increase with time. A two-tailed test is carried out above; this corresponds to the two-tailed p-value which is the default value produced by the package. A one-tailed test would mean that only the area to the left of -3.17 would be included in the p-value, giving $p=0.0125$.

used, then for a two-tailed test the critical values for $n-2=5$ degrees of freedom are ± 2.57. Recall that the appropriate t-distribution is that with the same number of degrees of freedom as the s that appears in the standard error formula. Since the observed t-statistic is less than -2.57 the hypothesis that the slope is zero is rejected. The estimated slope coefficient of -0.168 is said to be statistically significantly different from zero. The conclusion from the regression analysis is that potency decreases with time.

Note that the package automatically provides a p-value associated with the test statistic. The interpretation is the same as for previously encountered p-values: the p-value is a measure of how unlikely the observed t-statistic would be if the null hypothesis were true. A small p-value is taken as a signal that the null hypothesis should be rejected. The correspondence with the use of the critical values for making the decision is that $p < 0.05$ corresponds to a t-statistic in the rejection region. Here, $p = 0.025$ indicates that the null hypothesis should be rejected.

A confidence interval for β_1 will provide a more realistic assessment of the change in potency with time than does the regression coefficient $\hat{\beta}_1$ on its own. This interval is given by

$$\hat{\beta}_1 \pm t\hat{SE}(\hat{\beta}_1) \tag{6.8}$$

where the t value has $n-2$ degrees of freedom. For the data of Figure 6.1 a 95% confidence interval for β_1 is given by:

$$-0.16826 \pm 2.57(0.05302)$$

$$-0.16826 \pm 0.13626$$

The interpretation of this interval is that batch potency decreases by between 0.03 and 0.30% of the label claim for every month elapsed from manufacture; this estimate has an associated confidence level of 95%. The width of the interval could have been made narrower by increasing the number of measurements. This might have been done by making several independent potency determinations at each time point* or by increasing the number of time points at which measurements were made, or both.

Similar tests can be carried out and a confidence interval calculated for the intercept parameter β_0 (see Exercise 6.1). The test of the hypothesis

*Since measurements at different time points belong, necessarily, to different runs, any replicates should also be in different runs. Within-run replicates should be averaged before carrying out the regression analysis. The situation is directly analogous to drawing X-bar charts. See Section 2.6.

that β_0 equals zero, which is automatically calculated by the package ($p = 0.000$ in Table 6.1, p 251), is clearly nonsensical in the current context – it asks if the mean batch potency is zero at time of manufacture! Such a test will, however, often be meaningful when calibration relationships are being studied – they will be discussed in Section 6.3.

Exercises

6.1 Tests may be carried out and confidence intervals calculated for the intercept parameter β_0 using the summary statistics of Table 6.1, p 251. Test the hypothesis that $\beta_0 = 100\%$ using a significance level of 0.05. Calculate a 95% confidence interval for β_0. Interpret and relate the results of the two calculations.

6.2 The data shown in Table 6.2 are the 24 hour percentage dissolution results for a drug product tested in a USP Type 1 dissolution apparatus. The analysis was carried out using a UV spectrophotometer. The data come from a stability monitoring programme – the product was stored at 25°C and 60% relative humidity and was measured at the six time points indicated in Table 6.2. Each test result is the mean of six replicates (though this is not relevant to your calculations).

- Draw a scatterplot of the data.
- Fit a least squares regression line to the data.
- Calculate a 95% confidence interval for the slope and interpret the interval.
- Is the slope statistically significantly different from zero?
- Calculate a 95% confidence interval for the intercept and interpret the interval.
- Is the intercept statistically significantly different from 100%?

Table 6.2 *Stability monitoring results*

Month	0	3	6	9	12	18
% Dissolution	82.7	82.55	73.8	70.52	64.28	57.82

Predicting Mean Batch Potency. If it is desired to estimate the mean batch potency at any particular time point (X_c) then equation (6.2) may be used to obtain a point estimate, $\hat{Y}_c$. A $(1-\alpha)\%$ confidence interval for the mean batch potency at time X_c is given by:

$$\hat{Y}_c \pm t\sqrt{s^2\left(\frac{1}{n}+\frac{(X_c-\bar{X})^2}{\sum_{i=1}^{n}(X_i-\bar{X})^2}\right)} \tag{6.9}$$

where t is the critical value for the t-distribution with $n-2$ degrees of freedom which encloses $(1-\alpha)\%$ of the curve, and s estimates σ. For example, the estimated mean batch potency for the data of Figure 6.1 at fifteen months after manufacture is given by:

$$\hat{Y}_{15} = 101.159 - 0.16826(15) = 98.635$$

and the 95% confidence bounds are:

$$98.635 \pm 2.57\sqrt{1.207\left(\frac{1}{7}+\frac{(15-10.286)^2}{429.429}\right)}$$

giving an the interval estimate between 97.39 and 99.88% of label claim. Figure 6.3 shows a plot of the fitted line and the bounds generated by Expression (6.9).

Three comments are appropriate:

- Firstly, the bounds represent an interval estimate of the *mean* batch potency at any given time. There is, therefore, no cause for concern if *individual* results, *e.g.*, the observed value at time point 6 months,

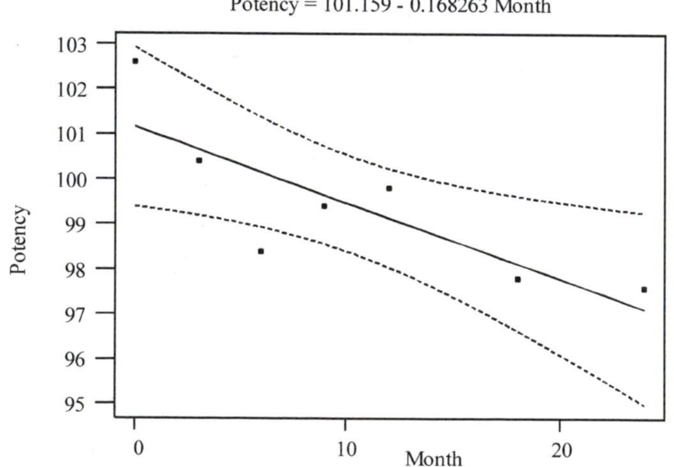

Figure 6.3 *Confidence bounds for mean batch potency at different time points*

lie outside the bounds. Individual test results are subject to more chance variation than predicted mean values.

- Secondly, the derivation of Expression (6.9) is based on the assumption that what is required is a single confidence interval at X_c. The bounds shown do not, therefore, represent *simultaneous* bounds for all points on the regression line, *i.e.*, the bounds shown in Figure 6.3 are not bounds for the true line. If many predictions are to be made from a fitted regression line and it is required that an overall confidence level (*e.g.* 95%) should apply to them all, *simultaneously*, then different, somewhat wider bounds are required. One approach simply replaces the t-tmultiplier in Expression (6.9) by W, where $W^2 = 2F(1 - \alpha, 2, n-2)$; the resulting bounds are called 'Working–Hotelling bounds'. For the stability data $W = \sqrt{2(5.79)} = 3.40$, in contrast with a t-value of $t = 2.57$, for a 95% confidence level. Such bounds are discussed in the paper by Hunter[2] and in specialized regression texts, see for example Neter, Wasserman and Kutner.[3]

- Finally, the shape of the confidence bounds tells something interesting about the estimation error. Expression (6.9) shows that the confidence bounds on the estimated mean batch potency will be narrowest at $X_c = \bar{X}$. The greater the difference between X_c and $\bar{X}$, the bigger the term $(X_c - \bar{X})^2$ becomes, resulting in the bands becoming wider as they move outwards from $(\bar{X}, \bar{Y})$.

Exercise

6.3 Calculate a 95% confidence interval for the mean batch potency at 6 months and confirm that the measured result at 6 months is not within the interval, as shown in Figure 6.3.

Determining Shelf Life. The ICH guidelines define the shelf life of a product to be the time point at which the 95% lower confidence bound for the batch potency intersects the 'acceptable lower specification limit'. This means that the shelf life X_c is determined by solving Equation (6.10) which describes the point of intersection of the lower confidence bound* and the specification limit:

$$(\hat{\beta}_0 + \hat{\beta}_1 X_c) - t\sqrt{s^2\left(\frac{1}{n} + \frac{(X_c - \bar{X})^2}{\sum_{i=1}^{n}(X_i - \bar{X})^2}\right)} = Y_{LSL} \qquad (6.10)$$

*Note that to obtain a one-side 95% lower bound the critical t-value is that which has 5% of the area in the left-hand tail of the t-curve; this is the same value ($t = -2.015$ in the current example) that is required for a two-sided 90% confidence interval.

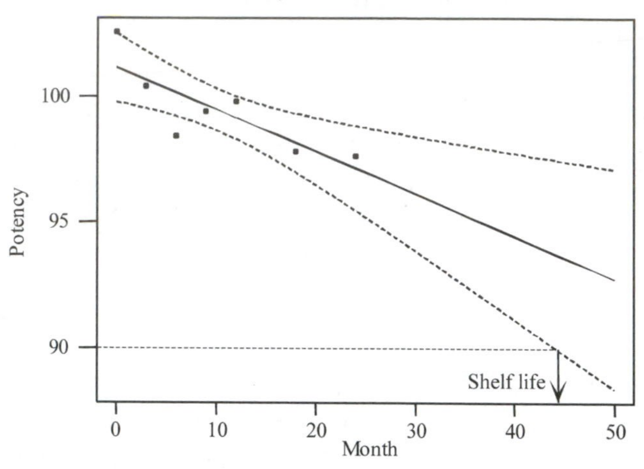

Figure 6.4 *Estimating product shelf life*

where Y_{LSL} is the 'acceptable lower specification limit'. This equation can be solved easily using a spreadsheet: the left-hand-side of the equation may be set up as a formula and then a series of values can be substituted for X_c until the result equals the right-hand-side, Y_{LSL}. Alternatively, the value can be read from a plot similar to Figure 6.4, when this is provided by the statistical software. For the current example, Figure 6.4 indicates that the shelf life will be approximately 45 months, if the acceptable lower specification limit is taken to be 90% of label claim. Minitab calculates a lower bound of 89.99 at 44.25 months.

Exercise

6.4 Use Equation (6.10) to obtain 95% lower confidence bounds at 44, 45 and 46 months and confirm that the shelf life is approximately 45 months. If you regularly use a spreadsheet, set the formula up in a worksheet and do the calculations there.

Monitoring Product Stability. Expression (6.9) gives confidence limits within which the *mean* batch potency is expected to lie at any given time point. Expression (6.11), below, gives what are known as 'prediction limits' – a single new potency measurement at any given time point would be expected to lie within these limits. Note that the prediction formula (6.11) is very similar to Expression (6.9). The difference, the insertion of a '1' under the square root sign, reflects the extra chance variation, due to sampling and measurement error, in predicting *individual* test results, rather than long-run averages.

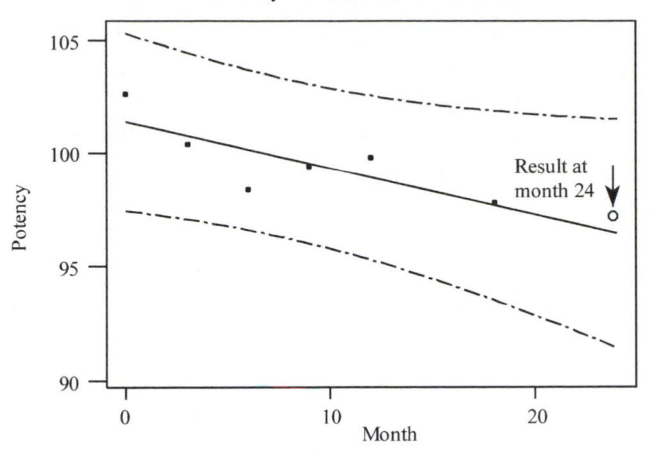

Figure 6.5 *Regression line and 95% prediction limits based on first six data points, i.e., data up to 18 months, only*

$$\hat{Y}_c \pm t\sqrt{s^2\left(1+\frac{1}{n}+\frac{(X_c-\bar{X})^2}{\sum_{i=1}^{n}(X_i-\bar{X})^2}\right)} \tag{6.11}$$

Figure 6.5 shows prediction limits for product potency, based only on the data up to 18 months after manufacture. Table 6.3 shows the regression summary statistics and includes both a confidence interval for the mean batch potency and a prediction interval for a new measurement at 24 months.

The observed test result of 97.6 at 24 months is well inside the 95% prediction limits (91.4, 101.5) derived from the data up to 18 months. This indicates that the 24 month test result is entirely consistent with

Table 6.3 *Regression based on data up to 18 months; predictions at 24 months*

Predictor	Coef	SE Coef	T	P
Constant	101.371	0.804	126.04	0.000
Month	−0.20476	0.08083	−2.53	0.064

S = 1.171 R−Sq = 61.6%

Predicted Values for New Observation at 24 months

New Obs	Fit	SE Fit	95.0% CI	95.0% PI
1	96.457	1.379	(92.629, 100.286)	(91.434, 101.481)

the prior data: the prediction limits bracket the test result values that might be expected based on the earlier measurements. Thus, there is no cause for concern – the new results do not call into question the established degradation relationship.

Using the regression line and prediction limits in this way is essentially equivalent to using control charts for monitoring system stability, as described in Chapter 2. If the 24 month test result had fallen below the lower prediction limit, this would be a cause for concern, as it would suggest either an acceleration in degradation or, perhaps, problems with the analysis of the test portion – an immediate investigation would be indicated.

Exercise

6.5 The data of Exercise 6.2, p 255, come from a stability monitoring study for the first 18 months after manufacture of a product. Within what bounds would you expect the measured dissolution rate to be at 24 months? Use a 95% prediction interval.

6.2.3 Analysis of Variance

Table 6.4 contains a block of output, labelled 'Analysis of Variance (ANOVA)', that is typically produced by statistical packages as part of a regression analysis. It is based on the full shelf life dataset, but was not included in Table 6.1 in order to focus attention on the estimated regression coefficients and their standard errors, which are the most important elements of any regression analysis. ANOVA leads into definitions of the coefficient of determination (r^2) and the correlation coefficient (r), one of which commonly appears in method validation reports; these are discussed below.

The Analysis of Variance (ANOVA) technique is a general method for decomposing statistical variation into different components and it will be encountered in different guises in this and the next two chapters. Variation in statistics is usually measured by sums of squares, which are

Table 6.4 *Analysis of Variance for the shelf life data*

Analysis of Variance

Source	DF	SS	MS	F	P
Regression	1	12.158	12.158	10.07	0.025
Residual Error	5	6.036	1.207		
Total	6	18.194			

then averaged to give mean squares. The most familiar example of this is the sample variance (the square of the sample standard deviation):

$$s^2 = \frac{\sum_{i=1}^{n} (Y_i - \bar{Y})^2}{n-1} \tag{6.12}$$

which is a ratio of the sum of the squared deviations of the data points from their own mean to the 'degrees of freedom'; this was introduced in Chapter 1. It measures the 'total' variation in the data.

The ANOVA table simply breaks both parts of this expression into two components, one associated with the regression line (assumed systematic) and the other, with variation around the regression line (assumed random). Figure 6.6 shows that the deviation of any point from the mean of all the data, *i.e.*, $Y_i - \bar{Y}$, is the sum of two parts, *viz.*, the deviation of any point from the corresponding fitted point on the regression line, $Y_i - \hat{Y}_i$, and the deviation, $\hat{Y}_i - \bar{Y}$, of that fitted point from the overall mean:

$$(Y_i - \bar{Y}) = (Y_i - \hat{Y}_i) + (\hat{Y}_i - \bar{Y}) \tag{6.13}$$

When the two sides of Equation (6.13) are squared and summed over all the data points, the cross-product of the two terms on the right-hand side sums to zero, resulting in Equation (6.14):

$$\sum_{i=1}^{n} (Y_i - \bar{Y})^2 = \sum_{i=1}^{n} (Y_i - \hat{Y}_i)^2 + \sum_{i=1}^{n} (\hat{Y}_i - \bar{Y})^2 \tag{6.14}$$

$$\text{SSTO} = \text{SSE} + \text{SSR}$$

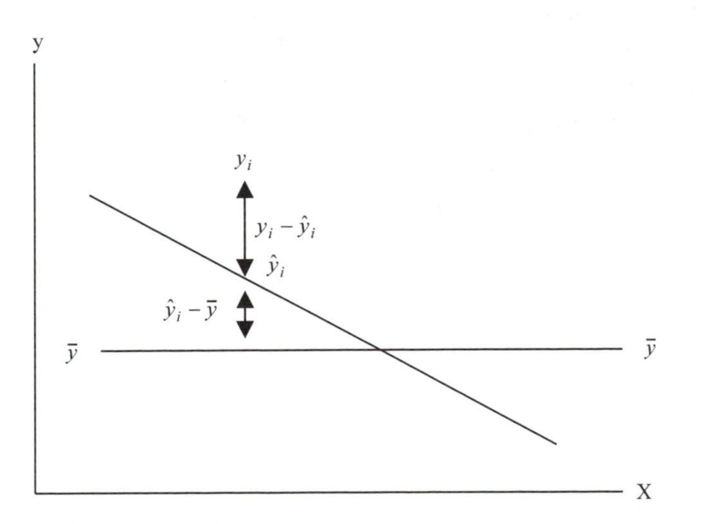

Figure 6.6 *Breaking $y_i - \bar{y}$ into two parts*

Thus, the total variation in Y, as measured by the 'total sum of squares', SSTO, is partitioned in two components, the 'error or residual sum of squares', SSE, and the 'regression sum of squares', SSR. The error sum of squares measures the chance variation of the data points around the regression line. The regression sum of squares is a measure of the extent of the variation in Y as X varies – if it is a large fraction of the total sum of squares then X is a good predictor of Y, if it is small then knowing X is not helpful in predicting Y.

Ideally, we would measure the variation in the data around the long-run mean, μ, but since this is unknown we use $\bar{Y}$ in its place; in doing so we use some of the information contained in the data and lose a degree of freedom in the process. Thus, SSTO has $n-1$ degrees of freedom. Similarly, SSE has $n-2$ degrees of freedom, since two degrees of freedom are lost in estimating both the intercept and slope that are used in calculating the fitted values, $\hat{Y}_i$. Consequently, SSR has only one degree of freedom.

Table 6.4 shows the ANOVA calculations for the stability testing data. There are 7 observations, giving 6 degrees of freedom for SSTO, 5 for SSE and 1 for SSR. The sum of squares for regression is SSR = 12.158, that for error is SSE = 6.036, which together sum to the total SSTO = 18.194. The column of mean squares is just the ratio of the sums of squares to the degrees of freedom: thus, MSE = SSE/$(n-2)$ = 1.207. The square root of the MSE is the standard deviation of the data about the regression line, $s = 1.099$, which was given in Table 6.1. The estimation of these quantities (s or MSE = s^2) is the main role of the ANOVA table, as one of the two appears in all statistical tests, confidence or prediction intervals associated with the regression line. The ANOVA table also contains an F-test (the F value is just the ratio MSR/MSE) which provides a test of the hypothesis that β_1 equals zero. However, since this is exactly equivalent to the t-test discussed earlier (check that the F value is just the square of the t value) it will not be considered further here.

Coefficients of Determination and Correlation. The ANOVA table leads to a measure of the closeness of agreement between the data points and the corresponding fitted values on the regression line; this is called the coefficient of determination (r^2, labelled R-sq in Table 6.1) and is defined as:

$$r^2 = \frac{SSR}{SSTO} = 1 - \frac{SSE}{SSTO} \tag{6.15}$$

This summary statistic measures the proportion of the total variation that is associated with the regression line, as opposed to the variation (presumed random) of the data points around the line. Since it is a proportion it must lie between zero and 1, though packages often multiply by 100 and quote the result as a percentage; thus, r^2 is given as 66.8% in the stability data output of Table 6.1. Values of r^2 close to 1 indicate good predictive ability. When all the data points lie on the fitted line, as shown in the top panel of Figure 6.7, the error sum of squares is zero – in such a case the line is a perfect predictor of Y, given an X value, and r^2 is equal to 1. On the other hand, if there is no relationship between Y and X, the fitted line will neither increase nor

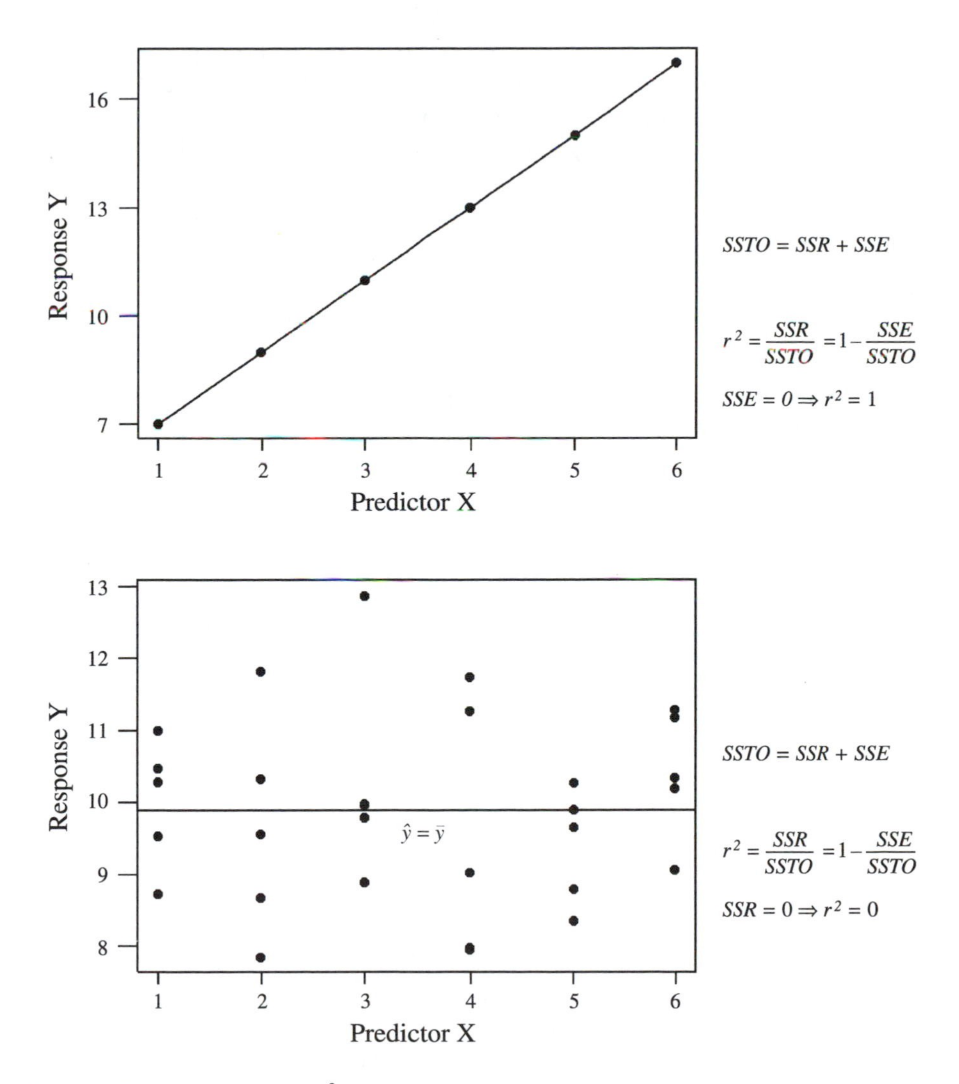

Figure 6.7 *The bounds for r^2*

decrease as X varies – $\hat{Y}$ will coincide with $\bar{Y}$, as shown in the bottom panel of Figure 6.7, and $SSR = \sum_{i=1}^{n} (\hat{Y} - \bar{Y})^2 = 0$, which means $r^2 = 0$. As r^2 increases from zero to 1 the data points lie correspondingly closer to the fitted line and the predictive ability of the regression equation increases accordingly.

The square root of the coefficient of determination is numerically equal to a summary measure called the correlation coefficient or, sometimes, Pearson's product-moment correlation coefficient. The correlation coefficient was devised as a measure of the extent of the association between two random variables (*e.g.*, the heights of fathers and sons), but because of its numerical equivalence to the square root of the coefficient of determination, it is very often quoted as a measure of the goodness of fit of a regression line to the data on which it is based. The positive square root is taken if the line slopes upwards to the right (a 'positive' relationship); the negative square root indicates Y decreases as X increases.

Either summary measure (r^2 or r) may be used to indicate the strength of a linear relationship – values close to 1 indicate good predictive ability. They should not, as is sometimes done, be quoted as measures of linearity;[4] an example will be shown later of a clearly non-linear relation between two variables, but which gives an r^2 value of 0.997 when a straight line is fitted to the data. A scatterplot is the simplest way of examining linearity; a formal statistical significance test for linearity will be discussed later.

Exercise

6.6 For the dissolution stability data of Exercise 6.2, p 255, compute an ANOVA table and from this calculate the coefficient of determination, r^2.

6.3 CALIBRATION

Modern measurement systems generally involve some kind of instrumental response (peak area, peak height, electric current, *etc.*) which is an indirect reflection of the magnitude of the quantity being measured. For example, when measuring dissolved reactive phosphorus in water a mixture of reagents is added to a filtered water sample. A blue colour, which absorbs light at 882 nm, develops in proportion to the concentration of phosphate ions. The colour intensity is then measured on a spectrophotometer which gives absorbance readings on a scale of zero (no colour) to about two (strong colour). In order to use the instrumental responses for measuring samples of unknown concentration, the instrument must be calibrated, *i.e.* a quantitative relationship must be established between the instrumental response (here absorbance) and the magnitude of the quantity being measured. This is done by preparing a set of standard concentrations whose values are considered known, determining the corresponding absorbances and deriving a statistical relationship between the responses and the standard values.

6.3.1 Example

The data shown in Figure 6.8 below come from a calibration exercise involving a spectrophotometer which was to be used for measuring phosphates (dissolved reactive phosphorus, DRP, *i.e.*, PO_4-P) as described above. In order to avoid long strings of leading zeros in the regression output, the absorbance was multiplied by 1000 to give the variable 'Response'. Using this as the response variable in place of absorbance makes no difference to the test results produced by the analytical system. Table 6.5 contains a regression analysis of the calibration data.

For the phosphate data the estimated calibration relation,

$$\hat{Y} = \hat{\beta}_0 + \hat{\beta}_1 X$$

is

$$\text{Response} = -0.500 + 0.608DRP$$

DRP	Absorbance	Response
0	0.000	0
10	0.005	5
25	0.015	15
50	0.029	29
75	0.046	46
100	0.060	60

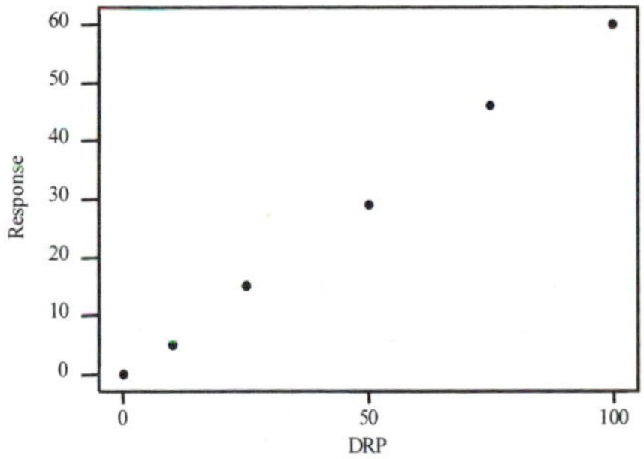

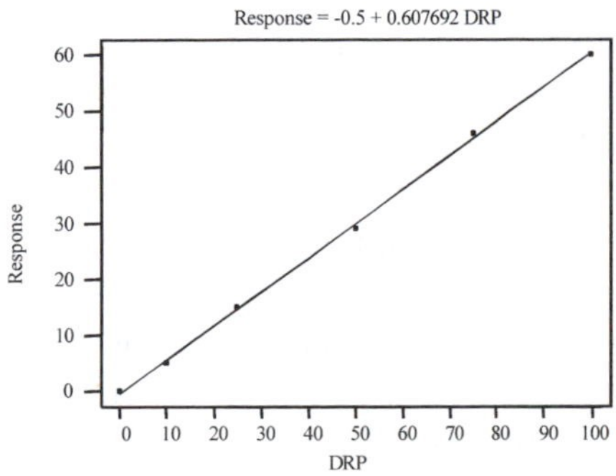

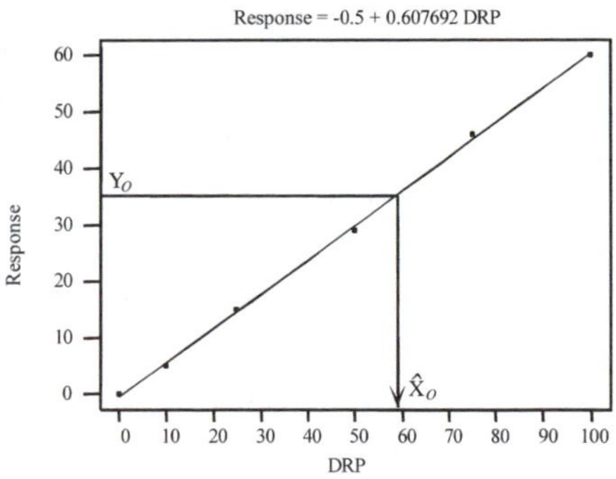

Response = -0.5 + 0.607692 DRP

Response = -0.5 + 0.607692 DRP

Figure 6.8 *The DRP data, scatterplot and fitted regression line*

Table 6.5 *Regression output for the DRP calibration data*

Predictor	Coef	SE Coef	T	P
Constant	−0.5000	0.4970	−1.01	0.371
DRP	0.607692	0.008866	68.54	0.000
S = 0.7721	R−Sq = 99.9%			

This equation indicates that for a change of one unit in DRP concentration the Response changes by $\hat{\beta}_1 = 0.608$ units, on average (the regression using Absorbance as the Y variable would have three zeros between the decimal point and the 6 in the slope coefficient and before the 5 in the intercept coefficient). Note that the fitted line does not go through the origin even though it would be expected that for zero phosphate concentration the response should be zero. This will be discussed in some detail later.

When a test portion is measured and a response, say Y_o, is obtained, the calibration equation may be used to estimate the unknown concentration of phosphate in the test portion. The bottom right-hand panel of Figure 6.8 shows how this is done: the response, Y_o, is projected across to the fitted line and then downwards to the concentration axis. This is equivalent to the calculation shown in Equation (6.16) where the predicted DRP concentration is labelled $\hat{X}_o$.

$$\hat{X}_o = \frac{Y_o - \hat{\beta}_0}{\hat{\beta}_1} \tag{6.16}$$

Thus, if a test portion gives a response of 35, *i.e.*, an absorbance reading of 0.035, the calibration line predicts a concentration of:

$$\hat{X}_o = \frac{35 - (-0.5)}{0.608} = 58.4\,\mu g/L$$

This calculation illustrates the most common use of calibration curves.

Note that the regression line may be used for prediction purposes in two distinct ways. Given any particular known DRP concentration, the regression line may be used to predict the system response that will be observed; this system response is subject to chance measurement variation. In the calibration context the direction of prediction is in the opposite direction and is often referred to as 'inverse prediction'. Using Equation (6.16), we move from an observed response, Y_o, which is subject to chance measurement variation, to estimate a fixed, but unknown, true concentration level in the test sample. Since the value Y_o from which we predict involves a random error component, the resulting prediction will, inevitably, be affected by this chance variation. To take account of this, methods have been developed to place error bounds on the predicted concentration – one approach is described below.

Exercise

6.7 Hunter[2] reports the data in Table 6.6 as coming from a calibration study of a HPLC system. The concentrations are for naphthionic acid (FD&C Red No. 2) and the responses are chromatographic peak area.

Fit the calibration line and obtain an inverse prediction of the concentration for a peak area response of 20 units.

Table 6.6 *HPLC calibration data*
(Reprinted from the Journal of AOAC INTERNATIONAL, 1981, Vol. 64, pp 574. Copyright, 1981, by AOAC INTERNATIONAL.)

Conc.	*Peak area*
0.180	26.666
0.350	50.651
0.055	9.628
0.022	4.634
0.290	40.206
0.150	21.369
0.044	5.948
0.028	4.245
0.044	4.786
0.073	11.321
0.130	18.456
0.088	12.865
0.260	35.186
0.160	24.245
0.100	14.175

6.3.2 Error Bounds for the Estimated Concentration

Figure 6.9 shows how error bounds may be placed on predictions of system responses at known DRP concentrations. These bounds are given by Expression (6.11) and correspond directly to those used for monitoring product stability in the shelf life example.

Error bounds may be placed on the inversely predicted concentration $\hat{X}_o$, also; they are a recognition of the fact that, given the inherent chance variation in the analytical system, any particular response Y_o might have been generated by a range of true concentration values. This is illustrated in Figure 6.10 below.

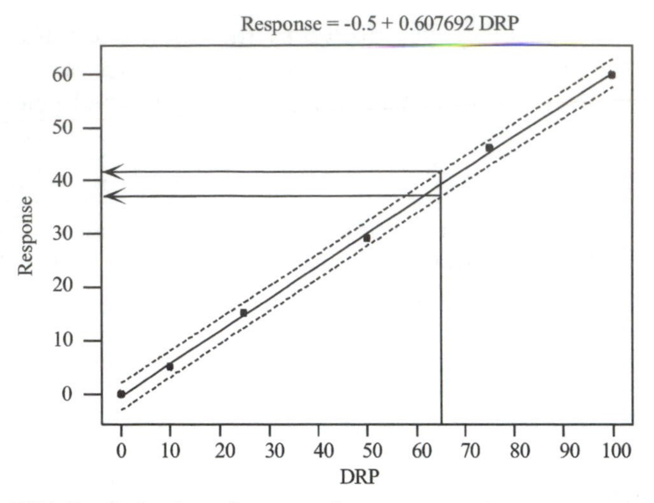

Figure 6.9 *95% Prediction bounds at any given concentration*

The error bounds on $\hat{X}_o$ can be obtained graphically by using Expression (6.11) to draw the prediction limits as shown in Figure 6.10 and then drawing a horizontal line from the observed Y_o and noting the intersection points on the prediction limit bounds as shown in the Figure. An approximate algebraic expression for these bounds is:

$$\hat{X}_o \pm \frac{t}{\hat{\beta}_1}\sqrt{s^2\left(1+\frac{1}{n}+\frac{(Y_o-\bar{Y})^2}{\hat{\beta}_1^2\sum_{i=1}^{n}(X_i-\bar{X})^2}\right)} \quad (6.17)$$

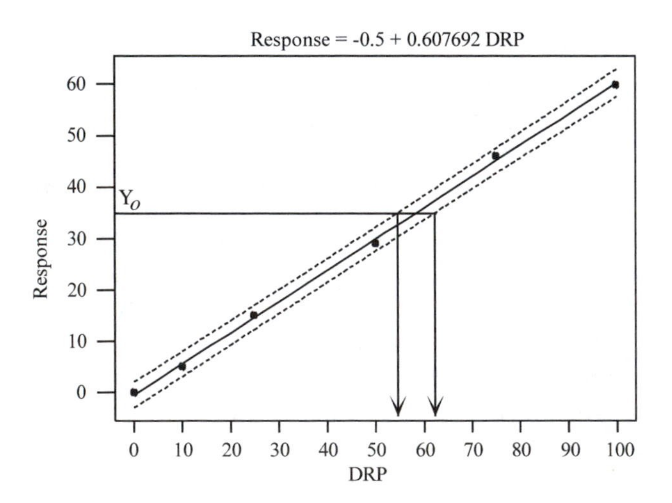

Figure 6.10 *Error bounds on an estimated concentration*

where $\hat{X}_o = \dfrac{Y_o - \hat{\beta}_0}{\hat{\beta}_1}$. The approximation involved will be good provided the fitted line is very close to the data, as would be expected in analytical calibrations.

For the DRP data, a test sample response of $Y_o = 35$ gives an estimated concentration of $\hat{X}_o = 58.42$; Expression (6.17) gives error bounds of 58.42 ± 3.86, so the estimated concentration is between 54.56 μg per L and 62.28 μg per L.

Exercise

6.8 For the HPLC calibration data of Exercise 6.7, calculate 95% prediction limit bounds for the inverse prediction of concentration when the peak area response is 20 units. Note that the sum of the squared deviations of the concentration values around their mean is 0.1431 and the MSE $= s^2 = 1.1941$.

Replicate Measurements. The error bound expressions used above assume that the measurement protocol for the test materials is the same as that for the calibration standards. This will not always be the case. Thus, each test result in the calibration study might be the average of several replicate measurements, while routine testing might involve single measurements or a different number of replicates.

Suppose first that in obtaining the calibration line n individual measurements are made and the calibration line is then fitted to these n data points. If a single measurement is subsequently made on the test material, the bounds, which reflect the variability that was encountered in the calibration study, will then apply to the test result, *i.e.*, the test result would be predicted to fall within the bounds. Next, if each of the n data points used in setting up the calibration line is the average of m replicates ($m = 2$ or 3 would be common), and if the test material is also to be the average of m replicate measurements, then Expressions (6.11) and (6.17) are still appropriate for calculating error bounds. The inherent random error will have been reduced, by averaging, but this will apply equally to the calibration study and the test material. If, however, the calibration and routine test protocols are different, then account must be taken of any differences when calculating the error bounds.

If the n calibration data points are single measurements and the test material result is to be the average of m replicate measurements, then Expression (6.11) changes to (6.18).

$$\hat{Y}_c \pm t\sqrt{s^2\left(\frac{1}{m} + \frac{1}{n} + \frac{(X_c - \bar{X})^2}{\sum_{i=1}^{n}(X_i - \bar{X})^2}\right)} \qquad (6.18)$$

Note the difference – '1' (*i.e.*, one s^2) has changed to '$1/m$'. This adjustment is directly equivalent to a similar adjustment for single sample studies encountered in Chapter 3. There, we used s as a measure of the variability of single measurements, while $s/\sqrt{m}$, *i.e.*, $\sqrt{s^2/m}$, was the appropriate measure of the variability of means based on m values.

For inverse prediction a similar modification is required. Expression (6.17) changes to (6.19) – again the only difference is that '1' (*i.e.*, one s^2) has been replaced by '$1/m$' to reflect the fact that Y_o is an average of m replicates in (6.19), whereas it represented a single measurement in (6.17).

$$\hat{X}_o \pm \frac{t_{.025}}{\hat{\beta}_1}\sqrt{s^2\left(\frac{1}{m}+\frac{1}{n}+\frac{(Y_o-\bar{Y})^2}{\hat{\beta}_1^2\sum_{i=1}^{n}(X_i-\bar{X})^2}\right)} \tag{6.19}$$

Note that averaging replicates before fitting the line does not change the resulting line. The least squares line fitted to six averages of three replicates will be identical to that fitted to the eighteen original data points. However, the variation of the data around the line will obviously change, since averages are less variable than individual data points. Unless the same number of replicates as are used in the calibration protocol are also measured during routine testing, the simplest approach will be to fit the calibration line to the individual data points and then use Equation (6.19) to allow for the effects of averaging in routine testing. Note also that these formulae are applicable where true replicates are obtained; refer to Section 2.5 for a discussion of the nature of replication.

6.3.3 Zero-intercept Calibration Lines

At first sight it might appear that zero system response would be expected where no analyte is present, and that the calibration line should go through the origin. In the DRP calibration, for example, this would imply a zero absorbance reading on the instrument if there are no phosphate ions in the water. A non-zero response might be obtained for several reasons; two are considered below. Firstly, random measurement error may result in a non-zero absorbance reading even though there are no phosphate ions in the water. Secondly, even though what is considered to be ion-free water is used in making up the standard calibration solutions, there may be, unknown to the analyst, some level of phosphate in the water. If this is the case then all the standards will contain a slightly higher level of phosphate than intended. Consequently, the responses will be increased and will result in a calibration line that does not go through the origin, *i.e.*, it will have a positive intercept on the vertical or response axis. The first case is simply a reflection of the random error that is a part of all measurement systems and is allowed for by the error bounds placed on estimates of unknown concentrations. The second reflects a constant upward bias in the system responses and may or may not have implications for the validity of the results, depending on the measurement context, the magnitude of the bias and the intended use of the results.

In this section we will examine first how the regression results can help us to decide if a non-zero intercept represents a random or a systematic effect and, following that, we will discuss how a regression line through the origin may be fitted. For convenience the phosphate calibration data and part of the regression output are shown again in Table 6.7.

Table 6.7 *The DRP data and regression summaries for the calibration line*

DRP μg L^{-1}	*Absorbance*	*Response*
0	0.000	0
10	0.005	5
25	0.015	15
50	0.029	29
75	0.046	46
100	0.060	60

Predictor	Coef	SE coef	T	P
Constant	−0.5000	0.4970	−1.01	0.371
DRP	0.60769	0.00887	68.54	0.000

Note that the calibration line does not go through the origin although a zero response was obtained for the blank standard. Is this due to chance variation or does it reflect a systematic bias? A statistical significance test of the hypothesis that the true intercept is zero distinguishes between the two possibilities. The fact that the intercept is negative suggests that it differs from zero by chance. Nevertheless, we will carry out the test to illustrate the procedure.

If the null hypothesis that $\beta_o = 0$ is true, then the sampling distribution of the test statistic

$$t = \frac{\hat{\beta}_o - 0}{S\hat{E}(\hat{\beta}_o)} = \frac{\hat{\beta}_o - 0}{\sqrt{s^2\left(1/n + \bar{X}^2 / \sum_{i=1}^{n}(X_i - \bar{X})^2\right)}}$$

is the t-distribution with $n-2=4$ degrees of freedom. Recall that two degrees of freedom are lost when both the slope and the intercept of the line are estimated from the data. The frequency curve of the t-distribution with 4 degrees of freedom has 95% of its area between −2.78 and 2.78, as shown in Figure 6.11.

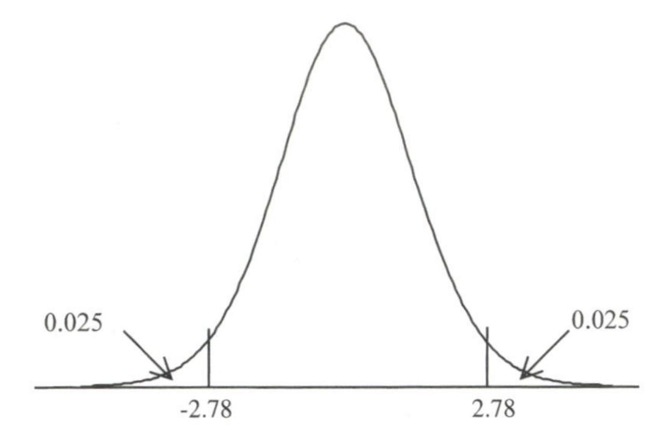

Figure 6.11 *Student's t-distribution with 4 degrees of freedom*

For the DRP dataset the calculated t-value is:

$$t = \frac{-0.0005000 - 0}{0.0004970} = -1.01$$

Since this lies between the critical values there is no basis for rejecting the hypothesis that the true intercept is zero. Note that the regression output gives the fitted intercept, its standard error, the t-ratio and also a p-value. The p-value indicates that 37% of the area under the distribution curve is further than 1.01 units from zero, *i.e.* the value of -1.01 is not at all unusual; the fitted intercept is not statistically significantly different from zero.

The p-value for the slope coefficient is given as $p = 0.000$ in the table, indicating rejection of the hypothesis that the slope β_1 is zero, *i.e.*, the hypothesis that there is no relationship between the response and the predictor variables. It would be a strange calibration study, indeed, if there were any likelihood of this being the case!

Exercise

6.9 Test the hypothesis that the calibration line for the HPLC data of Table 6.6, p 267, goes through the origin, *i.e.*, that the intercept is zero.

Fitting the Line Through the Origin. In many cases analysts will choose to fit the calibration line through the origin where no response is expected at zero analyte concentration. The slope should not differ very much from that obtained by the usual regression method, provided the intercept term found by the latter method is not statistically significant. If the fitted intercept is statistically significant (indicating a reproducible non-zero response at zero analyte concentration) the line should not be fitted through the origin. If, on technical grounds, it is believed that the line *should* go through the origin then the source of the bias must be identified and corrected before using a line through the origin.

Fitting through the origin may be carried out routinely by some automated analytical systems (*e.g.* chromatographs), especially when single point calibrations are used, *i.e.*, when only one standard is measured and the line is drawn from the origin to this point (there isn't much choice in this case!). Single point calibrations are undesirable – since the line is determined by a single measurement it will be influenced in a major way by the chance component in that measurement. Technically, this will be reflected in the standard errors of the slope and of predictions based on the line having large values. Where the calibration line is based on several standards (as in the DRP example) the chance components tend to be averaged out, but with only one result there is no opportunity for this to happen. Making several replicate measurements at the end-point of the working range will improve matters considerably. However, it will still provide no information on the linearity of the system response. Accordingly, if a single point calibration is to be used it is essential that the linearity of the response be investigated carefully when the system is first validated.

The results from fitting a regression through the origin for the DRP data are shown in Table 6.8 below. These results are based on the model:

$$Y_i = \beta X_i + \varepsilon_i \tag{6.20}$$

Table 6.8 *Regression analysis for DRP data: no intercept fitted*[a]

Predictor	Coef	SE Coef	T	P
Noconstant				
DRP	0.600796	0.005631	106.70	0.000

S = 0.7731

Analysis of Variance

Source	DF	SS	MS	F	P
Regression	1	6804.0	6804.0	11385.32	0.000
Residual Error	5	3.0	0.6		
Total	6	6807.0			

[a]Note that when fitting through the origin the ANOVA table changes somewhat, for technical reasons. In Table 6.8 $SSTO = \sum_{i=1}^{n} Y_i^2$, instead of the usual $\sum_{i=1}^{n}(Y_i - \bar{Y})^2$, and its degrees of freedom are n rather than $n-1$. SSE has $n-1$ degrees of freedom, since only one parameter β_1, is estimated from the data. For the same technical reasons, the coefficient of determination, r^2, is not calculated by Minitab, as under certain circumstances (not likely to be encountered in analytical calibrations) it can lie outside the range zero to one.

where i labels the observations. This equation states that each response Y_i is a simple multiple of the concentration of phosphate in the standard, plus a measurement error ε_i. The multiplier is the slope of the calibration line and this is estimated by the package using least squares, as before.

The slope of the line differs only marginally from that obtained by regression analysis, when an intercept term was included: 0.600796 *versus* 0.607692. To estimate the concentration of an unknown sample from its response Y_o, the regression equation

$$\hat{Y} = \hat{\beta}X$$

is inverted to give

$$\hat{X}_o = \frac{Y_o}{\hat{\beta}} \tag{6.21}$$

The formula for error bounds on this estimate is modified slightly from that used previously to give:

$$\hat{X}_o \pm \frac{t_{.025}}{\hat{\beta}} \sqrt{s^2\left(1 + \frac{Y_o^2}{\hat{\beta}^2 \sum_{i=1}^{n} X_i^2}\right)} \tag{6.22}$$

where $t_{.025}$ is based on $n - 1 = 5$ degrees of freedom.

If this equation is used to estimate the phosphate ion concentration for a response of 35, *i.e.*, an absorbance of 0.035, we obtain:*

$$\hat{X}_o \pm \frac{2.57}{0.600796}\sqrt{0.597613\left(1 + \frac{35^2}{(0.600796)^2 18850}\right)}$$

$$58.26 \pm 3.59\,\mu g L^{-1}$$

The results obtained here are different from those obtained from the calibration line that includes an intercept term. The error bounds were ± 3.86 previously, whereas they are $\pm$ 3.59 here. The main reason for this is that the t-value is smaller, 2.57 instead of 2.78. This reflects increased precision in the estimate of σ: only one regression coefficient is estimated here, rather than the two that were estimated when the intercept was included, leaving one more degree of freedom for estimating σ, 5 rather than 4. For larger sample sizes this difference would be smaller. In many cases, however, calibration lines are based on quite small sample sizes.

As before, if the calibration study is based on n individual measurements while the test material will be measured several times

*Note that all these calculations were done on a computer and rounded afterwards for presentation: accordingly minor differences may arise if the calculations are carried out using the summary statistics presented above.

and the results then averaged, account must be taken of the different measurement protocol. The modification to Expression (6.22) involves replacing 1 by $1/m$, as before. The error bounds are then given by Expression (6.23).

$$\hat{X}_o \pm \frac{t_{.025}}{\hat{\beta}} \sqrt{s^2 \left(\frac{1}{m} + \frac{Y_o^2}{\hat{\beta}^2 \sum_{i=1}^{n} X_i^2} \right)} \tag{6.23}$$

Note that Y_o is the average of m readings in (6.23), while it represents a single reading in (6.22).

Exercise

6.10 Re-fit the data of Table 6.6, p 267, through the origin. Calculate the inverse prediction limits corresponding to a peak area of 20 units and compare your results to those obtained when the intercept was included in the fitted line.

Concluding Remarks. The regression model underlying the calibration example discussed in this section assumed that the chance variation around the regression line was constant at all concentration levels, or at least approximately so. This is likely to be the case in examples such as the DRP data, involving only a quite limited response range; note that the absorbances ranged from zero to 0.06. Where much wider analytical ranges are involved in the study, it will very often be found that the variability of the system responses will increase with the concentrations of the standards or tests materials. In such cases, using ordinary least squares regression, as we have done, would not be the best statistical approach to the problem. While the fitted line will be an unbiased estimator of the true line, one with better statistical properties (smaller standard errors for the regression coefficients) will be obtained if weighted least squares is used in the fitting process. Also, the ordinary least squares confidence and prediction intervals, discussed earlier, will be inappropriate, as they take no account of the changing variability. The use of weighted regression is discussed in Section 6.6.

6.4 DETECTION LIMIT

The concept of a detection limit is important in trace analysis in that it provides a measure of the extent to which an analytical system can or cannot detect very low levels of an analyte. Many definitions of such a limit exist and the nomenclature in this area can be confusing: different names have been attached to the same quantity and different quantities

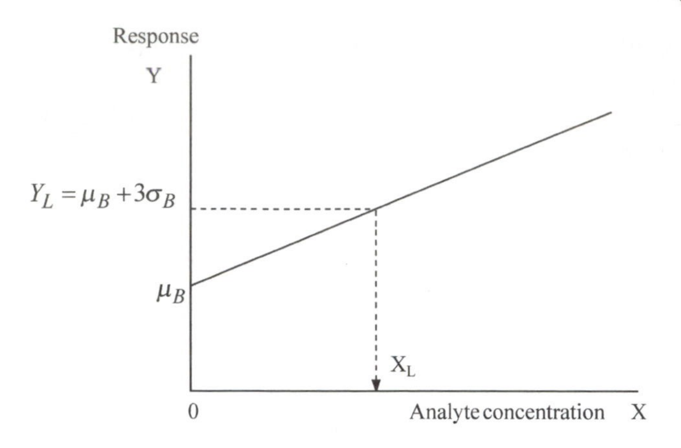

Figure 6.12 *Detection limits in the response and concentration domains*

have been given the same name by different authors. Here, the recommendation of the Analytical Methods Committee (AMC) of the Royal Society of Chemistry[5] will be adopted; this is in agreement with the IUPAC definition of detection limit.*

The AMC defined the detection limit of an analytical system to be the concentration or amount corresponding to a measurement level $3\sigma_B$ units above the value for zero analyte. The quantity σ_B is the standard deviation of responses of field blanks. A field blank is a sample containing zero concentration of analyte. It recommended that "other usages, definitions and named limits be discontinued. The only qualification to the detection limit should be the specification of the analytical system."

Figure 6.12 shows a schematic calibration line in which the detection limit in the response or signal domain Y_L is mapped onto the detection limit X_L in the concentration domain. The logic of the definition can be seen in Figure 6.13, where it is assumed that replicate measurements on a field blank will produce a Normal curve with mean μ_B and standard deviation σ_B.

This diagram indicates that if 10 000 measurements were made on a field blank only 13 would be expected to exceed Y_L. Accordingly, if an analysis results in a value that does exceed Y_L it can safely be taken to indicate the presence of the analyte. Values less than Y_L could have been generated by a test sample from which the analyte is absent. Often it will be appropriate to report such values as 'less than the detection limit', but

*Note that the AMC subsequently discussed the possibility of using the concept of uncertainty of measurement in the definition of detection limit (see Chapter 8). The older approach is discussed here since at the time of writing the use of measurement uncertainties is still not widespread in routine laboratory practice. Also, it provides an opportunity to discuss negative test results.

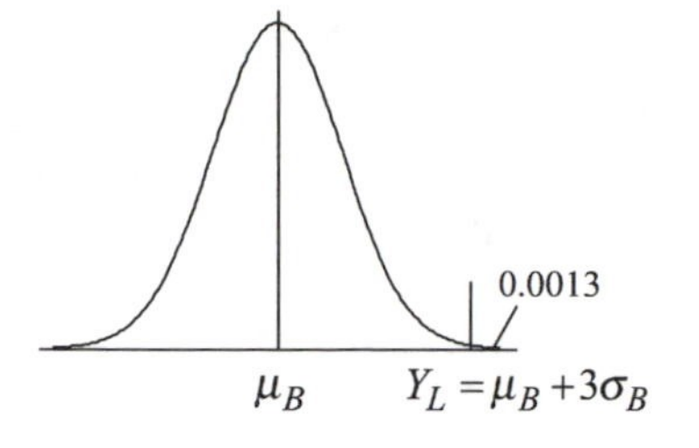

Figure 6.13 *Distribution of responses to a field blank*

if the results are to be incorporated into subsequent statistical calculations the actual test results must be reported.

To measure the Limit of Detection of an analytical system what is required is an estimate of σ_B, the standard deviation of replicate measurements of a field blank. A useful way to accumulate such information would be to insert one or more field blanks into each analytical run (see recommendations of the 'Harmonized Guidelines on Internal Quality Control in Analytical Laboratories' in Chapter 2, p 65) as part of the routine QC procedures. In this way the determinations will have been subjected to all the random influences that affect the measurement process over an extended time period. They will, therefore, be more representative of routine operating conditions than data derived from a special short-term study.

It is important in this context to note the implications of the random errors in the response domain for the possible measured results in the concentration domain. If the mean of the responses is mapped onto zero concentration, as shown in Figure 6.12, this implies an assumption of zero bias. Responses greater than the intercept value will correspond to positive results in the concentration domain and responses below the point of intersection of the calibration line on the response axis will correspond to *negative* values. When making replicate measurements on field blanks we should expect negative test results about half of the time! While this may seem odd (perhaps even nonsensical) Figure 6.12 shows that it is a simple consequence of the measurement process – the response is an indirect measure of concentration and random variation around the average response translates, through the calibration line, into random variation around zero, when measuring blanks. For the purposes of calculating the standard deviation of the replicate results when measuring the blanks, it is essential that the negative values be retained in the calculations. If they are truncated to zero, then the magnitude of the standard deviation will be seriously underestimated and an over-optimistic estimate of the Limit of Detection will be obtained.

The reader is referred to the AMC paper for a careful and lucid discussion of the theoretical and practical issues relevant to the definition and estimation of the Limit of Detection, including what to do when a realistic field blank is not available.

Exercise

6.11 In a validation study of the method for measuring total fat in foods and feeds, data from which were discussed in Chapters 3 and 4, Pendl *et al.*[6] measured 19 replicate method blanks. They obtained the results shown in Table 6.9 (units are mg).

They obtained the Limit of Detection (LOD) as the mean of the results, plus three standard deviations. Calculate the LOD and express it as a percentage of the sample weight, 5 mg.

Table 6.9 *Method blank replicate results*
(Reprinted from the Journal of AOAC INTERNATIONAL 1998, Vol. 81, pp 907. copyright, 1998, by AOAC INTERNATIONAL.)

4.00	1.83	2.00	2.17	3.00	2.83	2.67
2.17	2.50	2.00	3.00	2.83	2.83	
2.67	2.00	2.83	2.67	3.00	3.50	

6.5 RESIDUAL ANALYSIS

As with all statistical methods, the conclusions drawn from a regression analysis may be wrong if the assumptions underlying the model do not hold. It is worthwhile, therefore, to carry out some simple checks on these assumptions. This would be quite tedious without the aid of a computer but requires little extra effort if suitable statistical software is available.

The statistical model that underlies linear regression assumes that at any given level of the predictor X, the response Y has a mean value given by the true regression line. Replicate observations of Y at a given X will vary around the mean and will follow a Normal distribution with some standard deviation. The standard deviation of the distribution of Y values around the line will be the same for all values of X. These assumptions are illustrated in Figure 6.2, page 251.

This model implies that all the systematic variation is embodied in the straight line and, consequently, no further systematic variation should be discernable in the data. This means that if a scatterplot of the data that

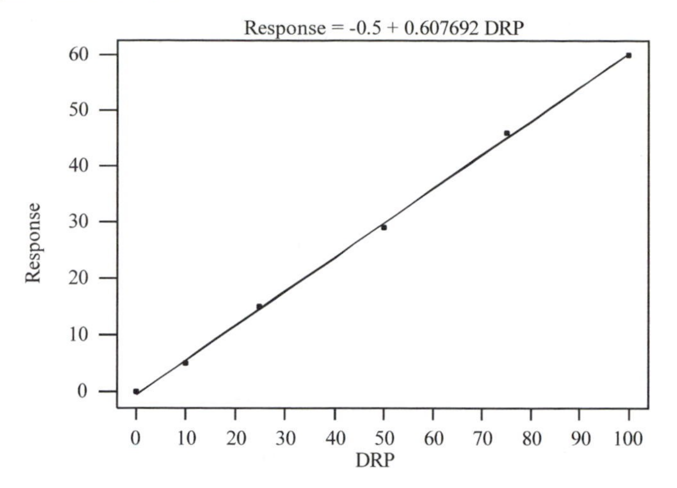

Figure 6.14 *A scatterplot of the DRP data with fitted regression line*

includes the fitted regression line is inspected, the data should vary randomly around the fitted line.

Example 1: the DRP Data. Figure 6.14 shows such a scatterplot for the DRP calibration data. While no very obvious non-random pattern is discernable, it is fair to say that the closeness of the data points to the fitted line would make it difficult to detect any trend that might be present.

Our ability to detect departures from random scatter will be much improved if, instead of plotting the raw data, we use residual plots for model validation. The residuals

$$e_i = Y_i - \hat{Y}_i = Y_i - (\hat{\beta}_0 + \hat{\beta}_1 X_i)$$

are the vertical deviations, e_i, of the data points from the fitted regression line. These residuals are estimates of the error terms ε_i and should have approximately the same properties. Accordingly, they should show no systematic trends or outliers – if they do, they call into question the assumptions underlying the regression model.

Experience suggests that departures from assumptions frequently involve some form of relationship between residuals and fitted values. Examples include error variation increasing with the magnitude of the responses or the existence of a non-linear response relationship with the predictor variable, X. Such relationships are likely to be evident in a scatterplot of residuals *versus* fitted values, as will be illustrated below.

Figure 6.15 shows a scatterplot of the residuals *versus* the fitted values for the DRP data. The residuals appear to be randomly scattered about zero with no discernable relationship with the predicted values. Thus, no

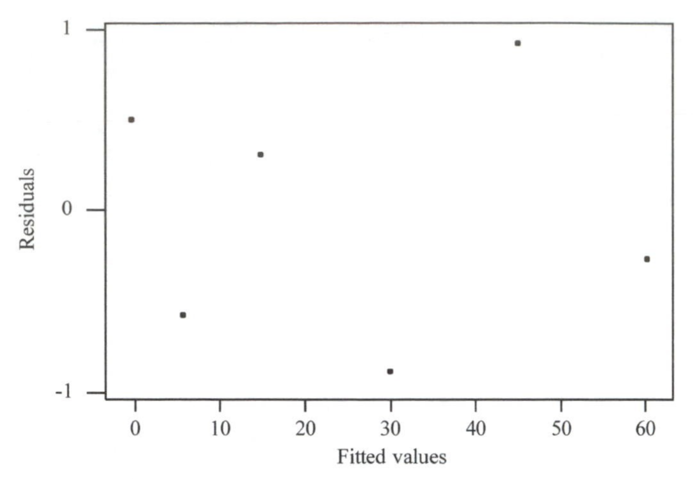

Figure 6.15 *Residuals plotted against fitted values for the DRP regression*

problems are indicated here. Since the data are assumed Normally distributed around the regression line the residuals should have the same properties as a sample from a Normal distribution with zero mean. Thus, if a Normal plot of residuals is drawn it should give an approximately straight line. Figure 6.16 shows such a plot; since the data approximate a straight line the Normality assumption is not called into question. Note, however, that Figures 6.15 and 6.16 are based on only six observations, hardly a satisfactory basis for assessing model assumptions. These data come from a routine calibration of the analytical system and so were not

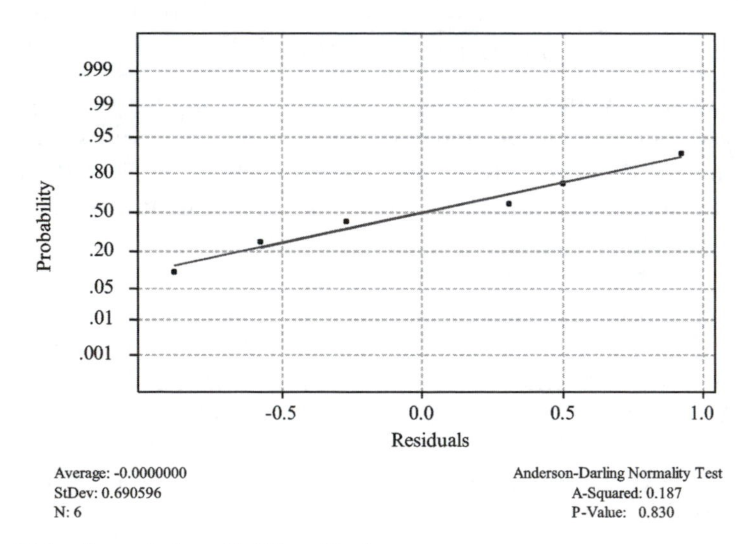

Figure 6.16 *Normal plot of DRP residuals*

collected for the purpose of model validation. Much more data might be expected to be available at the method development and validation stages. Using these, the model assumptions could be challenged in a more exacting way than was possible here.

Example 2: a Bio-analytical Assay. The data of Table 6.10 come from a bio-analytical assay validation study in which five replicate measurements were made at each of five concentration levels. The standards are spiked blood plasma samples (μg L^{-1}). The responses are 1000 times the peak height ratios, *i.e.*, the ratios of the heights of analyte peaks to the peak height for the internal standard. Figure 6.17 shows a regression line fitted to the calibration data. It is evident from Figure 6.17 that the scatter in the data is greater at the higher concentration levels.

Table 6.10 *Bio-analytical calibration data*

10	*20*	*50*	*100*	*150*
69.2	134.5	302.0	495.7	773.0
64.5	136.8	246.8	537.9	813.6
50.0	142.8	257.5	474.5	681.6
52.2	85.4	239.0	516.0	763.9
40.3	87.2	198.4	348.5	629.9

Figure 6.18 is the corresponding residuals *versus* fitted values plot. Figure 6.18 shows even more clearly the tendency for the variability of the data to increase with the response, or equivalently, with the concentration. The usual assumption of constant response variability is

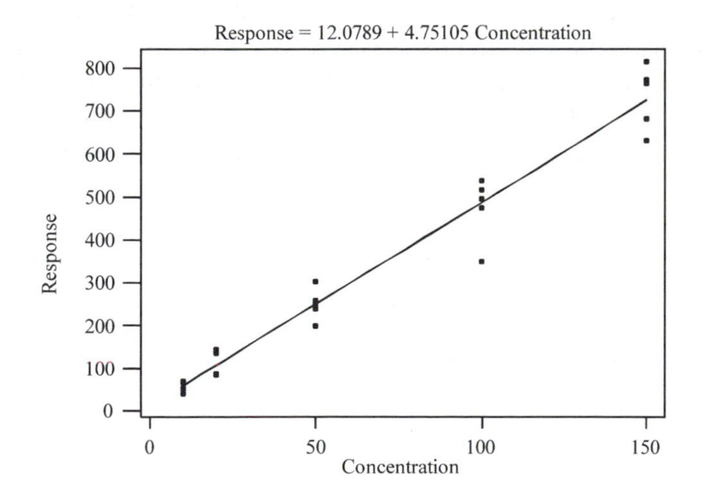

Figure 6.17 *A calibration line for the bio-analytical assay*

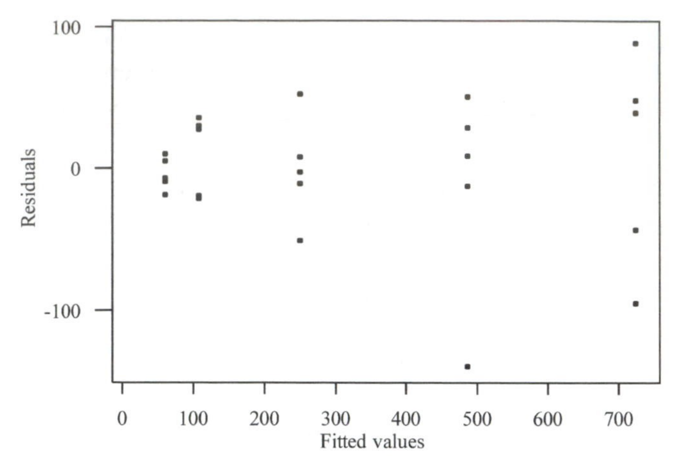

Figure 6.18 *Residuals versus fitted values for the bio-analytical data*

invalid in this case. Section 6.6 discusses how 'weighted' least squares can be used in place of 'ordinary' least squares to take account of the changing variability.

The pattern shown in Figures 6.17 and 6.18 is quite common for analytical calibrations, particularly where the concentrations or amounts measured vary over a wide analytical range. In Chapter 1 it was pointed out that where variability changes with the magnitude of the measured values a single standard deviation is inadequate as a measure of precision. In such cases the relative standard deviation or coefficient of variation will often be a more appropriate system performance measure.

Example 3: a Second Bio-analytical Calibration. Table 6.11 contains data from a second bio-analytical calibration study. The data comprise seven replicate measurements at each of six levels of concentration. Again, the standards are spiked blood plasma samples (μg L^{-1}) and the responses are 1000 times the peak height ratios, *i.e.*, the ratios of the heights of analyte peaks to the peak height for the internal standard.

Table 6.11 *Bio-analytical calibration data*

1	*2*	*4*	*40*	*80*	*160*
10.6	15.6	34.8	241.2	490.6	872.6
11.3	13.4	34.0	229.7	497.0	891.5
9.1	21.5	27.0	262.1	487.4	907.9
16.9	16.3	29.2	242.0	502.4	870.3
13.1	21.4	30.4	250.2	462.3	857.0
11.0	24.6	32.0	251.0	471.6	918.8
13.6	15.2	34.9	247.9	487.4	864.4

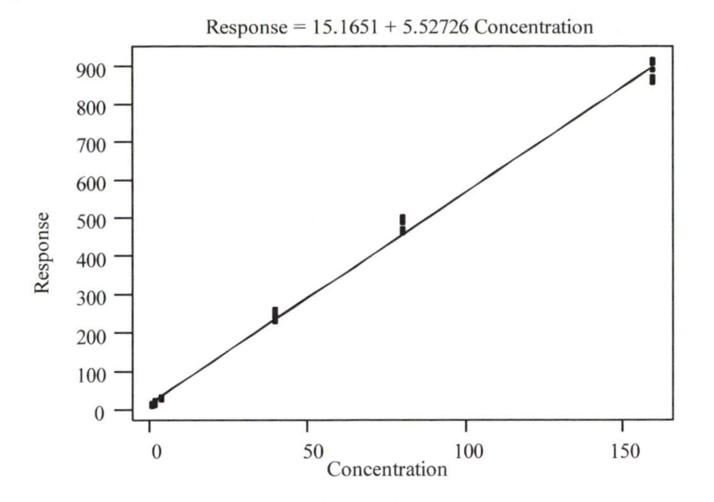

Figure 6.19 *The calibration line for the bio-analytical data*

Figure 6.19 shows a regression line fitted to the data; Figure 6.20 is the corresponding residuals *versus* fitted values plot. The data lie close to the calibration line but, as can be seen in Figure 6.19 and more clearly in Figure 6.20, there is evidence of two types of departure from the standard assumptions. The relationship between response and concentration appears to be non-linear, as there is a curvilinear pattern in the residuals *versus* fitted values plot. Also, the variability increases with the magnitude of the response values. These data furnish a good example of the occurrence of a very high r^2 value, 0.997, despite the relationship

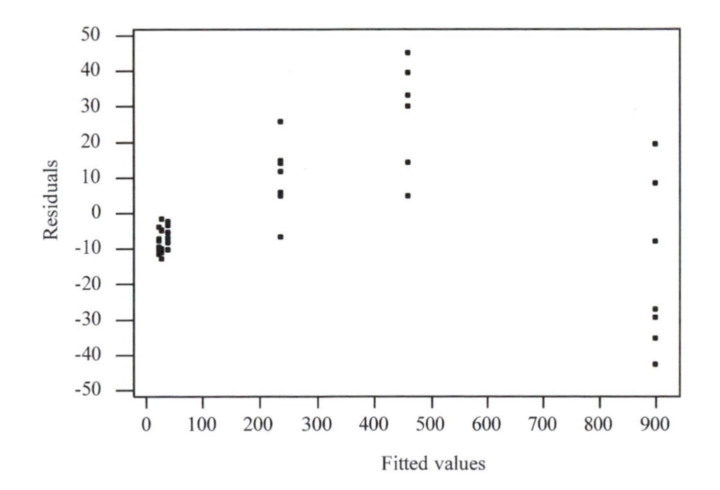

Figure 6.20 *Residuals versus fitted values from the calibration*

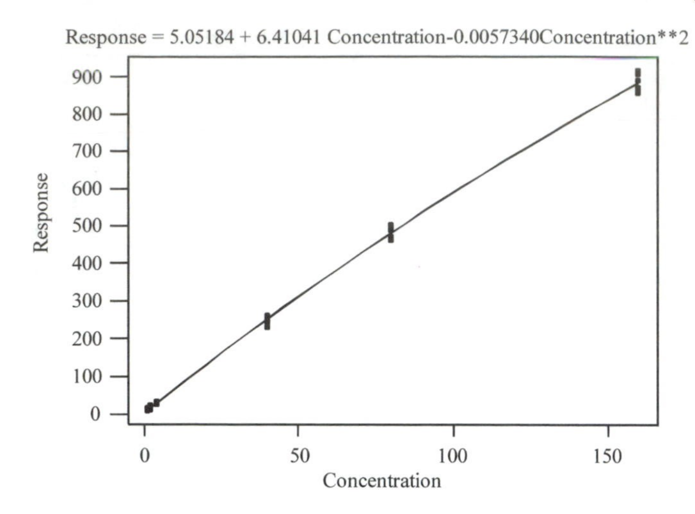

Figure 6.21 *A curved response function for the bio-analytical calibration*

being non-linear. As a summary statistic, r^2 measures closeness of fit of the line to the data, not linearity, as is commonly assumed.

The curvature can be allowed for by fitting a non-linear calibration line to the response data. How this is done is discussed later. Figure 6.21 shows the results of the fitting process: the fitted equation is given above the regression plot (the notation concentration**2 means concentration squared). Figure 6.22 shows the residuals *versus* the fitted values plot. The residuals are the deviations of the data points from the curved line of Figure 6.21.

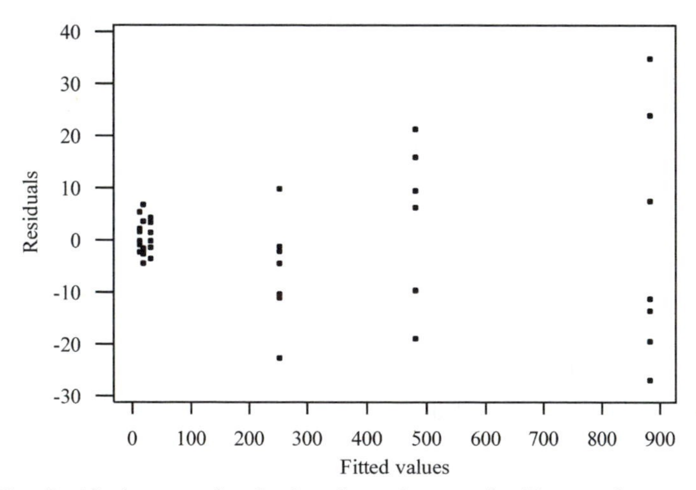

Figure 6.22 *Residuals versus fitted values from the curved calibration line*

In Figure 6.22 the curvature that was evident in Figure 6.20 has disappeared. This means that the curved line involving the square of the concentration is adequate to take account of the curvature in the data. In general, there is no guarantee that such a simple curve will be an adequate model and that is why is important to carry out residual analysis. The tendency for the variability to increase with the size of the response is still striking, however; fitting the curved line would not be expected to change this.

Calibration lines whose slopes decrease at higher response levels are commonplace. It is sometimes recommended that the 'linear range' of a method should be found by progressively deleting the largest values until curvature disappears; see *e.g.*, the NCCLS document EP-6.[7] This could be done based on visual inspection of plots such as Figures 6.19 and 6.20 or a statistical test for linearity might be employed in making the cut-off decision. Such a test is discussed later.

Exercises

6.12 Carry out a residual analysis for the HPLC calibration data of Table 6.6, p 267.

6.13 The data shown in Table 6.12 come from a paper published by the Analytical Methods Committee (AMC) of the Royal Society of Chemistry[8] with the title 'Is my calibration linear?' They are calibration data for a spectrophotometric method used for the determination of iron. The concentrations are in $\mu g\ mL^{-1}$.

- Draw a scatterplot of the data.
- Fit the least squares calibration line.
- Obtain residuals and fitted values and plot them against each other.
- What do you conclude?

Table 6.12 *Spectrophotometric calibration data*
(Reprinted from the Analyst, 1994, Vol. 119, pp 2363. Copyright, 1994, Royal Society of Chemistry.)

Conc.	*Absorbance*	
64	0.138	0.142
128	0.280	0.282
192	0.423	0.423
256	0.565	0.571
320	0.720	0.718
384	0.870	0.866

6.14 The paper cited above contains a second dataset, which is shown in Table 6.13. These are calibration data for ^{239}Pu determined by electrothermal vaporization inductively coupled plasma mass spectrometry (ETV-ICP-MS). The concentration units are pg mL^{-1}. Responses were measured in triplicate in a stratified random order. Fit a linear regression model to the data and carry out a residual analysis. What do you conclude?

Table 6.13 *ETV-ICP-MS calibration data*
(Reprinted from the *Analyst*, 1994, Vol. 119, pp 2363. Copyright, 1994, Royal Society of Chemistry.)

Conc.	*Responses*		
0	13	18	14
10	710	709	768
20	1494	1453	1482
30	2117	2207	2136
40	3035	2939	3160
50	3604	3677	3497
60	4348	4352	4321
70	5045	5374	5187

6.6 WEIGHTED REGRESSION

We have seen that ordinary least squares regression calculates the regression coefficients by minimizing the sum of the squared vertical deviations about the fitted line:

$$Q_{OLS} = \sum_{i=1}^{n} (Y_i - \hat{Y}_i)^2 = \sum_{i=1}^{n} (Y_i - \hat{\beta}_o - \hat{\beta}_1 X_i)^2 \tag{6.24}$$

i.e., the values of $\hat{\beta}_o$ and $\hat{\beta}_1$ are chosen such that the sum Q_{OLS} is a minimum. This calculation treats all observations equally. However, for situations such as that illustrated by Figures 6.18 and 6.20, in which the data on the left-hand side of the plot are subject to less chance variation than those on the right-hand side, it seems appropriate to allow the observations subject to less chance variation a greater influence in determining the values of the regression coefficients. One way to achieve this is to use Weighted Least Squares (WLS).

The WLS regression line is found by minimizing:

$$Q_{WLS} = \sum_{i=1}^{n} w_i(Y_i - \hat{Y}_i)^2 = \sum_{i=1}^{n} w_i(Y_i - \hat{\beta}_o - \hat{\beta}_1 X_i)^2 \tag{6.25}$$

in which the weight attached to each squared deviation is the reciprocal of the variance of the corresponding observation (*i.e.*, $w_i = 1/Var(Y_i)$). Because the reciprocal is used, observations with large variance will have relatively little weight in determining the fitted line, while those subject to smaller chance variability will have greater influence.*

6.6.1 Fitting a Calibration Line by WLS

The fitting of a weighted least squares line will be illustrated using the data of Example 2 of Section 6.5 (Table 6.10, p 281). Figure 6.17 is a scatterplot of the data and Figure 6.18 is the corresponding residuals *versus* fitted values plot, based on an ordinary least squares fit. They both suggest that WLS rather than OLS is appropriate in fitting the calibration line. In order to carry out a weighted least squares analysis a set of weights is required; unless they are known in advance, the weights must be estimated from the calibration data. Accordingly, fitting the regression line takes place in two stages:

- we use the replicated data of Table 6.10 to obtain appropriate weights;
- we use the resulting weights in a WLS fit of the calibration line.

Estimating Weights. An obvious set of estimated weights may be calculated directly from the standard deviations of the response data shown in Table 6.10 ($w_i = 1/s_i^2$ for each group of observations). Being based on the data, these weights are subject to chance variation. However, we can reduce the effect of chance variation by taking advantage of the relationship that exists between response standard deviation and response magnitude.

It is often found that where response variability changes, the standard deviations of responses tend to be linearly related to the response magnitudes, which, in turn, are linearly related to the concentration. Figure 6.23 shows the standard deviations of the response data of Table 6.10 plotted against the concentrations. A regression line has been added to the plot to guide the eye. The plotted standard deviations are evidently subject to chance variation; the line smoothes this chance variation. Hence, weights calculated from the standard deviations that

*Note that when each deviation, $Y_i - \hat{Y}_i$, is scaled by dividing by the standard deviation, σ_i, of the responses at that level of X, we get $(Y_i - \hat{Y}_i)/\sigma_i$. If these quantities are squared and summed, the result is Q_{WLS} as shown in equation (6.25). This is a simpler way of viewing what is involved, but since computer packages tend to describe the weighting in terms of the reciprocals of the variances, using Equation (6.25), WLS has been presented in this way in the text.

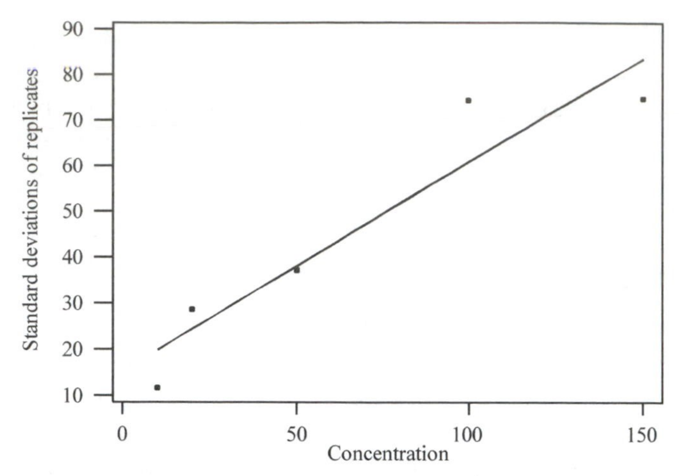

Figure 6.23 *Smoothing the standard deviations (SDs) by regression*

are predicted by the regression line will also be smoother than weights based directly on the standard deviations of the observed data.

Weighted least squares is used here in fitting a regression line to the standard deviations since the variability of sample standard deviations is known to be proportional to the corresponding true standard deviations that are to be estimated. Thus, for Normal variables the standard error of the sample standard deviation, s_i, is $\sigma_i/\sqrt{2n_i}$, where the sample standard deviation, s_i, is based on n_{i} results, and σ_i is the long-run or true standard deviation of the response, Y_i. This may be estimated by $s_i/\sqrt{2n_i}$ which gives weights $w_i = 2n/s_i^2$ for fitting a regression line of the form:

$$\hat{\sigma}_i = a_0 + a_1 X_i \tag{6.26}$$

From Equation (6.26) we obtain smoothed estimates, $\hat{\sigma}_i$, of the standard deviations at the different concentration levels. These are used to calculate new weights which are then used in place of the previous set of weights in fitting a second regression line. This process is iterated until the fitted line is only marginally different from the previous line. The final line provides a set of weights that can be used to fit the calibration line using WLS. Table 6.14 shows the results of four iterations of this fitting process. The regression coefficients for fit number 4 are only marginally different from those for fit number 3; the fitting process is halted as a result.

Fitting the Calibration Line. The equation generated by fit number 4 of Table 6.14 is:

Table 6.14 *Results of regression analyses to determine weights*

Fit number	*Weights*	*Regression of SD on X to obtain weights*		
		a_o	a_1	*r-sq*
1	initial-wts	7.92592	0.54446	0.889
2	fit-1	9.48394	0.57247	0.845
3	fit-2	9.67892	0.56703	0.850
4	fit-3	9.72458	0.56578	0.851

$$\hat{\sigma}_i = 9.724 + 0.566X_i \tag{6.27}$$

From this the weights for the calibration line regression are given by $w_i = 1/\hat{\sigma}_i^2$. These are plotted in Figure 6.24, in which the weights have been multiplied by 10^4 to eliminate the leading zeros.

Figure 6.24 shows that responses at the higher concentration levels are given very little weight, relatively, in determining the regression coefficients, and hence, the fitted line. Note that the weights are often rescaled such that they sum either to one or to n, the total number of observations in the dataset. Since it is only the relative magnitudes of the weights that are important, rescaling has no effect on the results.

When these weights are used in fitting the line the following calibration equation is obtained:

$$\hat{Y} = 12.14 + 4.76X$$

with an r^2 value of 0.956.

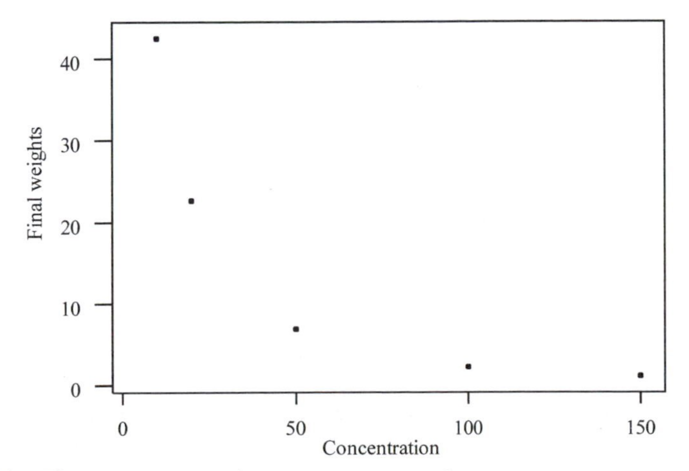

Figure 6.24 *The regression weights versus concentration*

6.6.2 Is Weighting Worthwhile?

To decide if using weighted least squares in place of the simpler ordinary least squares has been worthwhile, we examine first the difference it makes to the fitted calibration line and then consider the implications for using the calibration line for prediction.

A Comparison of the OLS and WLS Calibration Lines. Table 6.15 shows the regression coefficients and their standard errors for both ordinary and weighted least squares calibration lines for the bio-analytical data. The results indicate that the fitted lines are only marginally different from each other – this is often the case. The major difference is in the estimated standard error of the intercept, for which the WLS estimate is less than half that of the OLS estimate. The reason for this is that OLS regression takes no account of the fact that the variation of the data close to the origin is much smaller than at higher concentrations.

Table 6.15 *Ordinary and weighted least squares regression summaries*

	$\hat{\beta}_o$	$S\hat{E}(\hat{\beta}_o)$	$\hat{\beta}_1$	$S\hat{E}(\hat{\beta}_1)$
OLS	12.08	15.80	4.75	0.19
WLS	12.14	6.87	4.76	0.21

Prediction. The influence of the weighting will be important when determining prediction error. A prediction interval for the concentration corresponding to a response of Y_o is given by:

$$\frac{Y_o - \hat{\beta}_o}{\hat{\beta}_1} \pm \frac{t}{\hat{\beta}_1}\sqrt{MSE\left(\frac{1}{w_o} + \frac{1}{\sum_{i=1}^{n} w_i} + \frac{(Y_o - \bar{Y}_w)^2}{\hat{\beta}_1^2 \sum_{i=1}^{n} w_i(X_i - \bar{X}_w)^2}\right)} \qquad (6.28)$$

The means that appear in the formula are weighted means; thus, $\bar{X}_w = \sum_{i=1}^{n} w_i X_i / \sum_{i=1}^{n} w_i$ and the weighted mean of the Y values is similarly defined. The MSE that appears in the formula is that derived from the weighted regression.* The formula for the bounds includes a weight, w_o, for the

*The weighted ANOVA decomposition of sums of squares is given by: $\sum_{i=1}^{n} w_i(Y_i - \bar{Y}_w)^2 = \sum_{i=1}^{n} w_i(Y_i - \hat{Y}_w)^2 + \sum_{i=1}^{n} w_i(\hat{Y}_i - \bar{Y}_w)^2$. Accordingly, MSE is given by the first term on the right-hand side, divided by the degrees of freedom $n-2$.

response, Y_o. To find an appropriate weight we insert the corresponding point predictor

$$\hat{X}_o = \frac{Y_o - \hat{\beta}_o}{\hat{\beta}_1} \tag{6.29}$$

into the equation that describes the relationship between the standard deviation of responses and the concentration:

$$\hat{\sigma}_o = 9.724 + 0.566\hat{X}_0$$

to obtain the standard deviation, which in turn gives the corresponding weight $w_o = 1/\hat{\sigma}_o^2$.

Figure 6.25 is a schematic representation of the weighted prediction bounds. The bounds spread out as we move away from the weighted means of the data. The relative magnitudes of the weights result in the weighted means, $\bar{Y}_w$ and $\bar{X}_w$, being closer to the origin than are the corresponding unweighted means. Because of this, the inverse predictions, $\hat{X}_o$, corresponding to large response values, Y_o, being further removed from the means, have much wider prediction errors, as illustrated in the Figure.

Table 6.16 shows prediction intervals corresponding to two response values, one at low and one at high concentration. The WLS bounds are nearly four times as wide for the predicted concentration corresponding to a response of 600 than they are for the lower value of 100; this is consistent with the general shape illustrated in Figure 6.25. The two sets of OLS bounds, on the other hand, are virtually the same. The error bounds from the WLS fit are much narrower for $Y_o = 100$ and wider for $Y_o = 600$ than are the corresponding OLS bounds. This is in line with

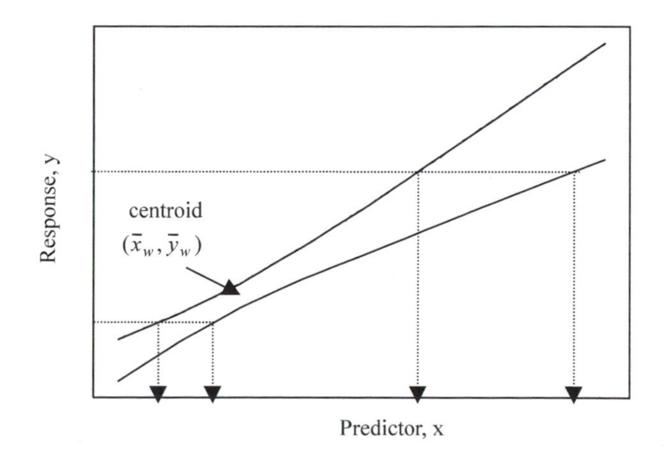

Figure 6.25 *A schematic illustration of weighted prediction bounds*

Table 6.16 *WLS and OLS prediction bounds for the bio-analytical data*

	OLS regression		*WLS regression*	
Y_o	$\hat{X}_o$	*OLS bounds*	$\hat{X}_o$	*WLS bounds*
100	18.5	±22.1	18.5	±9.0
600	123.8	±22.3	123.5	±35.8

what might be expected intuitively, since WLS takes account of the smaller chance variation at low concentration and the greater variation at high concentration values.

For both the weighted and unweighted regressions the prediction bounds fan outwards symmetrically from the means of the X and Y values (the centroid), where the means are weighted means in the WLS case and unweighted means for the OLS regression. For the unweighted analysis, $\bar{Y} = 325.7$, which is almost exactly half way between the two responses, 100 and 600, and, accordingly, the widths of the prediction bounds are very similar. For the weighted analysis, $\bar{Y}_w = 114.3$ is very much closer to 100 than to 600 and the prediction bounds at the lower response are correspondingly narrower, as illustrated in Figure 6.25.

Residual Analysis. Is the weighted least squares line a good fit? We carry out residual analysis as we did earlier, but plot *weighted* residuals instead of ordinary or *unweighted* residuals. A weighted residual is given by $\sqrt{w_i}(Y_i - \hat{Y}_i)$. Note that OLS minimizes the sum of the squared unweighted residuals, $\sum_{i=1}^{n}(Y_i - \hat{Y}_i)^2$, while WLS minimizes the sum of the squared weighted residuals $\sum_{i=1}^{n} w_i(Y_i - \hat{Y}_i)^2$, so there is a direct correspondence to what we did before.

Figure 6.26 shows a plot of the weighted residuals *versus* the fitted values. The tendency for the variability of the residuals to increase with the fitted values that was apparent in Figure 6.18 has disappeared. Figure 6.27 is a Normal probability plot of the weighted residuals. Figures 6.26 and 6.27 suggest that, after adjusting for the changing variability by using the weights, the residuals may be regarded as coming from a Normal distribution with mean zero and constant standard deviation, thus validating the model on which the regression analysis is based.

Comments. Although the calculations required for weighted regression are somewhat more complicated than those for ordinary least squares, conceptually there is little difference between the two situations and the

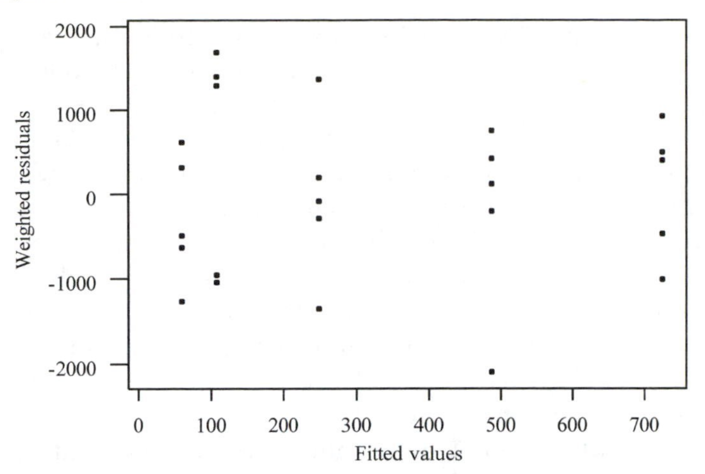

Figure 6.26 *A plot of weighted residuals versus fitted values*

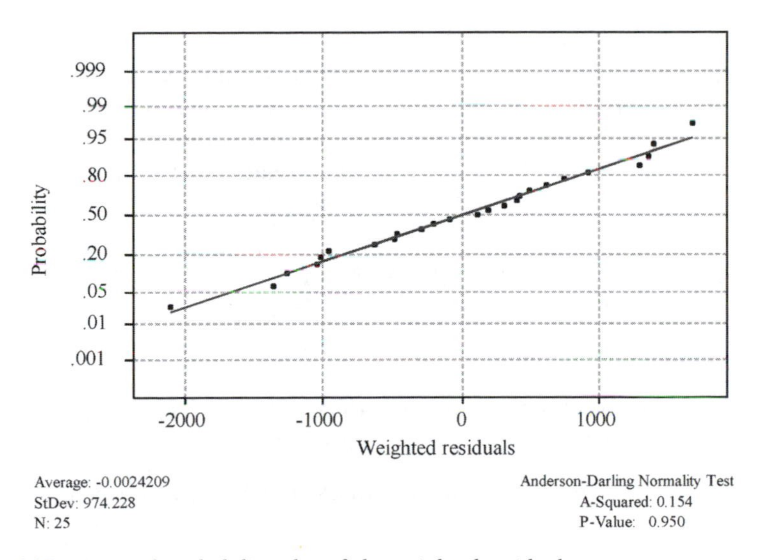

Figure 6.27 *Normal probability plot of the weighted residuals*

reader is urged not to be deterred by the extra complications in the calculations. Good statistics software will take care of these.

What are of more concern, though, are the analytical considerations that are implied by having to deal with data having such properties. The key implication is that any statement about analytical precision must refer to the concentration level or amount of analyte being measured. The reader is referred back to Section 1.5, where the ideas are discussed briefly and to the references cited there, for further discussion.

In the case study of Section 1.6 an approach to accumulating information on repeatability precision from routine data collection was suggested, as an alternative to special validation studies. The same considerations apply here: by collecting small amounts of data regularly at different levels of X, we can build up information on the relationship between response variability and concentration or analyte amount. Such information may be collected by inserting replicate standards, at different concentration/amount levels, into routine analytical runs. Alternatively, where test materials are measured at least in duplicate, then the same information can be obtained by modelling the relationship between the standard deviations of the replicates at different concentration levels and the concentration levels. The resulting information can then be used in fitting routine calibration lines and in providing measures of precision at different analyte levels.

Exercises

6.15 To simplify the procedure described above, suppose we decide that when finding weights we will regress the sample standard deviations on the concentrations, once only by OLS, instead of iteratively by WLS. Will this have a major impact on the results? To investigate, carry out the following analyses.

- Calculate the sample standard deviation of responses at each of the concentration levels (c_i) of Table 6.10, p 281.
- Regress the SDs on the concentrations using OLS.
- Calculate weights for the different concentration levels (c_i) using $W_\text{i} = 1/$(fitted value (c_i))^2.
- Do a weighted regression of the responses on the concentrations using these weights.
- From the results extract MSE, the regression coefficients, and their standard errors.
- Compare your results to those in Table 6.15.
- Use a spreadsheet to set up formulae to carry out each of the following calculations.
- Calculate weighted means for X and Y (concentration and response, respectively) and the weighted sum of squared deviations of the X values.
- Calculate the inverse predictor, $\hat{X}_o$, (see Equation (6.29) and the discussion immediately following) and the weights corresponding to $Y_\text{o} = 100$ and 600.

- Use Equation (6.28) to calculate error bounds for the inverse predictor and compare your results to those of Table 6.16.

6.16 Exercise 6.14 was concerned with calibration data for ^{239}Pu determined by electrothermal vaporization inductively coupled plasma mass spectrometry (ETV–ICP–MS). The concentration units are pg mL^{-1}. Responses were measured in triplicate in a stratified random order.

- Calculate the sample standard deviations at each concentration level and plot them against the concentrations (X).
- You will find that one standard deviation is anomalously low.
- Regress the sample standard deviations on X, using OLS and find the fitted standard deviations. Do this once, *i.e.*, follow the same procedure as in the last Exercise. Disregard the anomalously low standard deviation when doing this regression.
- You will find that the fitted standard deviation corresponding to $X = 0$ is negative. Replace it with the observed value, 2.646, in what follows.
- Use the fitted standard deviations to obtain weights.
- Carry out the WLS analysis and obtain residuals and fitted values.
- Plot both the residuals and weighted residuals (*i.e.*, residuals multiplied by the square root of the corresponding weights) against the fitted values. What do you conclude?
- Obtain weighted inverse prediction intervals for $Y_o = 1000$ and $Y_o = 5000$.

6.7 NON-LINEAR RELATIONSHIPS

Apart from one example illustrating residual analysis in Section 6.5, our discussion of regression analysis up to this has been concerned with fitting a linear relationship between a single predictor variable, X, and a response variable, Y. It is obvious that in very many cases the relationships between variables encountered in the analytical laboratory will not be linear and that more complex methods will be required for their analysis. Often, the relationship can be transformed to linearity, and the analysis then proceeds as described earlier. An alternative approach is to fit a non-linear model using multiple regression. Only a short introduction to what is a very wide subject area will be attempted here. The reader will then be referred to more specialized works for further details.

6.7.1 A Single Predictor Variable

As a first example, consider the data of Table 6.17. The data come from a study that investigated optimal conditions for the extraction of tobacco alkaloids using ultrasonic and microwave extraction methods.[9] The first column shows the extraction temperature (°C), while the second column gives the percentage nicotine extracted from tobacco.

Figure 6.28 shows nicotine concentration plotted against extraction temperature. The curve superimposed on the scatterplot was fitted by multiple regression: the fitted curve was obtained by choosing values for the coefficients of the quadratic equation

$$\hat{Y}_i = \hat{\beta}_o + \hat{\beta}_1 X_i + \hat{\beta}_2 X_i^2 \tag{6.30}$$

such that the sum of the squared deviations of the observations from the corresponding points on the curve was minimized, *i.e.*, by ordinary least squares. A regression equation is described as a 'multiple regression' equation if it contains more than one predictor variable. Where the terms in the model are powers of one or more variables then the equation is described as a 'polynomial regression model'. Every general-purpose statistical package will allow the fitting of such curves.

Table 6.17 *The nicotine study data*
(Reprinted form the Journal of AOAC INTERNATIONAL, 2001, Vol. 84, pp 309. Copyright, 2001, by AOAC INTERNATIONAL.)

Temperature/°C	*%Nicotine*
41.0	3.279
41.0	3.401
41.0	3.750
41.0	3.745
74.0	3.973
74.0	4.319
74.0	4.498
74.0	4.483
30.0	3.145
85.0	4.595
57.5	3.945
57.5	4.243
57.5	4.217
57.5	4.164
57.5	4.393
57.5	4.023
57.5	4.174
57.5	4.126

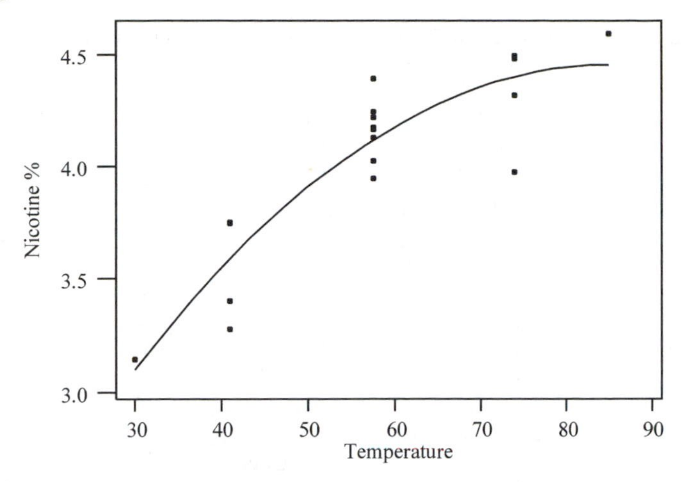

Figure 6.28 *Nicotine versus temperature (°C) with a superimposed quadratic curve*

It is clear from Figure 6.28 that larger values of temperature give higher extraction rates and that a temperature close to 80 °C would be a suitable system setting if, as here, the influences of the other variables are ignored.*

Table 6.18 shows two sets of Minitab output; the first is from a straight-line regression of nicotine on temperature and the second is from the curved model of Equation (6.30). The r^2 value is higher for the curved model as the curvature term, X^2, brings the fitted line closer to the data. This must happen since the objective of the least squares procedure is to find the line or curve that is closest to the data; if the X^2 term did not help to achieve this, then the fitting algorithm would automatically set its coefficient to zero or very close to zero. Note that the degrees of freedom in the ANOVA tables differ because, in general, those for error are $n-p$ and those for regression are $p-1$, where n is the total number of observations and p is the number of parameters estimated in the fitting procedure; p is 2 for the simple linear regression, while it is 3 for the polynomial regression.

Is the Curvature Term Worth Including? Table 6.18 shows that fitting the X^2 term increases the fraction of the total sum of squares that is associated with the regression model by approximately 7%. Is this improvement in the fit of the model to the data an indication that the curved model is a better representation of the underlying relationship, or

*The variation between the 'replicates' at three levels of temperature is not pure chance error, in that not all other system conditions were held constant for these runs. The data were extracted from a more complex experimental design (a response surface design of the type discussed below). However, the complications are ignored here – the data presented above are sufficiently well-behaved for illustrating the use of a polynomial regression.

Table 6.18 *Linear and quadratic regression models for the nicotine data*

Regression Analysis: Nicotine% *versus* Temp

Predictor	Coef	SE Coef	T	P	
Constant	2.6086	0.2121	12.30	0.000	
Temp	0.024656	0.003579	6.89	0.000	
S = 0.2174	R-Sq = 74.8%				
Analysis of Variance					
Source	DF	SS	MS	F	P
Regression	1	2.2435	2.2435	47.46	0.000
Residual Error	16	0.7563	0.0473		
Total	17	2.9998			

Regression Analysis: Nicotine% *versus* Temp, Temp-sqrd

Predictor	Coef	SE Coef	T	P	
Constant	1.2041	0.6312	1.91	0.076	
Temp	0.07674	0.02257	3.40	0.004	
Temp-sqr	−0.0004529	0.0001943	−2.33	0.034	
S = 0.1924	R-Sq = 81.5%				
Analysis of Variance					
Source	DF	SS	MS	F	P
Regression	2	2.4445	1.2223	33.02	0.000
Residual Error	15	0.5553	0.0370		
Total	17	2.9998			

could the improved r^2 be the result of chance variation? A statistical test will answer this question.

The t-statistic associated with the X^2 term, $t = -2.33$, tests the null hypothesis that $\beta_2 = 0$ against the alternative that $\beta_2 \neq 0$, *i.e.*, it compares the two models:

$$H_o : Y_i = \beta_o + \beta_1 X_i + \varepsilon_i \tag{6.31}$$

$$H_1 : Y_i = \beta_o + \beta_1 X_i + \beta_2 X_i^2 + \varepsilon_i$$

Since the p-value corresponding to $t = -2.33$ is less than 0.05, we reject the null hypothesis and conclude that the curvature term is statistically significant – it is unlikely to be an artefact of the chance variation in the system.

It is worth noting, here, that although multiple regression models are easy to fit using appropriate software, their interpretation and use can present many difficulties. Note, for example, that the coefficient of temperature in the linear model is $\hat{\beta}_1 = 0.025$ while it is $\hat{\beta}_1 = 0.077$ for the polynomial model. It will often be the case that the regression coefficient for any one variable, and its statistical significance or lack of it, will depend on which other variables are present in a multiple regression model. This, of course, makes the resulting equations difficult to interpret in terms of the separate influences of individual variables on the response. However, while the individual effects may be difficult to disentangle, the combined predictive power of the set of predictors can be useful in optimizing system performance. This is illustrated in the response surface example described later in this section; further information will be found in the regression textbooks cited in the references.

Exercise

6.17 Exercise 6.13 was concerned with calibration data for a spectrophotometric method used for the determination of iron. The residual analysis carried out in that exercise suggested a non-linear response. Fit a polynomial model with both X and X^2 and assess whether the squared term is worth including in the model. The data are given in Table 6.12, p 285.

6.7.2 A 'Lack-of-fit' Test

The statistical test for curvature described above provides a useful approach to the assessment of linearity, which is a common requirement of validation protocols. The test is specific in that it assesses linearity against the specific alternative of a quadratic response. It is, however, likely to be useful in very many cases where a calibration line is being fitted and there is the possibility that the slope of the response will decrease as the analyte concentration or amount level increases.

A more general-purpose 'lack-of-fit' test may be carried out when the data contain replicate responses at some of the levels of the predictor variable. Consider for example the spectrophotometric calibration data of Exercise 6.13, that relate to a method used for the determination of iron. The responses shown in Table 6.19 are the absorbances multiplied by 10,000.

Table 6.19 *Calibration data for iron*
(Reprinted from the Analyst, 1994, Vol. 119, pp 2363. Copyright, 1994, Royal Society of Chemistry.)

Conc.	*Response*	
64	1380	1420
128	2800	2820
192	4230	4230
256	5650	5710
320	7200	7180
384	8700	8660

If the variances (s^2) of the duplicates at each of the six concentration levels are calculated and averaged, then the resultant is a mean square for 'pure error' of 633.33 (with 6 degrees of freedom, one from each duplicate). This is shown in the Minitab output in Table 6.20. The corresponding sum of squares for pure error is 3880.

If the straight-line model is appropriate for the calibration data, then the residual or error sum of squares will reflect only chance variation around the regression line. If, on the other hand, the straight line is inappropriate as a description of the variation in mean response with concentration level, then the residual sum of squares will contain a bias, or lack-of-fit, component, since the residuals represent the differences between the observations and the fitted line. Thus, if the fitted line is biased, the bias will automatically become part of the residuals and therefore a component in their sum of squares. A sum of squares to estimate the magnitude of this possible lack-of-fit component can be obtained by subtraction.

$$\text{SS(lack-of-fit)} = \text{SS(residual error)} - \text{SS(pure error)} = 14621$$
$$\text{df(lack-of-fit)} = \text{df(residual error)} - \text{df(pure error)} = 4$$

The ratio MSLF/MSPE = 5.77 provides a formal test for lack of fit. The null hypothesis is that the model fitted to the data (here a simple linear regression model) is an adequate description of the relationship between the response means and the concentration levels. This hypothesis is rejected since the p-value is 0.03. The alternative hypothesis for the test simply negates the null hypothesis of a straight-line relationship. Accordingly, it is non-specific, unlike the test carried out above,

Table 6.20 *Minitab regression analysis of iron data, including lack-of-fit test*

The regression equation is
Response = − 101 + 22.8 Conc.

Predictor	Coef	SE Coef	T	P
Constant	−100.67	28.25	−3.56	0.005
Conc.	22.7634	0.1134	200.81	0.000

S = 42.92 R−Sq = 100.0% R-Sq(adj) = 100.0%

Analysis of Variance

Source	DF	SS	MS	F	P
Regression	1	74285146	74285146	40326.44	0.000
Residual Error	10	18421	1842		
Lack of Fit	4	14621	3655	5.77	0.030
Pure Error	6	3800	633		
Total	11	74303567			

which postulated a quadratic response as an alternative to the linear relationship. This test can be used to assess the adequacy of non-linear models in exactly the same way, as discussed in the exercise below.

Exercise

6.18 When a polynomial model with both X and X^2 is fitted to the iron data the residual sum of squares is 4699, with 9 degrees of freedom. The pure error sum of squares and degrees of freedom are the same as above. Calculate the sum of squares for lack of fit and the corresponding degrees of freedom and carry out an F-test to assess the goodness of fit of the quadratic model.

6.7.3 Response Surface Modelling

Danaher *et al.*[10] describe a study concerned with the development of a derivitization method for the determination of drug residues in food; the data reported in the paper refer to the determination of avermectins and milbemycins in animal livers. The part of the study described below involved examination of the effects of varying three derivatization conditions on the peak areas produced in a HPLC system, with fluorescence detection. The factors were: temperature, X_1, time, X_2, and acid volume, X_3. Chromatographic responses were obtained using the design shown in Table 6.21; this is discussed below. Multiple linear regression

Table 6.21 *The central composite design matrix for the study*

A	*B*	*C*
−1	−1	−1
1	−1	−1
−1	1	−1
1	1	−1
−1	−1	1
1	−1	1
−1	1	1
1	1	1
α	0	0
$-\alpha$	0	0
0	α	0
0	$-\alpha$	0
0	0	α
0	0	$-\alpha$
0	0	0
0	0	0
0	0	0
0	0	0
0	0	0

was then used to fit the full quadratic model for eprinomectin peak area (Equation (6.32)); this equation contains linear terms, X, squared terms, X^2, and cross-product terms, X_i,X_j, for each variable.

$$\hat{Y} = \hat{\beta}_o + \hat{\beta}_1 X_1 + \hat{\beta}_2 X_2 + \hat{\beta}_3 X_3 + \hat{\beta}_{11} X_1^2 + \hat{\beta}_{22} X_2^2 + \hat{\beta}_{33} X_3^2 + \hat{\beta}_{12} X_1 X_2 + \hat{\beta}_{13} X_1 X_3 + \hat{\beta}_{23} X_2 X_3 \quad (6.32)$$

Without applying some mathematical tricks, it is virtually impossible to 'read' such an equation in a way that will allow the nature of the system response to be understood easily. However, with the aid of appropriate software the system response can be represented graphically. Equation (6.32) can be interpreted as a 'response surface' for which the response Y is the vertical deviation above the plane formed by pairs of the predictor variables. Figure 6.29, which was presented in the original paper, shows how peak area for eprinomectin changed as a function of pairs of the experimental variables, at fixed levels for the third variable. The authors drew the following conclusions from the Figure.

"The plot of eprinomectin peak area as a function of reaction time and volume of acid indicated that the optimum reaction time was between 27 and 47 min and that the optimum volume of acid was 33–67 μL. The plot of eprinomectin peak area as a function of acid and temperature indicated similar optimum values for volume of acid. The results indicated that a temperature higher than 65°C would give a larger eprinomectin peak area, but this temperature was chosen as an upper limit to avoid potential problems of solvent evaporation. From the response surface plots, suitable derivatization conditions were established to be a derivatization time of 30 min, 50 μL of acid added and a reaction temperature of 65°C."

Experimental Designs for Fitting Response Surfaces. In order to generate data that will capture the curvature involved in quadratic models, several levels of each experimental variable need to be studied. A full factorial design for three factors, each with five levels, would require $5^3 = 125$ runs. The design used in the eprinomectin study was a 'central composite design' (CCD). The structure of the CCD design is shown in Table 6.21; note that it contains five levels for each of the design variables, but only 19 runs. A 3-factor CCD design involves 8 runs corresponding to the 2^3 factorial design points (these are the first eight runs of Table 6.21), 6 runs corresponding to axial points stretching out from the centres of the 6 faces of the cube (points 9–14 of Table 6.21) and several points, here 5, at the centre of the design (points 15–19 of Table 6.21).

The factorial levels are coded $\pm$ 1 as usual, the centre points as zero, and the axial points as $\pm \alpha_i$. A commonly used value is $\alpha = 1.682$; this

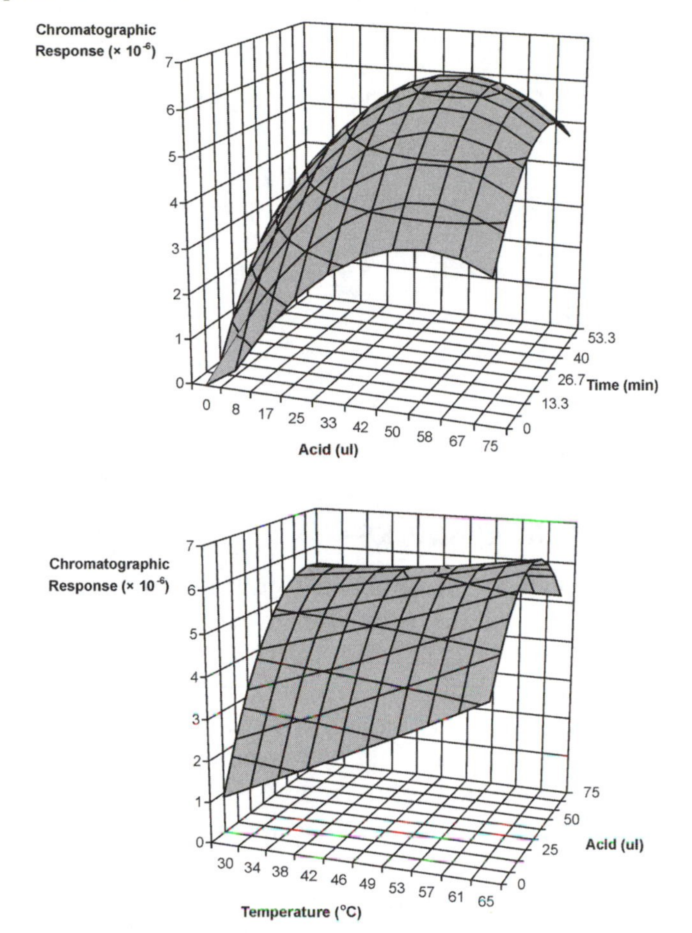

Figure 6.29 *Response surface plots based on the full quadratic model* (Reprinted from the *Analyst*, 2001, Vol. 126, pp 576. Copyright, 2001, Royal Society of Chemistry.)

means that if the distance from the centre to a factorial point is 1 unit, the distance from the centre to the axial point is 1.682 units. The replicates at the centre of the design and the axial points increase the number of levels of each variable to allow information on the curvature of the response surface to be obtained. The number of centre points and the axial distances, α_i (i = 1, 2, 3), may be varied to optimize the statistical properties of the design. The replicates at the centre of the design provide information on pure chance variation and this may be used to check on the goodness of fit of the final model. The geometry of the CCD design is illustrated in Figure 6.30. For an authoritative account of the methodology see Box and Draper[11] and for an excellent introductory example see Box, Hunter and Hunter.[12]

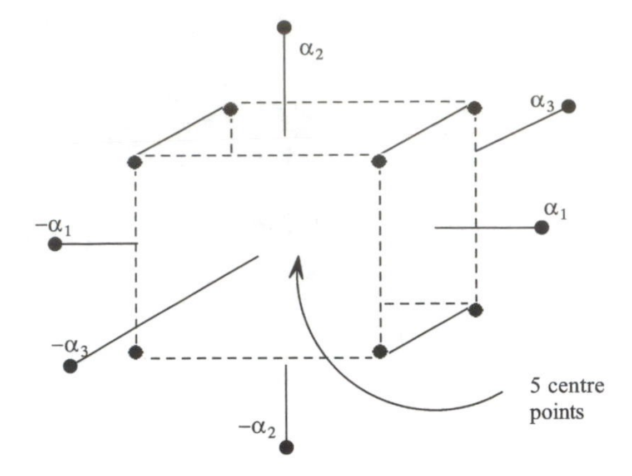

Figure 6.30 *The geometry of the central composite design*

6.8 CONCLUDING REMARKS

This chapter provided an introduction to fitting equations to data using regression analysis and to the uses to which the resulting equations might be put. Regression analysis is a very broad area within statistics, with many variants and complications. Its apparent simplicity and the widespread availability of regression software have meant that it has often been applied inappropriately. One example is given below.

One of the key assumptions of the regression model is that the X values are known constants, or at least subject to much less chance variation than the responses, Y. This assumption will very often not hold when two methods are compared by measuring a number of materials by each method and then regressing the responses of one method on those of the other. Such an approach to method comparison is widely used. Since the standard deviations of replicate measurements will often be similar for the two methods, the regression assumption is violated. Use of standard statistical tests, based on the regression output, for assessing relative biases between the methods can lead to erroneous conclusions. What is required for a valid analysis is a method that treats both the X and Y variables symmetrically, allowing for the chance variation in both sets of results. This is a well-known problem and is discussed in more advanced statistics textbooks. For a simple introduction and relevant software see the Analytical Methods Committee's technical brief 'Fitting a linear functional relationship to data with error on both variables'.[13]

For further discussion of this and other aspects of regression analysis in an analytical chemistry context, see the papers by Thompson[14] and Miller[15] and the books by Massart *et al.*[16] and Strike.[17] The general books on applied regression analysis, cited earlier,[3,11] will provide the reader with stepping stones from what was intended as a simple introduction to the basic ideas of regression analysis, to the more challenging topics that may need to be addressed.

6.9 REVIEW EXERCISES

6.19 A stability monitoring programme for a pharmaceutical product involved measuring the 'related substances' in the product at the time periods (months) shown in Table 6.22. The product was stored at 25 °C and 60% relative humidity. The measurements were made by HPLC with a UV spectrophotometer detector.

Table 6.22 *Stability monitoring data for related substances*

Time (months)	*0*	*3*	*6*	*9*	*12*
Rel. subs. (%)	0.149	0.185	0.245	0.297	0.320

- Draw a scatterplot of the data – does the relationship appear linear?
- Fit an OLS regression line
- By how much do related substances increase per month? Obtain a 95% confidence interval.
- Use the regression line to obtain a 95% confidence interval estimate of the percentage of related substances in the product at the beginning of the stability programme (which was close to time of manufacture).
- Obtain a 95% prediction interval for related substances at 18 months.
- The measured amount at 18 months was 0.673. What do you conclude?

6.20 The data in Table 6.23 come from a calibration study of an X-ray fluorescence spectrometer: fifteen certified reference materials with different silicon contents (Si %) were measured; the response is counts per second (CPS).

- Plot the cps against the Si % and verify the existence of a close linear relationship.
- Fit an OLS regression line; obtain residuals and fitted values. Plot them against each other and note the changing variability.

Table 6.23 *Silicon calibration data*

Si%	*CPS*
0.004	57.9
0.043	112.4
0.080	165.6
0.180	311.7
0.240	380.0
0.270	420.0
0.310	489.2
0.380	601.0
0.540	795.0
0.590	867.7
0.690	1021.6
1.000	1424.6
1.040	1529.9
1.070	1540.2
1.380	2005.9

- There are two reasons why this might be expected for these data. Firstly, the responses vary from around 20 to 2000 CPS; where data are spread over a very wide response range it is usually found that the precision varies over that range. Secondly, the responses are counts and very often standard deviations for counts are related to the magnitude of the counts. Because the data are counts statistical theory can guide us in choosing a set of weights. Where counts fluctuate purely randomly about a given mean level, a Poisson distribution will often describe the statistical properties of the resulting data.* The variance for a Poisson variable is equal to the mean. Since the fitted values from the OLS regression predict the mean CPS response at each %Si level, we can use their reciprocals as weights for a WLS analysis.

- Use the reciprocals of the fitted values at each %Si level as weights, obtain a WLS regression line and the corresponding residuals and fitted values. Calculate weighted residuals by multiplying the residuals by the square roots of the weights. Plot both the weighted and unweighted residuals against the fitted values. What do you conclude?

- Prepare a table of regression coefficients and their estimated standard errors for both the OLS and WLS calibration lines.

- Obtain predictions and prediction limits for the %Si corresponding to responses of 100 and 1500 CPS. Comment on the results.

*The Poisson distribution gives the probability of observing r counts as $p(r) = \frac{\lambda^r}{r!}e^{-\lambda}$ where λ is the average count level and r takes values $r = 0,1,2,3...$

6.10 REFERENCES

1. International Conference on Harmonization of Technical Requirements for Registration of Pharmaceuticals for Human Use (ICH), *Guidelines for Industry: Validation of Analytical Procedures*, ICH Q2A, ICH, Geneva, 1994.
2. J.S. Hunter, *J. AOAC Int.*, 1981, **64**, 574.
3. J. Neter, W. Wasserman, M.H. Kutner, *Applied Linear Regression Models*, Irwin, Homewood, Illinois, 1983.
4. Analytical Methods Committee, *Analyst*, 1988, **113**, 1469.
5. Analytical Methods Committee, *Analyst*, 1987, **112**, 199.
6. R. Pendl, M. Bauer, R. Caviezel, P. Schulthess, *J. AOAC Int.*, 1998, **81**, 907.
7. *Evaluation of the Linearity of Quantitative Analytical Methods; Proposed Guideline*, NCCLS-EP6-P, National Committee for Clinical Laboratory Standards, Villanova, PA, 1986.
8. Analytical Methods Committee, *Analyst*, 1994, **119**, 2363.
9. N.M. Jones, M.G. Bernardo-Gil, M.G. Lourenco, *J. AOAC Int.*, 2001, **84**, 309.
10. M. Danaher, M. O'Keeffe, J.D. Glennon, L. Howells, *Analyst*, 2001, **126**, 576.
11. G.E.P. Box, N.R. Draper, *Empirical Model-building and Response Surfaces*, Wiley, New York, 1987.
12. G.E.P. Box, W.G. Hunter, J.S. Hunter, *Statistics for Experimenters*, Wiley, New York, 1978.
13. Analytical Methods Committee, *AMC Technical Brief*, 2002, No. 10 (available on RSC website, www.rsc.org).
14. M. Thompson, *Analyst*, 1982, **107**, 1169.
15. J.N. Miller, *Analyst*, 1991, **116**, 3.
16. D.L. Massart, B.G.M. Vandeginste, S.N. Deming, Y. Michotte, L. Kaufman, *Chemometrics: a Textbook*, Elsevier, Amsterdam, 1998.
17. P.W. Strike, *Statistical Methods in Laboratory Medicine*, 2nd edn, Butterworth–Heinemann, London, 1991.

CHAPTER 7

The Design and Analysis of Laboratory Studies Re-visited

7.1 INTRODUCTION

Two sets of ideas will be discussed in this chapter. The first is concerned principally with the extension of the experimental designs discussed in Chapter 5 to situations where more than two factor levels need to be studied. The second relates to the estimation of the magnitudes of the components of measurement variability. What ties the two topics together is the use of analysis of variance (ANOVA) for the data analysis in both cases.

In Chapter 4, a two-sample *t*-test was used to analyze the data arising out of the simplest comparative design, which involves two levels of a single factor. In Chapter 5, this was extended to multi-factor designs, involving several factors each with two levels. Sections 7.2–7.4 extend the discussion further by allowing for many levels for each of the factors involved in the experiment. What is of interest in such studies is to compare the mean responses of the system for the different factor level combinations, usually with a view to selecting the best.

Nested designs are discussed in Section 7.5. Nested or hierarchical data structures have already been encountered in Chapter 2, where several replicate measurements on a control material were obtained for many analytical runs, for the purpose of drawing control charts; the replicates are said to be 'nested' within the runs. The US National Committee for Clinical Laboratory Standards (NCCLS)[1] recommends a three-level hierarchy, in which replicates are nested within runs, which in turn are nested within days, for studying the components of analytical variability. We will see how ANOVA can be used to analyze such data structures to produce measures, in the form of standard deviations or variances, that describe how the total analytical variation can be decomposed into components that characterize the contributions arising at each level of the hierarchy. The same approach will be used in Chapter 8 to analyze the results of collaborative trials carried out to validate new analytical methods and to characterize their performance in terms of the repeatability and reproducibility measures that were discussed in Chapter 1.

7.2 COMPARING SEVERAL MEANS

7.2.1 Example 1: A Laboratory Comparison Study

A multi-national corporation makes adhesive products in which boron is an important trace element at the parts per million (ppm) level. Concerns had been expressed about the comparability of the analytical results for boron produced by different laboratories in the corporation. The laboratories all use ICP–AES systems for the analysis, but these systems are not all identical in their configurations.

An inter-laboratory study was conducted which involved four laboratories each measuring several products over a period of weeks. Table 7.1 shows results for one product; the six replicates were measured on different days, so the variation between them reflects intermediate precision within the laboratories. Before engaging in a formal statistical analysis it is usually a good idea to plot the data and calculate some summary statistics. This will give an initial impression of the structure of the study results and will often identify unusual observations and signal potential problems with the formal analysis. When only a small number of data points are available dotplots are probably the most informative graphical summaries: dotplots (Figure 7.1) and summary statistics are shown below.

What is of interest here is to compare the average results obtained in the four laboratories, to determine if the laboratories are biased relative to each other. The intermediate precision might also vary from

Table 7.1 *The GC method development study data (peak areas)*

Injection temperature/°C	*Split ratio*	
	1:30	*1:25*
220	39.6	56.7
	37.6	54.6
	42.4	51.2
200	50.6	54.2
	49.9	55.9
	46.8	56.6

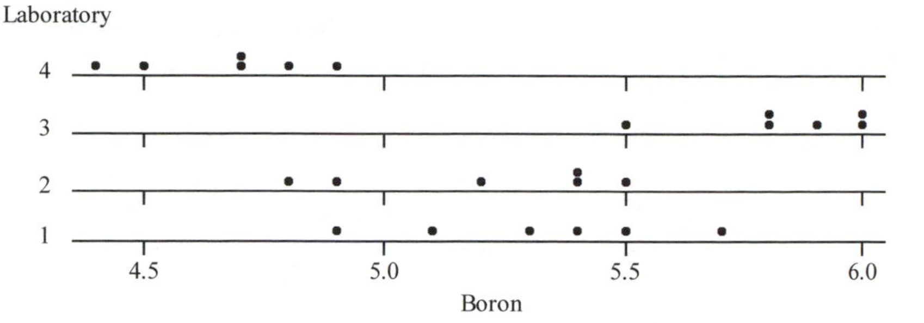

Figure 7.1 *Dotplots of the laboratory data*

laboratory to laboratory, but we will assume for the moment that this is not the case; we will return to this question later. The dotplots and summary statistics suggest that differences exist between the laboratory means, but there is always the possibility that the observed differences are simply the result of chance variation. Statistical tests are intended to distinguish between chance and systematic variation, so it is natural to look for a test which will address our question.

If there were only two laboratories then the question could be addressed using a t-test for the difference between the means, followed by a confidence interval to measure the relative bias should the test indicate the presence of bias. The simplest form of Analysis of Variance generalizes the t-test to more than two groups, and addresses the question: do the data cause us to reject the (null) hypothesis that the analytical systems in the four laboratories vary around the same mean response when the same material is measured?

Measuring Variation. Variation between several measurements is usually measured by the standard deviation or, alternatively, by its square, the variance. Thus, suppose we obtain a random sample of n observations, $y_1, y_2, y_3, \ldots, y_n$, from a fixed population or process whose mean is μ and whose variance is σ^2; the sample variance is defined as:

$$s^2 = \frac{\sum_{i=1}^{n} (y_i - \bar{y})^2}{n - 1}$$

This quantity is the average of the squared deviations of the data, y_i, from the overall mean, $\bar{y}$, where the divisor is $n - 1$, the 'degrees of freedom'. The term 'degrees of freedom' is used because only $n - 1$ of the deviations $y_i - \bar{y}$ are free to vary: they are constrained by the fact that they must sum to zero.* Thus, if $n - 1$ deviations sum to some arbitrary value, the last deviation is, necessarily, equal to minus that value.

When the data collection mechanism is more complicated than that assumed above, the method known as Analysis of Variance (ANOVA) may be used to break up both the total sum of squares and the total degrees of freedom into components associated with the structure of the data collection process. In Chapter 6 we saw how the total sum of squares and degrees of freedom could be decomposed into components associated with the regression line and variation around the regression line. Here, again, the totals will be decomposed into two components: one associated with random variation within groups (laboratories) and one associated with variation between groups.

Before beginning our discussion of ANOVA, it is convenient to introduce here the idea of 'expectation', which we will find useful later. When we calculate the sample mean, $\bar{y}$, based on our sample of n measurements from a fixed population or process, we do so to estimate the population or process mean, μ. Statistical theory tells us that the sample mean, $\bar{y}$, is an unbiased estimator of μ, *i.e.*, if we had an infinite number of observations then $\bar{y}$ would coincide with the long-run mean, μ. We say that μ is the 'expected value' of $\bar{y}$ or:

$$E(\bar{y}) = \mu$$

Similarly, the sample variance s^2 is an unbiased estimator of the population or process variance σ^2:

$$E(s^2) = \sigma^2$$

The fact that a sample value is an unbiased estimator of a population or process parameter is considered a useful property in statistics: it is directly analogous to the idea of an analytical method being unbiased in analytical science. In each case it means that we are aiming at the right target. Expected values will be used later to guide us in choosing suitable estimators of the different components of total measurement variance.

Analysis of Variance. A slightly more elaborate notation, which reflects the within-group/between-group data structure, is helpful when dealing with data from several groups. Thus, for the data of Table 7.1, each observation y_{ij} may be uniquely labelled by two subscripts: i refers to the group (laboratory) in which the observation was generated ($i = 1,2,3,4$) while j labels the replicates within the laboratory ($j = 1,2,..6$). Figure 7.2 shows schematically how the deviation of any observation y_{ij} from the

*This was discussed in Chapter 1, p 15.

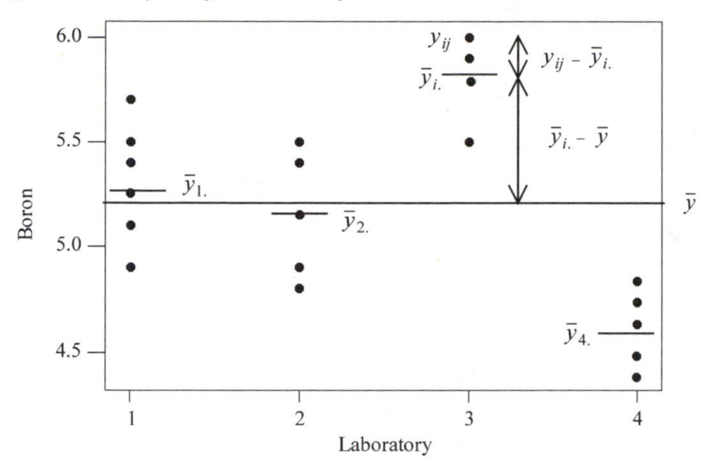

Figure 7.2 *The ANOVA decomposition*

overall mean $\bar{y}$ can be regarded as the sum of two components – its deviation from its own group mean* $y_{ij} - \bar{y}_{i.}$ and the deviation of its group mean from the overall mean $\bar{y}_{i.} - \bar{y}$. Thus we can write:

$$y_{ij} - \bar{y} = (y_{ij} - \bar{y}_{i.}) + (\bar{y}_{i.} - \bar{y}) \tag{7.1}$$

When both sides of Equation (7.1) are squared and summed over all the data points, the cross-product of the two terms on the right-hand side sums to zero and we get:

$$\sum_{\substack{all\\data}} (y_{ij} - \bar{y})^2 = \sum_{\substack{all\\data}} (y_{ij} - \bar{y}_{i.})^2 + \sum_{\substack{all\\data}} (\bar{y}_{i.} - \bar{y})^2 \tag{7.2}$$

The left-hand side of Equation (7.2) is the sum of the squared deviations of all the observations from the overall mean. Accordingly, it measures the total variation in the data and is called the 'total sum of squares', SS(total). The first term on the right-hand side measures the variation of the observations around their respective group means; it is, therefore, the within-group sum of squares, SS(within group). The second term on the right-hand side of (7.2) measures the variation of the group means around the overall mean; it is labelled SS(between groups).

*The mean of the six results in laboratory i is $\bar{y}_{i}.$; the dot indicates that we have summed over the j sub-script; the bar that we have averaged, *i.e.*, $\bar{y}_{i}. = \frac{1}{6}\sum_{j=1}^{6} y_{ij}$. In general there are I groups and J replicates within each group. It is not necessary that the number of replicates should be the same for each group, but adjustments to the calculations are required if this is not the case; the statistical software will take care of these.

Equation (7.2) therefore may be re-written more simply as:

$$\text{SS(total)} = \text{SS(within groups}^{\dagger}) + \text{SS(between groups)} \quad (7.3)$$

For the laboratory study the decomposition of sums of squares is:

$$5.2996 = 1.1750 + 4.1246$$

The degrees of freedom can be decomposed in a similar manner. Degrees of freedom (df) in ANOVA are typically calculated as the number of objects being considered minus one. Thus, the df(total) is $IJ - 1$, where there are J replicate results in each of I groups. For the laboratory study this is $24 - 1 = 23$. There are $I = 4$ groups, so there are $I - 1 = 4 - 1 = 3$ degrees of freedom for the SS(between groups). Within each group there are $J = 6$ observations, so there are $J - 1 = 6 - 1 = 5$ degrees of freedom. When the within-group degrees of freedom are combined for the $I = 4$ groups, we get df(within groups) $= I(J - 1) = 4(6 - 1) = 20$.

In summary, for the degrees of freedom decomposition we have:

$$\begin{aligned} \text{df(total)} &= \text{df(within groups)} + \text{df(between groups)} \quad (7.4) \\ IJ - 1 &= I(J - 1) + (I - 1) \\ 24 - 1 &= 4(6 - 1) + (4 - 1) \end{aligned}$$

The Statistical Model. The decomposition of sums of squares discussed above reflects an underlying statistical model for the data. This model assumes that the data are generated independently within the four laboratories by measurement processes with some standard deviation, say σ, which is common to all the laboratories, while the process means, μ_i, may vary from laboratory to laboratory. The assumption of a common standard deviation within all laboratories is reflected in the sum of squares decomposition, in that the within-group sums of squares have been combined – it would not make sense to do this unless this assumption were valid. It is also usual to assume that measurement variation is Normally distributed. Thus, we can write:

$$y_{ij} \sim N(\mu_i, \sigma^2) \quad (7.5)$$

which is a concise way of writing the assumptions discussed above. An alternative way of writing the model:

$$y_{ij} \sim N(\mu + \alpha_i, \sigma^2) \quad (7.6)$$

†Note that statistics packages almost invariably refer to the component associated with purely random variation as the 'error' component. In the current context this is the within-group component, so 'within-group' and 'error' sums of squares and degrees of freedom will be used interchangeably.

expresses each laboratory mean, μ_i, as a deviation from their average, here labelled μ:

$$\mu_i = \mu + \alpha_i \tag{7.7}$$

Thus, α_1 is the amount by which the measurement process mean for Laboratory 1 is above or below the overall mean for the four laboratories. Because they are deviations, the α_i terms sum to zero. When a formal test of the equality of the laboratory measurement process means is required, the null hypothesis of equal means will be expressed either by specifying $\mu_i = \mu$ or, equivalently, $\alpha_i = 0$ for all laboratories.

The ANOVA Table and F-test. It is usual to collect the various quantities that have been calculated into what is called an ANOVA table. Such a table was encountered in Chapter 6 in a regression context; the ANOVA table for the current example is shown as Table 7.2 below. Note that the sums of squares and degrees of freedom are additive, so that once any two rows of the table are calculated, the third can be obtained by addition or subtraction, as appropriate; alternatively, the additivity can be used as a check on the correctness of the calculations.

The ratios of the sums of squares to the degrees of freedom give mean squares. For example, the ratio 1.1750/20 is the mean square within groups, or more commonly, the mean square error (MSE). These mean squares are sample variances and the fifth column of the table shows what their 'expectations' are. As discussed earlier, expectation is the statistical terminology for long-run average values. Thus, the expression $E(y_{ij}) = \mu_i$ indicates that the observations in laboratory i vary around a measurement process mean μ_i. Similarly, $E(MSE) = \sigma^2$ states that the mean square error (*i.e.*, within-group mean square) has as its long-run average the intermediate precision variance σ^2. This suggests that we should use MSE when we want to estimate σ^2, which is assumed to be

Table 7.2 *The ANOVA table for the laboratory data*

Source of variation	*Sums of squares*	*Degrees of freedom*	*Mean squares*	*Expected mean squares*	*F-value*	*p-value*
Between groups	4.1246	3	1.3749	$\sigma^2 + J\frac{\sum_{i=1}^{I}\alpha_i^2}{I-1}$	23.40	0.000
Within-groups or error	1.1750	20	0.0588	σ^2		
Total	5.2996	23				

the same for all four laboratories. Similarly, the mean square between groups estimates $\sigma^2 + \frac{6\sum_{i=1}^{4}\alpha_i^2}{4-1}$. Recall that the α_i terms measure the deviations of the individual laboratory means from their average value μ.

The last two columns of Table 7.2 provide us with a test of the null hypothesis of no differences between the long-run group means, *i.e.*, that there are no relative biases between the laboratories:

$$H_o : \mu_1 = \mu_2 = \mu_3 = \mu_4 = \mu \text{ (or, equivalently, } \alpha_1 = \alpha_2 = \alpha_3 = \alpha_4 = 0)$$
$$H_1 : \text{not all laboratory means } (\mu_i) \text{ are the same}$$

The appropriate test statistic is the *F*-ratio:

$$F = \frac{MS(between - groups)}{MS(within - groups)} = \frac{MS(between - groups)}{MS(error)}$$

If the null hypothesis is true, then both the numerator and denominator of the *F*-ratio have the same expectation, σ^2, (since all the α_i terms are zero when the null hypothesis is true) and the ratio should be about one. In practice, the ratio is subject to chance variation, which is described by an *F*-distribution with degrees of freedom corresponding to those of the numerator and the denominator, *i.e.*, 3 and 20, respectively.

A large *F*-ratio suggests that the term $\sum_{i=1}^{4}\alpha_i^2$ is non-zero, *i.e.*, that not all the laboratory deviations α_i are zero and, hence, that not all the laboratory measurement systems vary around the same mean. Accordingly, the null hypothesis should be rejected. If a significance level of $\alpha = 0.05$ is selected, then the critical value will be $F_c = 3.10$, as shown in

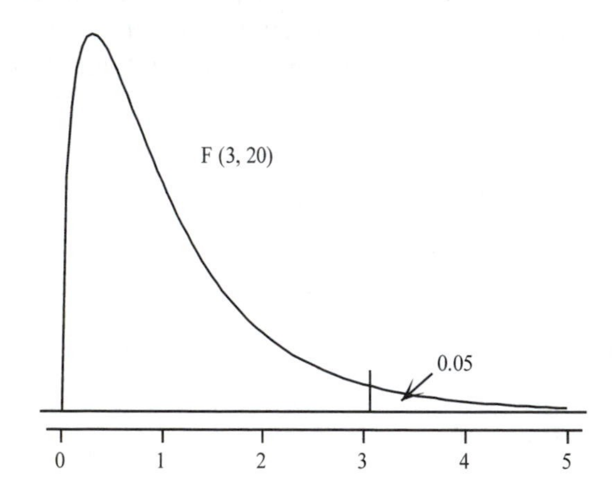

Figure 7.3 *The F-distribution with 3 and 20 degrees of freedom*

Figure 7.3. Since the F-value in the ANOVA table is larger than this, we reject the hypothesis of no relative biases and conclude that not all the laboratories produce comparable results. Statistical packages usually produce a p-value associated with the F-value – the p-value corresponding to our F-ratio of 23.40 is less than 0.0005, which indicates a highly statistically significant result. Note that the F-test in this case is inherently one-tailed, *i.e.*, the rejection region is in the right-hand tail only. A small value of F would be regarded a purely random event.

Model Validation. Before proceeding to examine further the differences between the laboratory means we will first assess the validity of the assumptions on which our analyses are based. Our assessment is based on the residuals. When the laboratory means (the fitted values) are subtracted from the corresponding data points within each laboratory, we are left with the residuals. If our model assumptions of Normality and constant within-laboratory intermediate precision are correct, these residuals should have the properties that would be expected from four random samples from a single Normal distribution, with zero mean.

Figure 7.4 shows the residuals plotted against the fitted values. The spread of values is not identical for all laboratories, but this could be the result of purely chance variation; here, there is no reason to believe that this variation is other than random. Note, in particular, that there is no tendency for the spread of values to increase with the magnitude of the results.

Figure 7.5 shows a Normal probability plot of the residuals; this plot is consistent with our assumption of data Normality. Figures 7.4 and 7.5,

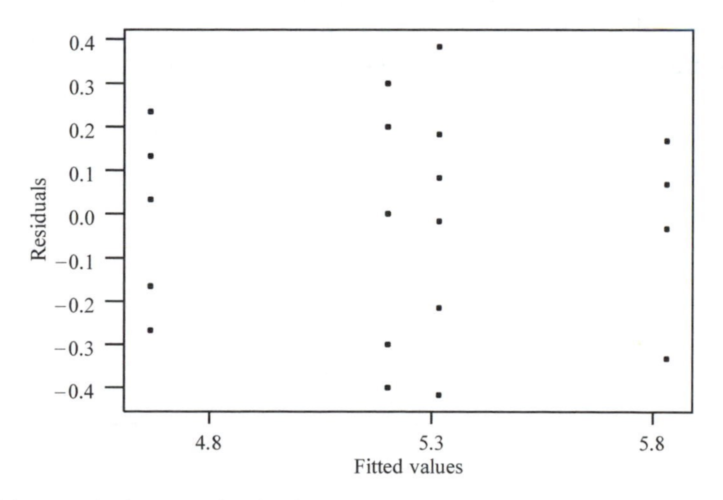

Figure 7.4 *Residuals versus fitted values*

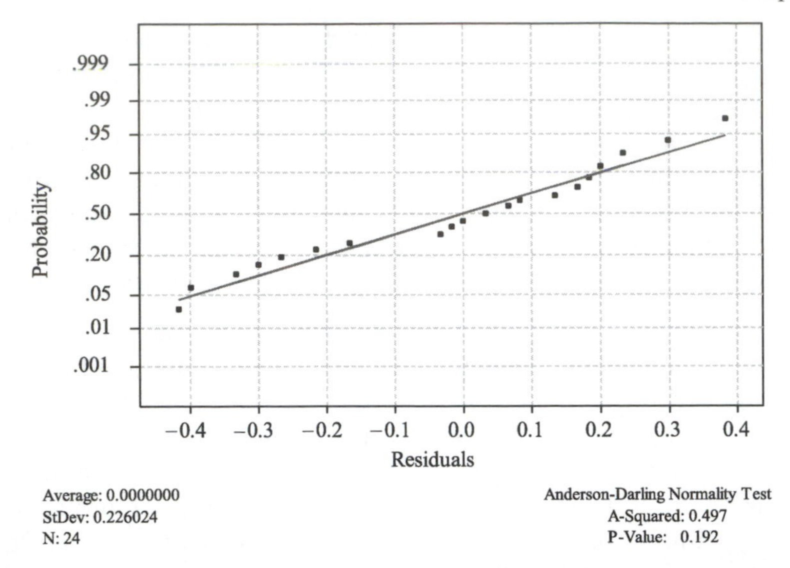

Figure 7.5 *A Normal probability plot for the laboratory residuals*

although based on only relatively small numbers of data points, give us some assurance that the statistical model on which our ANOVA F-test was based is acceptable. The same assumptions are required for the tests and confidence intervals in the next section.

Comparing Means. The F-test is a global test, which in this case indicates that not all the measurement process means are the same. To investigate the pattern of differences, we need to carry out some further analysis. The average results for the four laboratories are shown in Table 7.3 below, in ascending order of the means.

A natural approach to comparing any pair of sample means is to carry out a t-test of the hypothesis that their long-run values do not differ. We saw in Chapter 4 that the difference between any pair of means $\bar{y}_i$ and $\bar{y}_j$ is statistically significant if:

$$t = \frac{\bar{y}_i - \bar{y}_j}{\sqrt{\frac{2s^2}{n}}} > t_{0.025}$$

Table 7.3 *The laboratory means ranked by size*

Lab 4	*Lab 2*	*Lab 1*	*Lab 3*
4.67	5.20	5.32	5.83

that is if

$$\bar{y}_i - \bar{y}_j > t_{0.025}\sqrt{\frac{2s^2}{n}} \tag{7.8}$$

where the sample size is n in each case, s^2 is the combined within-group variance, $t_{0.025}$ is the critical value for a two-tailed test with a significance level of $\alpha = 0.05$ and $\bar{y}_i > \bar{y}_j$. In our laboratory example each mean is based on $J = 6$ observations and we have a combined within-laboratory variance of $s^2 = \text{MSE} = 0.0588$, which is based on all $I = 4$ laboratories.* The degrees of freedom for the t-distribution are the same as those for $s^2 = \text{MSE}$, and so they are 20 in this case.

The quantity on the right-hand side of Expression (7.8) is the smallest difference between two sample means which will lead us to conclude that the laboratory measurement process means should be considered different. Accordingly, this quantity is known as the 'Least Significant Difference' (LSD):

$$\text{Least Significant Difference (LSD)} = t_{.025}\sqrt{\frac{2MSE}{J}} = 2.09\sqrt{\frac{2(0.0588)}{6}} = 0.29 \tag{7.9}$$

The six possible comparisons between the four laboratory means can be carried out very quickly using the LSD. In Figure 7.6, the results are shown in ascending order of magnitude and a line is drawn under pairs of means that do not differ by at least 0.29, *i.e.*, that are not statistically significantly different.

Lab 4	Lab 2	Lab 1	Lab 3
4.67	5.20	5.32	5.83
	________	________	

Figure 7.6 *Comparisons of laboratory means*

Figure 7.6 indicates that the results from Laboratory 4 are statistically significantly lower and those from Laboratory 3 are higher than those from the other laboratories. The difference between Laboratories 1 and 2 is not statistically significant. This picture summarizes concisely the results of the inter-laboratory study.

A 95% confidence interval for the difference between any pair of laboratory measurement process means can be obtained in the usual way using:

$$\bar{y}_i - \bar{y}_j \pm t_{.025}\sqrt{\frac{2MSE}{J}} \tag{7.10}$$

*If s_i^2 is calculated for each laboratory then $s^2 = \text{MSE}$ is the simple average of the four values. Thus, ANOVA may be viewed as a general approach to the combining within-group variances where there are more than two groups.

Filling in the various quantities gives:

$$\bar{y}_i - \bar{y}_j \pm 2.09\sqrt{\frac{2(0.0588)}{6}}$$

$$\bar{y}_i - \bar{y}_j \pm 0.29$$

Thus, the error bounds on any difference $\bar{y}_i - \bar{y}_j$ are obtained by adding and subtracting 0.29 from the calculated difference.

7.2.2. Multiple Comparisons

Our investigation of the pattern of differences between the laboratory means essentially involved carrying out six *t*-tests. The number of possible comparisons would grow rapidly with the number of groups involved. Thus, six groups would involve $6 \times 5/1 \times 2 = 15$ comparisons, while ten groups would allow $10 \times 9/1 \times 2 = 45$ comparisons to be made. Multiple comparisons present us with a difficulty: the statistical properties of *t*-tests, as discussed in Chapter 3, hold for a single test, but will change radically if many tests are carried out simultaneously. This is very often ignored when only a small number of tests are carried out, but it becomes increasingly important when many comparisons are made.

To see where the problem lies, note that for any one test the significance level is the probability of rejecting the hypothesis of no difference between the long-run means, when, in fact, there is no difference, *i.e.*, when the null hypothesis is true. However, the probability is much higher than 0.05 that one or more of the six comparisons will produce a statistically significant result, when in fact there are no systematic differences. To understand why this is so, consider a simple coin tossing experiment. If a fair coin is tossed, the probability of the result being a head is 1/2. However, the probability of one or more heads in six tosses is considerably higher. Thus, the probability that all six results are tails is $(1/2)^6 = 1/64$, so that the probability of at least one head is $1 - (1/2)^6 = 63/64$. The calculations for the comparisons between the laboratories are more complicated (since the pairs of means are not all independent of each other, as was assumed for the coin tossing experiment) but the underlying ideas are the same. Accordingly, use of multiple statistical tests simultaneously can lead to much higher type 1 error rates (here, concluding that laboratory differences exist when, in fact, they do not) than the individual significance levels would suggest.

Various strategies are adopted to deal with this problem, but they all involve making the individual tests less powerful, *i.e.*, less likely to detect a small but real difference between the means. One simple approach is to reduce the significance level of the individual tests (say to $\alpha = 0.01$), so that the error rate for the family of comparisons is reduced to a more acceptable level. Minitab offers other multiple comparison alternatives. One of these is Tukey's 'Honestly Significant Difference' (HSD) method. The derivation of the HSD interval is based on the range of a set of means, *i.e.*, the difference between the largest and smallest values in the set. All other differences between pairs of means are, necessarily, smaller. Hence, the significance level chosen applies to the whole family of possible comparisons. The application of the HSD method is very similar to that of the LSD method – it simply requires the replacement of the critical t-value by another (T), which is given by:

$$T = \frac{1}{\sqrt{2}} q(1 - \alpha, I, I(J-1)) \tag{7.11}$$

where q is the Studentized range distribution, α is the significance level for the family of comparisons, I means are being compared and MSE has $I(J-1)$ degrees of freedom. For the laboratory study, Table A9 (Appendix) shows that requiring a family significance level of $\alpha = 0.05$ will lead to a critical T value of:

$$T = \frac{1}{\sqrt{2}} q(1 - \alpha, I, I(J-1)) = \frac{1}{\sqrt{2}} q(0.95, 4, 20) = \frac{1}{\sqrt{2}}(3.96) = 2.80$$

This means that HSD is 0.39 whereas LSD was 0.29. Accordingly, the HSD method is more conservative – results need to be further apart before they are declared statistically significantly different.

In this case the conclusions drawn, based on the HSD method, are exactly the same as those shown in Figure 7.4, which was based on the LSD method: Laboratories 1 and 2 are not significantly different from each other, but taking these as a pair they are significantly different from both 3 and 4, which are also significantly different from each other. We will see in the next example that the two methods do not always lead to the same conclusions.

7.2.3 Example 2: A Method Development Study

Four different configurations of a GC were investigated when developing a method for the analysis of trace compounds in distilled spirits. The objective was to maximize peak area. Each configuration (A, B, C, D) was run three times; the twelve runs were carried out in a randomized order. The results for the analysis of the propanol content of test portions from a single bottle of spirit are shown in Table 7.4 below.

The dotplots of the data, Figure 7.7, clearly suggest that not all the system means are the same. Configuration A appears to give smaller

Table 7.4 *Peak areas (arbitrary units) for the GC study*

	Configuration			
	A	*B*	*C*	*D*
	39.6	50.6	56.7	54.2
	37.6	49.9	54.6	55.9
	42.4	46.8	51.2	56.6
Mean	39.87	49.1	54.17	55.57
SD	2.41	2.02	2.78	1.23

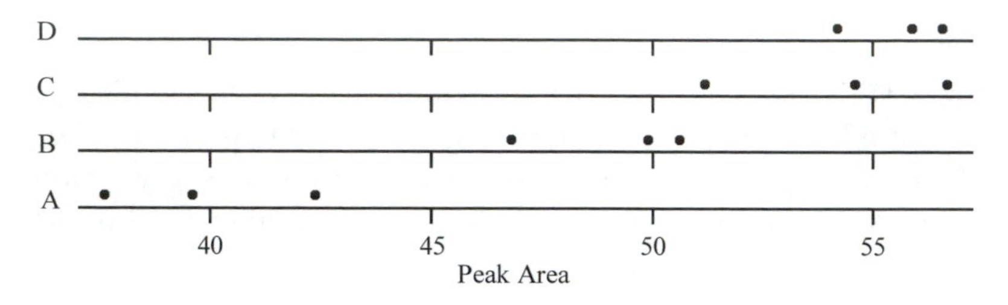

Figure 7.7 *Dotplots and means for the GC peak areas study*

peak areas than the others; the differences between configurations B, C and D are less marked and will require further analysis. The ANOVA table, Table 7.5, confirms that the long-run configuration means are unlikely to be the same; the F-value of 31.66 is very highly statistically significant ($p < 0.0005$).

Figure 7.8 shows the results of LSD and HSD analyses. The individual significance levels were 0.05 for the LSD analysis, while the family significance level was set as 0.05 for the HSD analysis. The values used in the analysis were $\mathrm{LSD} = 2.31\sqrt{\frac{2(4.78)}{3}} = 4.12$ and $\mathrm{HSD} = \frac{1}{\sqrt{2}}4.53\sqrt{\frac{2(4.78)}{3}} = 5.71$.

The LSD analysis suggests that configurations C and D give about the same peak areas, on average, and this average is greater than that of

Table 7.5 *The ANOVA table for the GC study*

Analysis of Variance for Peak Area

Source	DF	SS	MS	F	P
GC-config	3	454.26	151.42	31.66	0.000
Error	8	38.26	4.78		
Total	11	492.52			

LSD analysis:

A	B	C	D
39.87	49.10	54.17	55.57

HSD analysis:

A	B	C	D
39.87	49.10	54.17	55.57

Figure 7.8 *Comparisons of configuration means*

configuration B, which in turn is greater than that for A. The HSD comparisons also show A to be the smallest and indicate that D is statistically significantly bigger than B. However, D is not statistically significantly bigger than C, and similarly, C is not statistically significantly bigger than B. On purely logical grounds, this set of conclusions might appear odd. However, what we have here is a summary of the statistical evidence rather than a set of logical propositions: although B and D can be separated, the random variation in the data is such that B and C cannot be separated, and neither can C and D. Such ambiguities are commonplace when many pairwise comparisons are carried out. It is also not unusual that the more conservative HSD method should fail to distinguish between pairs of means that are reported as statistically significantly different when the LSD method is used.

The data of Table 7.4 are, in fact, the same data that appeared in Table 5.1, p 190, of Chapter 5; they were used there to illustrate the analysis of a 2^2 study design. Configurations A and B corresponded there to a split-ratio of 1:30, while C and D involved a split-ratio of 1:25. The injection temperature was 220°C for A and C, while it was 200°C for B and D. The combined s^2 value which was used in Section 5.3 to calculate the standard error of the various effects was $s^2 = 4.78$, which is identical to the MSE of the ANOVA table given above. In Section 5.3 residual analysis was carried out to investigate the model assumptions – there is, therefore, no need to repeat that exercise here.

The analysis of variance described above ignored the factorial structure of the design. This was done to illustrate how within-laboratory studies with an arbitrary number of levels for one factor can be carried out and analyzed using 'one-way' analysis of variance. Of course, if the study design involves a factorial structure, this should be recognized in the data analysis and this will be done for the GC data in the next section.

When regarded as an unstructured set of system configurations, the four configurations for the GC study could be extended to any number – for example, a study might involve comparing several different columns or several different carrier gases. No new conceptual issues arise as the number of levels increases, though, as discussed earlier, the multiple comparisons problem increases rapidly.

Exercises

7.1 During development of a multi-residue supercritical fluid extraction method for the extraction of benzimidazole residues from animal liver, Danaher *et al.*[2] investigated the influence on recovery rates of varying several method parameters, each over two levels. This was a factorial study, but here we look at four different variants of the method, A–D, without taking account of the factorial structure. We will return to the data later and take account of the structure. The percentage recovery results for one residue, triclabendazole (TCB) are shown in Table 7.6. What can we learn from the study?

Table 7.6 *Recoveries (%) of triclabendazole (TCB)*

A	*B*	*C*	*D*
41	8	71	0
60	6	81	15
43	25	83	24
66	8	67	10

7.3 MULTI-FACTOR STUDIES

Analysis of variance is particularly suited to the analysis of data arising from structured data collection exercises of the type discussed in Chapter 5, when the number of factor levels is greater than two. We will discuss now how the 'one-way' analysis of the previous section (*i.e.*, the analysis of a set of means corresponding to a number of levels of a single factor) can be modified to take account of the 2^2 factorial structure implicit in the GC study design. Following this discussion, an example of a study involving more than two levels of each factor will be presented.

7.3.1 Example 1: The GC Study Re-visited

In Section 7.2 we saw that the total variation in a set of data arising from J replicates within each of I groups could be broken up into a sum of squares within groups and a sum of squares between groups. Equations

(7.12) and (7.13) repeat Equations (7.2) and (7.3) from Section 7.2; in (7.12) the subscript i labels the groups and the subscript j labels the replicates within group.

$$\sum_{\substack{all\\data}}(y_{ij}-\bar{y})^2=\sum_{\substack{all\\data}}(y_{ij}-\bar{y}_{i.})^2+\sum_{\substack{all\\data}}(\bar{y}_{i.}-\bar{y})^2 \tag{7.12}$$

$$\text{SS(total)} = \text{SS(within groups)} + \text{SS(between groups)} \tag{7.13}$$

If, instead of having a single string of groups, labelled with a single subscript, i, we have a factorial structure, then it may be helpful to re-label the data to take account of the structure. The GC study data are shown again in Table 7.7 in the same form as they first appeared in Chapter 5, p 190. The two factors Injection Temperature and Split Ratio are said to be 'crossed' with each other, as each level of Injection Temperature appears with each level of Split Ratio, and *vice versa*.

Each measurement in Table 7.7 can be uniquely identified using three sub-scripts i, j, k, where $i=1, 2$ labels the Injection Temperature level, $j=1, 2$ labels the Split Ratio level and $k=1, 2, 3$ labels the individual data points within the four experimental groups. Using this labelling Equation (7.12) can be re-written as:

$$\sum_{\substack{all\\data}}(y_{ijk}-\bar{y})^2=\sum_{\substack{all\\data}}(y_{ijk}-\bar{y}_{ij.})^2+\sum_{\substack{all\\data}}(\bar{y}_{ij.}-\bar{y})^2 \tag{7.14}$$

$$\text{SS(total)} = \text{SS(within groups)} + \text{SS(between groups)}$$

In Equation (7.14) $\bar{y}_{ij.}$ is the mean of cell i, j of Table 7.7 and $\bar{y}$ is, again, the mean of all 12 data points. The between-group component can be

Table 7.7 *The GC method development study data (peak areas)*

Injection temperature/°C	*Split ratio*	
	1:30	*1:25*
220	39.6	56.7
	37.6	54.6
	42.4	51.2
200	50.6	54.2
	49.9	55.9
	46.8	56.6

broken down further to reflect the data structure:

$$(\bar{y}_{ij.} - \bar{y}) = (\bar{y}_{i..} - \bar{y}) + (\bar{y}_{.j.} - \bar{y}) + [(\bar{y}_{ij.} - \bar{y}) - (\bar{y}_{i..} - \bar{y}) + (\bar{y}_{.j.} - \bar{y})]$$

which gives

$$(\bar{y}_{ij.} - \bar{y}) = (\bar{y}_{i..} - \bar{y}) + (\bar{y}_{.j.} - \bar{y}) + (\bar{y}_{ij.} - \bar{y}_{i..} - \bar{y}_{.j.} + \bar{y}) \tag{7.15}$$

In Equation (7.15) $\bar{y}_{i..}$ represents the row means of the six data points at each of the two levels of Injection Temperature, *i.e.*, the data have been averaged over both the j and k subscripts; thus, $\bar{y}_{1..} = 52.33$ is the mean of the six measurements made at 200°C. Similarly, $\bar{y}_{.j.}$ represents the column means of the six data points at each of the two Split Ratio levels; thus, $\bar{y}_{.1.} = 44.48$ is the mean of the six measurements made when the Split Ratio was set at 1:30. The first term on the right-hand side (RHS) of Equation (7.15), $(\bar{y}_{i..} - \bar{y})$, measures the deviations of the Injection Temperature means from the overall mean, *i.e.*, it measures the 'main effect' of Injection Temperature. Similarly, the term $(\bar{y}_{.j.} - \bar{y})$ measures the deviations of the Split Ratio means from the overall mean, *i.e.*, the main effect of Split Ratio. Since the left-hand side (LHS) of Equation (7.15) measures the deviations of the cell means from the overall mean, the third term on the RHS, which is just the difference between the LHS term and the sum of the two main effects, is a 'non-additivity' component. This corresponds directly to what was labelled 'interaction' in Chapter 5. Where the mean response can be predicted from the separate (main) effects of the two factors, the factors act independently. Where this is not the case the factors are said to interact.

When the two sides of Equation (7.15) are squared, the three terms on the RHS will give three squared terms and three cross-product terms. However, when these are summed over all the data points the cross-product terms sum to zero and we are left with Equation (7.16). This shows that the between-group sum of squares can be decomposed into three terms, two main effects and the two-factor interaction.

$$\sum_{\substack{all\\data}} (\bar{y}_{ij.} - \bar{y})^2 = \sum_{\substack{all\\data}} (\bar{y}_{i..} - \bar{y})^2 + \sum_{\substack{all\\data}} (\bar{y}_{.j.} - \bar{y})^2 + \sum_{\substack{all\\data}} (\bar{y}_{ij.} - \bar{y}_{i..} - \bar{y}_{.j.} + \bar{y})^2 \tag{7.16}$$

$$\text{SS(between groups)} = \text{SS(Injection Temp)} + \text{SS(Split Ratio)} + \text{SS(interaction)} \tag{7.17}$$

When Equation (7.16) is substituted into (7.14) we obtain the ANOVA decomposition of sums of squares (7.18) which expresses the total sum of squares as a sum of four components: two main effects, the two-factor interaction and the within-group or error sum of squares.

Table 7.8 *ANOVA table for the GC study*

Analysis of Variance for Peak Area

Source	DF	SS	MS	F	P
Injection Temp	1	84.80	84.80	17.73	0.003
Split Ratio	1	323.44	323.44	67.63	0.000
Inj-Tempo*Split Ratio	1	46.02	46.02	9.62	0.015
Error	8	38.26	4.78		
Total	11	492.52			

$$\sum_{\substack{all\\data}}(y_{ijk}-\bar{y})^2 = \sum_{\substack{all\\data}}(\bar{y}_{i..}-\bar{y})^2 + \sum_{\substack{all\\data}}(\bar{y}_{.j.}-\bar{y})^2 + \sum_{\substack{all\\data}}(\bar{y}_{ij.}-\bar{y}_{i..}-\bar{y}_{.j.}+\bar{y})^2 + \sum_{\substack{all\\data}}(y_{ijk}-\bar{y}_{ij.})^2 \quad (7.18)$$

$$\text{SS(total)} = \text{SS(Injection Temp)} + \text{SS(Split Ratio)} + \text{SS(interaction)} + \text{SS(within-groups/error)} \quad (7.19)$$

The total degrees of freedom are 12 − 1 = 11. These can be partitioned into 4 − 1 = 3 between groups and 4 (3 − 1) = 8 degrees of freedom within groups, as before. The between-groups degrees of freedom can be further partitioned into 2 − 1 = 1 for Injection Temperature, 2 − 1 = 1 for Split Ratio, which leaves the third for the interaction.* Table 7.8 shows the ANOVA decomposition for the GC study. Note that its last two rows are identical to those in Table 7.5, p 322. The sums of squares and degrees of freedom of the first three rows, when summed, give the corresponding values in the first row of Table 7.5.

Table 7.8 provides *F*-tests of the hypotheses of null effects corresponding to the three sources of variation for the first three rows of the table. In each case the *F*-ratio is the ratio of the corresponding Mean Square to the MSE. Thus, the hypothesis of no interaction between the two factors is tested by:

$$\frac{MS(Interaction)}{MSE} = \frac{46.02}{4.78} = 9.62$$

This test gives a statistically significant result ($p = 0.015$), as we would expect, since we have already tested this same hypothesis in Chapter 5. Since the two factors interact we will want to study them together – the

*In general, there will be IJ combinations of the I levels of the first factor and J levels of the second factor. These give $I-1$ degrees of freedom for the first factor, $J-1$ for the second and $(I-1)(J-1)$ for the interaction. If there are K replicates in each cell, there will be $IJ(K-1)$ degrees of freedom for error. Thus, there are 2.2.(3 − 1) = 8 degrees of freedom for error in the current example.

four cell means need to be examined simultaneously. As discussed in Chapter 5, there is no need to examine the main effects once we have established that the factors interact. When the interaction is not statistically significant the F-tests on the main effects should be carried out and appropriate comparisons made between main effect means. For the GC study, the main effects are both highly statistically significant, as indicated by their p-values.

ANOVA or Effects Analysis? There is, in fact, an intimate connection between the ANOVA analysis discussed above and the 'effects analysis' discussed in Chapter 5. If you compare the two analyses you will find that the ANOVA F-statistics are the squares of the corresponding t-statistics. For example, in Chapter 5 the Injection Temp–Split Ratio interaction effect (AB) was 3.92 and its estimated standard error was:

$$S\hat{E}(AB) = \sqrt{\frac{4(4.783)}{12}} = 1.263$$

which gives a t-statistic of $3.917/1.263 = 3.101$ for testing the interaction effect. The square of this is 9.62, the F-statistic for the interaction effect in Table 7.8. Similarly, the critical value for the F-test $F(0.95,1,8) = 5.318$ is the square of the two-tailed critical t-statistic $t = 2.306$. Accordingly, since the two approaches to the data analysis give the same results, nothing is gained by using ANOVA in this case – the 'effects analysis' is simpler and more intuitive. However, this correspondence only applies when all factors have two levels, as was the case in Chapter 5. When the number of levels for any of the factors exceeds two, then ANOVA is the appropriate tool for data analysis and it was convenient to introduce its use *via* this simple familiar example. An example involving factors with more than two levels is discussed below.

Comparison of Means. The results of the study can be summarized graphically using an interaction plot, such as that in Figure 7.9. This picture illustrates the nature of the effects and their interaction: the lower temperature gives greater peak areas but the differential is much greater for the 1:30 split ratio. To make formal comparisons between the four means either the LSD or the HSD approaches discussed in Section 7.2 may be used; since both methods have already been applied to these data there is no need for further analysis here. Note though that, if the results had been different and the interaction had not been significant, we might have required comparisons between either or both of the two temperature averages or split ratio averages, depending on which main effect, if either, was significant. In each case the means would be based on

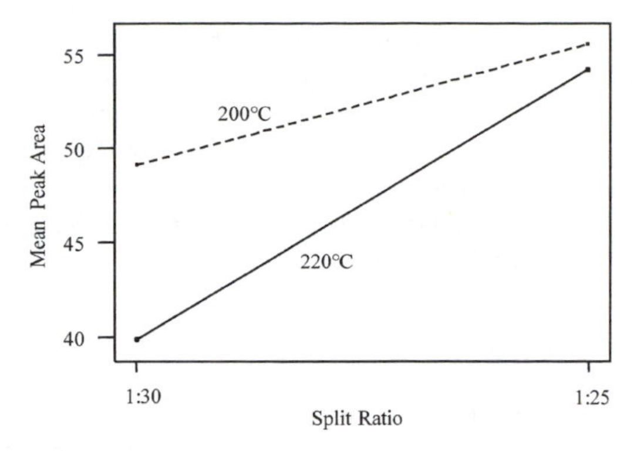

Figure 7.9 *The cell means represented as an interaction plot*

six observations and this would need to be taken into account in doing the calculations. Thus, the LSD for comparing the two means corresponding to either factor would be:

$$LSD = t_{\nu=8}^{0.025}\sqrt{\frac{2MSE}{6}} = 2.306\sqrt{\frac{2(4.78)}{6}} = 2.91$$

whereas the LSD for comparing cell means was 4.12; the smaller LSD value reflects the fact that the main effect means are based on 6 rather than 3 observations; hence, they are subject to less chance variation.

Exercise

7.2 During development of a multi-residue supercritical fluid extraction method for the extraction of benzimidazole residues from animal liver, Danaher *et al.*[2] investigated the influence on recovery rates of varying several method parameters, each over two levels. The percentage recovery results of a replicated 2^2 study are shown in Table 7.9 for triclabendazole (TCB). The factors varied

Table 7.9 *Recoveries (%) for triclabendazole (TCB)*

Pressure	*Carbon dioxide*	
	Low	*High*
High	41	71
	60	81
	43	83
	66	67
Low	8	0
	6	15
	25	24
	8	10

were CO_2 volume (40, 80 L) and Pressure (248, 543 bar). These data are, in fact, the same as those of Exercise 7.1, p 324, where the factorial structure was not explicitly recognized. They were also the basis for Exercise 5.3 (Table 5.7, p 201), where they were analyzed by 'effects analysis'. Analyze the data using two-way analysis of variance. Compare your results to those of Exercises 7.1 and 5.3. Verify that the ANOVA F-tests are the squares of the corresponding t-tests in the effects analysis of Exercise 5.3.

7.3.2 Example 2: A 3 × 3 Study

Matabudul *et al*[3]. reported the results of a study of a newly developed method for the determination of five anticoccidial drugs in animal livers and eggs by liquid chromatography linked with tandem mass spectrometry (LC–MS–MS). Residues of these drugs in animal tissue must be monitored as they are a potential consumer risk, if not properly controlled. Table 7.10 shows percentage recoveries for three matrices (poultry liver, sheep liver and egg) at each of three spiking levels 50, 100 and 200 $ng g^{-1}$ for lasalocid. Are the recovery rates matrix or spike level dependent? Note that the six replicates come from six different days (presumably one run per day), but, for convenience, I will ignore this aspect of the data structure here and treat the results as pure replicates. See Exercise 7.6, where a more complete analysis is outlined.

The ANOVA analysis carried out for the 2 × 2 table extends in a natural way to larger tables. No new ideas are required when the number of levels being studied increases. Equations (7.18) and (7.19) remain the same, symbolically, all that changes is the number of levels over which the

Table 7.10 *Residue recovery rates (%) for 3 × 3 study*
(Reprinted from the *Analyst*, 2002, Vol. 127, pp 760. Copyright, 2002, Royal Society of Chemistry.)

Matrix	*Spike level ng* g^{-1}					
	50		*100*		*200*	
	96	95	106	96	107	87
Poultry	90	83	101	99	95	96
liver	95	75	112	98	100	96
Sheep	93	104	106	99	84	81
liver	90	83	111	78	86	103
	79	79	110	82	97	105
	104	99	96	84	105	89
Egg	104	90	92	88	115	91
	98	87	97	97	94	79

summations take place, with consequent changes to the numbers of degrees of freedom. The statistical software allows for such changes automatically.

Table 7.11 shows the ANOVA summaries for the percentage recoveries of the residue, lasalocid. The are $3 - 1 = 2$ degrees of freedom for each main effect and $(3 - 1)(3 - 1) = 4$ degrees of freedom for the interaction term. Each of the nine cells of the data table contributes $6 - 1 = 5$ degrees of freedom to the 45 for error. The F-tests for the three effects are the ratios of their respective mean squares to the mean square error. In each case the null hypothesis is that the corresponding effect does not exist. Thus,

$$F = \frac{133.33}{88.67} = 1.50\ (p = 0.217)$$

tests for the existence of an interaction effect. The large p-value suggests that such an effect does not exist. The tests on the main effects are not statistically significant in either case. This suggests that the residue recovery rate is not matrix dependent (for these three matrix types) and does not depend on the spiking level (50–200 ng g^{-1}).

Figure 7.10 shows main effects plots for the residue recovery data. Despite the apparently large differences in the observed means, the interpretation of the F-tests in the ANOVA table is that the main effect means differ only by chance from the overall mean of all 54 recovery results. Why is this?

The standard deviation for replicate results is 9.42, the square root of the MSE, which is 88.67. Thus, the chance variation is quite large and hence, moderately large differences between means can be expected by chance, even where the means are based on 18 results. The minimum difference that would be required before any two levels for either main effect would be statistically significantly different is:

$$LSD = t\sqrt{\frac{2MSE}{18}} = 2.014\sqrt{\frac{2(88.67)}{18}} = 6.32$$

Table 7.11 *ANOVA analysis for lasalocid*

Analysis of Variance for %Recovery

Source	DF	SS	MS	F	P
Matrix	2	94.33	47.17	0.53	0.591
Spike	2	329.33	164.67	1.86	0.168
Matrix*Spike	4	533.33	133.33	1.50	0.217
Error	45	3990.33	88.67		
Total	53	4947.33			

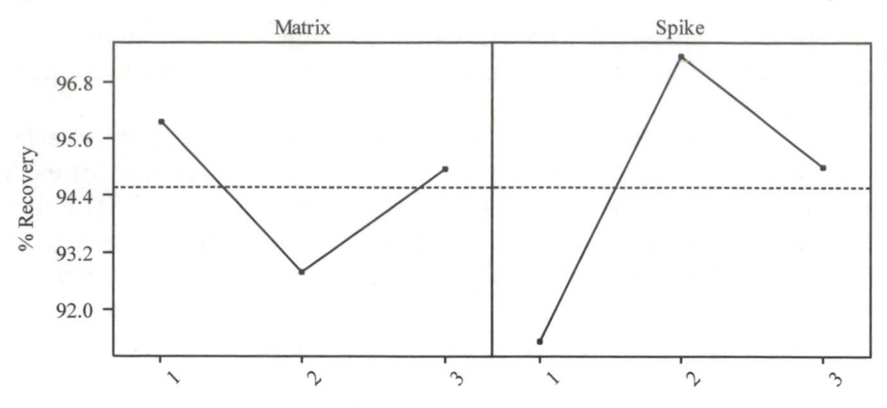

Figure 7.10 *Main effects plots of the matrix and spike level means*

where the degrees of freedom of the t-value are 45. Note that the HSD will be even larger. Since the smallest mean is 91.33 (spike level 1) and the largest is 97.33 (spike level 2) none of the differences between the main effect means are statistically significant.

Residual Plots. The ANOVA analysis assumes that the within-cell variability is constant for all nine cells, and that the variation is at least approximately Normal. Since the study involves spiking different matrices with a wide range of drug concentrations, it is important to verify these assumptions. Figures 7.11 and 7.12 show a residuals *versus* fitted values plot and a Normal probability plot, respectively. These plots

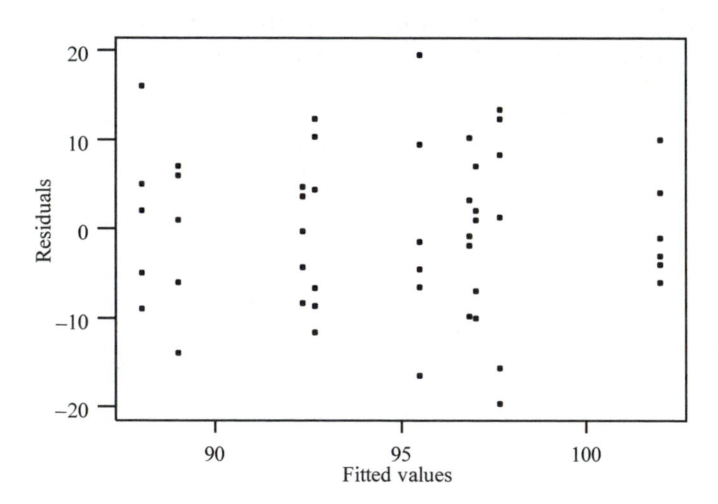

Figure 7.11 *Residuals versus fitted values for the drug residue study*

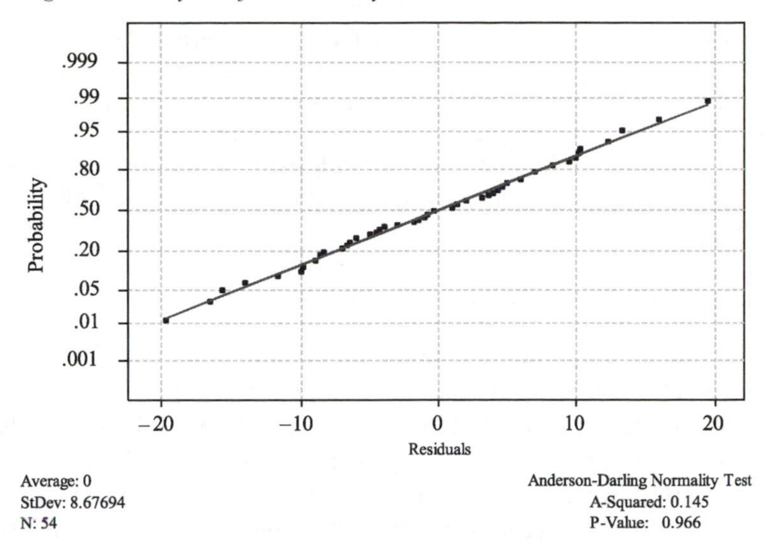

Figure 7.12 *Normal probability plot for drug residue study residuals*

do not raise doubts about the validity of the statistical assumptions underlying the analyses.

Exercise

7.3 The data shown in Table 7.12 are percentage recoveries for a second drug residue, narasin, from the study by Matabudul *et al.* described above.

Table 7.12 *Drug residue recoveries (ng g^{-1})*
(Reprinted from the *Analyst*, 2002, Vol. 127, pp 760. Copyright, 2002, Royal Society of Chemistry.)

Matrix	*Spike level/ng g^{-1}*					
	10		*20*		*40*	
	101	73	95	102	98	97
Poultry	79	86	88	99	81	88
liver	91	104	87	93	110	82
Sheep	87	98	92	100	98	109
liver	89	99	108	97	79	91
	79	91	104	89	89	90
	77	104	76	89	91	101
Egg	92	98	98	88	88	98
	84	74	95	101	94	98

The matrices are the same but the spiking levels are different. Again, the results should be treated as pure replicates, even though they were obtained on six different days (see Exercise 7.12 for a more sophisticated analysis). Since the drug is different and the spiking levels are much lower, we cannot assume that the results obtained for lasalocid will be replicated for narasin. Carry out the same set of analyses and determine whether recovery is affected by matrix or spike level.

7.3.3 Example 3: A 2^3 Study

The design and analysis of studies of more than two system parameters were discussed for two-level cases (2^k designs) in Chapter 5. If any of the factors in a multi-factorial study has more than two levels then ANOVA is used in carrying out the analysis. To illustrate the ANOVA analysis, we will return to our 2^3 example from Chapter 5. As we saw above, there are no conceptual differences between two-level and many-level studies as far as the ANOVA analysis is concerned – it simply involves comparisons between a greater number of means when multiple comparisons are carried out after the initial ANOVA analysis. Using a 2^3 example to illustrate the three-way analysis will allow comparison of the ANOVA approach and the effects analysis of Chapter 5.

A study was conducted to investigate the influence of various system parameters on the performance of a programmed temperature vaporizer (PTV) operated in the solvent split mode. The purpose was to develop a GC method for analyzing trace aroma compounds in wine, without the need for prior extraction. One phase of the investigation involved studying three factors each at two levels;

A: Injection Volume (100 or 200 μL),
B: Solvent Elimination Flow Rate (200 or 400 mL min^{-1}),
C: PTV Initial Temperature (20 or 40 °C).

The eight design points were replicated twice, giving a total of 16 observations. The order in which the 16 runs were carried out was randomized. One of the responses studied was the total peak area given by a synthetic mixture of aroma compounds in ethanol–water (12:88 v/v). The results (absolute peak areas) are shown in standard order in Table 7.13, which is a reproduction of Table 5.8 of Chapter 5.

Table 7.14 shows the ANOVA analysis for the PTV data. As always in factorial studies, we investigate the highest order interaction first. The *p*-value of 0.140 means that the three-factor interaction is not statistically significant. Only one of the two-factor interactions, Injection Volume-Flow Rate, is statistically significant ($p = 0.001$). This means

Table 7.13 *Design matrix and results for PTV development study (arbitrary units)*

Design point	*A*	*B*	*C*	*Results*	
1	−	−	−	37.5	24.6
2	+	−	−	149.4	146.5
3	−	+	−	24.9	28.8
4	+	+	−	117.3	114.0
5	−	−	+	27.5	32.2
6	+	−	+	129.5	136.6
7	−	+	+	23.7	22.4
8	+	+	+	111.0	112.6

that these two factors must be studied together to understand fully the nature of the system response. The main effect of the remaining factor, Initial Temperature, is also statistically significant ($p = 0.021$): the system gives a larger peak area at the lower Initial Temperature; in the study the mean response at the lower temperature was 80.38 as opposed to 74.44 at the higher temperature.

Figure 7.13 is an interaction plot. It shows that the main effect of Injection Volume is very large and also that the difference between the mean peak areas at the two Flow Rates depends on the Injection Volume level, the difference being greater at higher Injection Volume. Figure 7.13 is an alternative representation of the information shown in Figure 5.14 of Chapter 5. All the conclusions drawn above were arrived at in Chapter 5 by comparing the effects to their standard errors. The two analyses are equivalent, however, as the *F*-tests are just the squares of the corresponding *t*-tests of the effects. Thus, for the three-factor interaction we get:

Table 7.14 *ANOVA for 2^3 PTV study*

Analysis of Variance for Peak Area

Source	DF	SS	MS	F	P
Injection Volume	1	39531.4	39531.4	2277.89	0.000
Flow Rate	1	1041.7	1041.7	60.02	0.000
Initial Temp	1	141.0	141.0	8.13	0.021
Inject Vol*Flow Rate	1	452.6	452.6	26.08	0.001
Inject Vol*Initial Temp	1	47.3	47.3	2.72	0.137
Flow Rate*Initial Temp	1	17.9	17.9	1.03	0.340
Inject-Vol*Flow-R*Init-Temp	1	46.6	46.6	2.68	0.140
Error	8	138.8	17.4		
Total	15	41417.2			

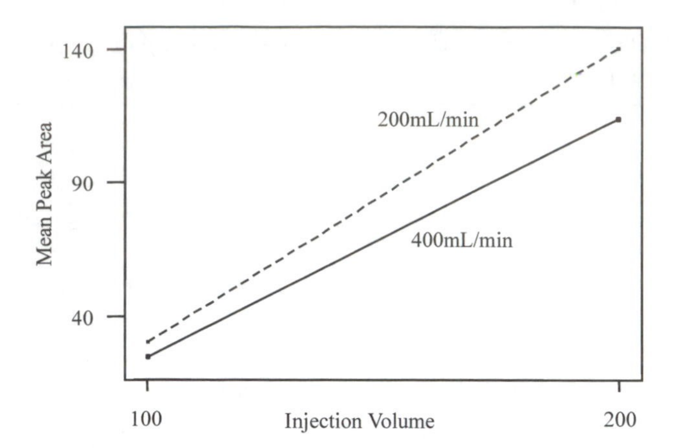

Figure 7.13 *The injection volume by flow rate interaction*

$$t^2 = \left(\frac{3.4}{2.08}\right)^2 = (1.6346)^2 = 2.67$$

which, apart from rounding error, is the same as the F-value for the three-factor interaction in Table 7.14. Formal comparisons between the mean responses may be carried out using LSD or HSD. The more conservative HSD approach is not usually used for 2^k designs (though in principle it should be used), since the number of comparisons implied by the designs is relatively small. In larger multi-level designs the possibility of spurious statistically significant differences arising by chance increases and use of the HSD approach may be advisable.

We investigated the assumptions underlying the analysis of the GC data in Chapter 5; we will not do so again here, as the analysis would be exactly the same as before.

The ANOVA analysis of the 2^3 study is quite general in that exactly the same approach can be applied irrespective of the number of levels for any of the factors studied. No new conceptual issues arise as the number of levels increases, but obviously the logistics of running the study will multiply with the levels and so also will the number of possible comparisons between the mean results at different design points.

Exercise

7.4 The data shown in Table 7.15 come from a study of the factors that influence the performance of a formic acid biosensor, which is used for air monitoring. For more information and a reference to the original study see Exercise 5.8, p 229; more factors are given there and the labelling here is

Table 7.15 *Biosensor development study (nA)*
(Reprinted from the Analyst, 2001, Vol. 126, pp 2008. Copyright, 2001, Royal Society of Chemistry.)

A	*C*	*E*	*Results*			
−1	−1	−1	213	217	238	263
1	−1	−1	296	266	250	314
−1	1	−1	285	302	264	244
1	1	−1	257	375	291	310
−1	−1	1	258	285	313	296
1	−1	1	278	362	277	320
−1	1	1	259	302	264	214
1	1	1	298	289	302	307

consistent with that in Exercise 5.8. For the factors listed below, the first level is coded −1 and the second level as +1 in the data table:

A: Concentration of Medola's blue (mg g^{-1}) 0.3, 1.5
C: Concentration of nicotinamide adenine dinucleotide (mg g^{-1}) 1.2, 4.8
E: Amount of solution placed on the electrode (μL) 8, 12

The response is the current (nA) at steady state when the biosensor was exposed to a formic acid vapour concentration of 1.4 mg m^{-3}. Large responses are desirable. There are four replicates at each of the 2^3 design points. Analyze the data, draw interaction plots, where appropriate, carry out a residual analysis and set out your conclusions as to what are the important features of the study results. These data were the basis for Exercise 5.4, p 210, in which the standard 'effects analysis' of a 2^3 design was required. Compare the results of the two analyses.

7.5 During development of a multi-residue supercritical fluid extraction method for the extraction of benzimidazole residues from animal liver, Danaher *et al.*[2] investigated the influence on recovery rates of varying several method

Table 7.16 *Recoveries (%) of albendazole sulfoxide (ABZ-SO)*

Press/bar	*Temp/°C*	*CO_2/L*	*ABZ-SO Recoveries*	
247.5	60	40	10	20
542.5	60	40	36	32
247.5	100	40	7	0
542.5	100	40	46	50
247.5	60	80	11	16
542.5	60	80	56	54
247.5	100	80	13	8
542.5	100	80	58	67

parameters, each over two levels. The percentage recovery results of a replicated 2^3 study are shown in Table 7.16 for albendazole sulfoxide (ABZ-SO). The factors varied were temperature (60, 100°C), CO_2 volume (40, 80 L) and Pressure (248, 543 bar).

These data were analyzed using an effects analysis in Exercise 5.5, p 212. Re-analyzed them now using a three-way ANOVA model. Compare your results with those obtained in Exercise 5.5 and verify the relationship between the F and t-tests for the two methods of analysis.

7.4 BLOCKING IN EXPERIMENTAL DESIGN

In Section 5.7 we encountered the idea of 'blocking' an experiment – there it involved running two half-fractions of a full factorial experiment at two different time periods. The rationale for this was that if a shift in system performance occurred between the two time periods, then the only effect that would be affected would be the five-factor interaction, which would be unlikely to exist in any case. Thus, a major source of variation – whatever caused the shift – would be excluded from the analysis of the effects of interest to the experimenters and the study would become more sensitive in the process. Just as the use of ANOVA allows generalization from two-level to many-level factorial experiments, so also its use allows the analysis of designs involving more than two blocks. The ideas will be introduced first through a simple example – we will return once more to our 2^2 GC development study – and this will be followed by a general discussion of the issues involved in blocking.

7.4.1 Example 1: The GC Development Study Re-visited, Again!

Table 7.17 shows the GC data presented in a layout which is slightly different from before. Each row corresponds to one set of experimental conditions: thus, row one has both factors at their low levels (Split Ratio 1:30 and Injection Temperature 200°C) – this corresponds to column B of Table 7.4, p 322 and the lower left-hand cell of Table 7.7, p 325.

An ANOVA analysis of the data in this table, treating the four rows as four groups each containing three replicate measurements, will give

Table 7.17 *The design matrix and GC study results*

Design points	*Split ratio*	*Inj. temp*	*Responses*			*Mean*	*Variance*
1	–	–	50.6	49.9	46.8	49.10	4.09
2	+	–	54.2	55.9	56.6	55.57	1.52
3	–	+	39.6	37.6	42.4	39.87	5.81
4	+	+	56.7	54.6	51.2	54.17	7.70

exactly the same results as those in Table 7.5, p 322. This is as expected since Table 7.17 is just a transposition of Table 7.4 where rows replace columns. The MSE will be the same as that found in both Tables 7.5, p 322 and 7.8, p 327, *i.e.*, MSE is 4.78. As we have seen, the MSE is just the combined s_i^2 from the within-group variances, *i.e.*, the average of the row variances shown in the last column of Table 7.17:

$$s^2 = \frac{\sum_{i-1}^{4} s_i^2}{4} = 4.78 = MSE$$

This is the key quantity required to measure the chance variation in the study and it is used in all the statistical tests and confidence intervals in the earlier sections of this chapter.

Each row of Table 7.18 contains the same values as the corresponding row of Table 7.17, but now the data have been grouped into columns corresponding to three laboratories L1, L2 and L3. In order to illustrate a blocked analysis, we pretend that the data were generated in three different laboratories, as shown in the table. The study might have been organized in this way in order to complete it very quickly – the three laboratories might be on one site, *e.g.*, a QC laboratory, a Technical Services laboratory and an R&D laboratory, all part of the same company. Note that each laboratory has carried out four runs, comprising a complete set of the four factorial design points.

The column of s_i^2 values on the right-hand-side of Table 7.18 is the same as that of Table 7.17, which it must be since the rows of the two tables contain the same numbers. However, the assumption underlying the analysis of Table 7.17 is that the s_i^2 values reflect purely chance variation. If the differences between the means of columns L1 – L3 of Table 7.18 reflect systematic differences between the results produced by the three laboratories, then each s_i^2 value will contain a component associated with laboratory-to-laboratory variation and combining the s_i^2 values will no longer provide a measure of pure chance variation.

Table 7.18 *The design matrix and grouped study results*

Design points	*Split ratio*	*Inj. temp*	*Responses*				
			L1	*L2*	*L3*	*Mean*	*Variance*
1	–	–	46.8	49.9	50.6	49.10	4.09
2	+	–	54.2	55.9	56.6	55.57	1.52
3	–	+	37.6	39.6	42.4	39.87	5.81
4	+	+	51.2	54.6	56.7	54.17	7.70
		Mean	47.45	50.00	51.58	49.68	

Table 7.19 *Blocked ANOVA analysis of GC study data*

Analysis of Variance for Peak-Area

Source	DF	SS	MS	F	P
Split ratio	1	323.441	323.441	539.82	0.000
Inj.-Temp.	1	84.801	84.801	141.53	0.000
Split ratio*Inj.-Temp	1	46.021	46.021	76.81	0.000
Blocks	2	34.665	17.333	28.93	0.001
Error	6	3.595	0.599		
Total	11	492.523			

The ANOVA analysis of Table 7.19 separates out the laboratory-to-laboratory variation as a distinct element in the analysis (called Blocks in the table); the p-value ($p = 0.001$) is strong evidence of systematic laboratory-to-laboratory variation. Note that the MSE has decreased markedly from its earlier value of 4.78 to a new value of 0.599. What has happened here is that the original SSE of 38.26 has been split into a component* for laboratories, SS(blocks) = 34.665, and a residual error of SSE = 3.595. What was originally seen as purely chance variation is now seen to be composed of mainly systematic laboratory-to-laboratory variation. The original 8 degrees of freedom for error have been broken into $3 - 1 = 2$ for blocks and $8 - 2 = 6$ for error. The ratios of the sums of squares to the degrees of freedom give the mean squares, as always. The experimental design can be thought of as a three-factor design which does not allow for any interactions between the third (blocking) factor and the two experimental factors, *i.e.*, between laboratories and either Split Ratio or Injection Temperature. Thus, the blocking factor appears only as a main effect in the ANOVA table.

The very large reduction in the MSE is a measure of the increased sensitivity of the blocked analysis. Since MSE appears in the denominator of all the F-tests, reducing MSE increases all the F-statistics and means that smaller effects may now be recognized as being statistically significant. This is not so important for the current example, as the interaction and two main effects were already statistically significant, despite the large MSE of 4.78. However, in other cases the presence of laboratory-to-laboratory variation could so inflate the MSE as to mask differences that would otherwise be statistically significant.

*The SS for blocks is $4\sum_{k=1}^{3}(\bar{y}_k - \bar{y})^2$, where $\bar{y}_k$ is the mean for laboratory k and $\bar{y}$ is the overall mean. The multiplier 4 is required as each $\bar{y}_k$ is the average of 4 values. Subtracting SS(blocks) from the original SSE gives the SSE for Table 7.19.

When it comes to making detailed comparisons between the mean results produced by the four GC configurations, though, the reduced MSE does have an impact. The Tukey HSD† now becomes 2.19, instead of 5.72 and the difference between 49.10 and 54.17, the mean results at design points 1 and 4, respectively (note these are labelled B and C in Figure 7.7), is now statistically significant. In fact, the HSD of 2.19 is smaller than the LSD of 4.12 calculated earlier. This means that the multiple comparisons of means are more precise, while still maintaining a significance level of $\alpha = 0.05$ for the whole family of comparisons.

The blocking of the 2^2 GC study was artificially created after the event by assigning the smallest of the three results for each design point to L1, the next smallest to L2 and the largest to L3. In a real experiment using this design, the order in which the four design points are run should be randomized separately within each laboratory. Such a design is known as a 'randomized block design', *i.e.*, the data are collected in several blocks and the runs within each block are carried out in a randomized order. Note that the three blocks, instead of corresponding to three different laboratories, might, for example, correspond to three sets of four runs carried out by three different analysts, or might involve the use of three different GC systems. Alternatively, they might have been carried out at three widely separated time periods.

Discussion. Note from Table 7.18 that each laboratory was required to carry out a full factorial design on the two factors. This means that exactly the same experimental runs were carried out in each laboratory and so a fair comparison between the laboratories could be made. If, instead of this, the laboratories had been assigned four of the twelve runs at random, then some laboratory might have a preponderance of design points that give low results (design pt 3), while another would be assigned design points that give high results (design pts 2, 4). It would not then be easy to disentangle the influences of the design variables (Split Ratio and Injection Temp) from the laboratory differences. When each laboratory runs the same set of design points the design is said to be 'balanced'.

The implications of blocking can be seen most clearly by focusing on just two design combinations, say design point 1 (−, −) and design point 2 (−, +). The difference between these is a measure of the effect of changing the Split Ratio from 1:30 to 1:25, at the low level of Injection Temperature, 200 °C. Because the two design points are measured within the same laboratory (block), it does not matter whether the laboratory tends to get high or low results – when the *difference* between the two results is

†HSD $= T\sqrt{\frac{2MSE}{3}} = 3.46\sqrt{\frac{2(0.599)}{3}} = 3.46(0.632) = 2.19$, where $T = \frac{1}{\sqrt{2}}q(0.95, 4, 6) = \frac{4.90}{\sqrt{2}} = 3.46$

calculated *within* any laboratory, the laboratory effect is eliminated from the comparison. The differences obtained in each of the three laboratories are then averaged. This is the essence of the blocking principle. Whenever there is a variable (*e.g.*, time, batches of materials/reagents, suppliers, people, machine, *etc.*) that may change the magnitudes of the results, then if the study is arranged such that all comparisons are carried out *within* each block (where each block corresponds to a single level of that variable), the comparisons will not be affected by changes in that variable *between* blocks. Accordingly, the study will become more sensitive, *i.e.*, it will be able to detect smaller systematic differences.

7.4.2 Example 2: Paired *t*-tests Revisited

In Chapter 4 we discussed a test for relative bias between two laboratories, using a paired *t*-test of the results obtained by each of the laboratories on six recent batches of product (see p 149–153). The data are reproduced in Table 7.20. The paired *t*-test gave a *t*-value of 6.586 with an associated *p*-value of 0.001. An alternative approach to the data analysis would be to regard the six batches as blocks within which the two laboratories are to be compared.

Table 7.21 shows an ANOVA analysis of the data. This is just a two-way analysis, which does not include an interaction between batches (blocks) and laboratories. The *F*-test statistic (43.376), which tests the null hypothesis of no difference between laboratory means, is just the square of the paired *t*-value and so the analyses are equivalent.

The ANOVA approach emphasizes one of the assumptions underlying the paired *t*-test: if there is a systematic difference between the average results from the two laboratories, then it is assumed that this difference is the same for all batches. In the ANOVA, this assumption is implemented by not fitting an interaction between laboratories and batches – an

Table 7.20 *Purity data with differences and summary statistics*

Batch	*Lab 1*	*Lab 2*	*Difference*
1	90.01	90.45	0.44
2	89.34	89.94	0.60
3	89.32	90.05	0.73
4	89.11	89.62	0.51
5	89.52	89.99	0.47
6	88.33	89.40	1.07
Mean	89.272	89.908	0.637
SD	0.553	0.364	0.237

Table 7.21 *ANOVA analysis of paired t-test data*

Analysis of variance for purity					
Source	DF	SS	MS	F	P
Laboratory	1	1.21603	1.21603	43.38	0.001
Batch	5	2.04920	0.40984	14.62	0.005
Error	5	0.14017	0.02803		
Total	11	3.40540			

interaction would imply that laboratory differences were batch dependent. In order to fit and test for an interaction effect it would be necessary for each laboratory to measure each batch of product more than once. Since this was not done here, the design of the study implicitly assumes the absence of a laboratory–product interaction.

In summary, blocking is a useful strategy for improving the sensitivity (power) of an experimental study whenever there is likely to be variation in the value of a factor which is not of direct interest to the study objectives, but which is likely to affect the system responses. Only the simplest type of blocking was introduced here. More sophisticated blocking strategies are discussed in specialist books on experimental design (see, for example, Box, Hunter and Hunter[4] or Montgomery[5]).

Exercise

7.6 Refer back to the drug residue data of Table 7.10, p 330. The six replicates came from six different days, presumably from a single run each day. The first column in each cell of the table represents days 1–3 and the second days 4–6. The earlier analysis ignored this important aspect of the data structure. A blocked ANOVA analysis allows for the day effect.

Regard day as a blocking factor, *i.e.*, consider the data as representing a 3×3 factorial study replicated on each of six days. Carry out an ANOVA analysis in which days appears as a block effect, *i.e.*, it appears as a main effect only (see text example above). Before carrying out the ANOVA using software, write down the degrees of freedom that you expect to appear in each row of the ANOVA table – this is a good way to check your understanding of how the analysis will turn out. Carry out appropriate residual analyses. Interpret the results of your analysis.

7.5 ESTIMATING COMPONENTS OF TEST RESULT VARIABILITY

Analysis of variance may be used to estimate the magnitudes of the standard deviations that describe the variation associated with different components of measurement error. We will apply it to two datasets generated within a single laboratory in this section; the same method of analysis will be applied to data generated in an inter-laboratory collaborative trial of a new analytical method in Chapter 8. Our first example involves the analysis of data routinely collected for the purposes of monitoring the stability of an analytical system using a control chart, as discussed in Chapter 2. The second is based on the US National Committee for Clinical Laboratory Standards (NCCLS) guidelines[1] for studying the components of measurement variability.

7.5.1 Example 1: Control Charts

Table 7.22, below, repeats potency data that were analyzed in the case study in Section 2.6 of Chapter 2. Five of the original runs have been excluded – three were excluded for the purposes of setting up the charts, but two extra runs, which gave suspiciously low mean values, have also been excluded here, leaving a total of twenty four runs. Each test result can be uniquely identified by two subscripts: thus y_{ij} is the jth replicate ($j = 1,2,3$) of the ith run ($i = 1,2,..24$). The data collection exercise can be represented schematically as shown in Figure 7.14. This is referred to as a hierarchical or nested design: the replicates are said to be 'nested' within the runs.

Table 7.22 *Twenty four sets of three replicate potency measurements (mg)* (Reprinted from the *Analyst*, 1999, Vol. 124, pp 433. Copyright, 1999, Royal Society of Chemistry.)

Run	*Potency*	*Run*	*Potency*	*Run*	*Potency*	*Run*	*Potency*
1	499.17	7	487.21	13	494.54	19	487.82
	492.52		485.35		493.99		489.23
	503.44		479.31		495.08		493.45
2	484.03	8	493.48	14	484.17	20	489.23
	494.50		496.37		490.72		491.11
	486.88		498.30		493.45		484.07
3	495.85	9	495.99	15	493.61	21	491.27
	493.48		499.36		488.20		488.90
	487.33		482.03		503.90		500.77
4	502.01	10	511.13	16	482.25	22	489.85
	496.80		504.37		475.75		488.42
	499.64		501.00		488.74		487.00
5	492.11	11	510.16	17	509.11	23	492.45
	485.58		498.59		510.18		484.96
	490.24		501.48		506.46		490.58
6	500.04	12	479.57	18	489.67	24	488.68
	499.11		462.64		487.77		476.01
	493.98		479.57		497.26		484.92

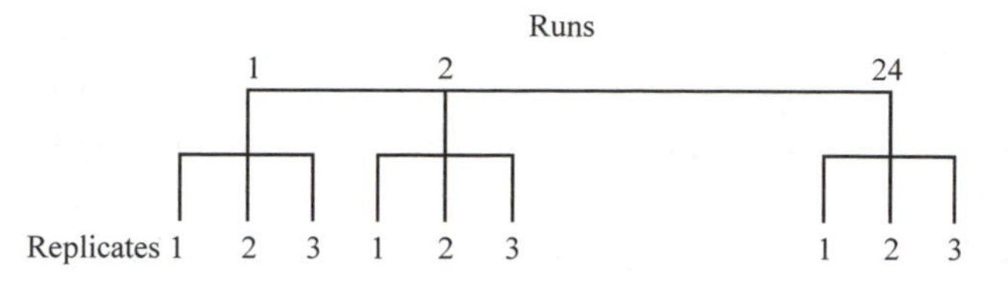

Figure 7.14 *The structure of the control chart data*

The following model describes the data structure and underlies the ANOVA analysis that follows:

$$y_{ij} = TV + R_i + W_{ij} \tag{7.20}$$

TV is the true value for the control material (assuming the analytical system is unbiased), R_i is the run bias for run i, and W_{ij} is the within-run or repeatability error for replicate j of run i. The run biases are assumed to be Normally distributed about a mean of zero and to have a standard deviation of σ_R. The within-run repeatability errors are assumed to be Normally distributed about a mean of zero and to have a standard deviation of σ_W. The two sources of chance variation are assumed to be independent of each other; these same assumptions were made for the laboratory comparison simulation study in Chapter 4. These assumptions imply that the variance of an observation y_{ij} is:

$$Var(y_{ij}) = \sigma_R^2 + \sigma_W^2 \tag{7.21}$$

since TV is a constant. This means that the observations y_{ij} come from a Normal distribution with mean TV and standard deviation $\sqrt{\sigma_R^2 + \sigma_W^2}$. The two elements on the right-hand side of Equation (7.21) are known as the variance components of the test results. ANOVA may be used to estimate these variance components.

It might be of interest to investigate the variance components for several reasons. In Chapter 1 we discussed a number of measures of the precision of an analytical system, which might be required as part of a method validation exercise. Thus the quantity σ_W is the repeatability standard deviation, while $\sqrt{\sigma_R^2 + \sigma_W^2}$ is an intermediate precision standard deviation; both of these quantities were discussed in Chapters 1 and 2. In Chapter 2 we carried out an analysis to estimate σ_W and σ_R with a view to assessing the quality of the analytical results. This led into a discussion of whether the replicates were best placed within a single analytical run, or whether single measurements in each of several runs would yield better information (see Section 2.8). There the calculations involved moving ranges (*i.e.*, the magnitudes of the differences between successive run means). The ANOVA calculations described below use the data in a statistically more efficient way in estimating σ_R. Accordingly, a good strategy would be to establish that the analytical system is stable, using a control chart, and then use ANOVA to estimate the variance components and, by taking square roots, the relevant standard deviations.

Analysis of Variance. The data structure here is exactly the same as for Example 1 of Section 7.2 at the beginning of the chapter and the analysis of the data proceeds as before: the total sum of squares and degrees of freedom are decomposed into components associated with the groups (runs) and the replicates within groups. The number of groups is different (there are 24 runs while there were only 4 laboratories) and the number of replicates is now 3 rather than 6, but the nested structure of the data collection is identical. The total sum of squares is given by:

$$\text{Total sum of squares} = \text{SS(total)} = \sum_{\substack{all \\ data}} (y_{ij} - \bar{y})^2 \tag{7.22}$$

where $\bar{y}$ is the mean of all the data. This has $IJ - 1 = 24(3) - 1 = 71$ degrees of freedom. The within-run/repeatability/error sum of squares is:

$$\text{Error sum of squares} = \text{SS(error)} = \sum_{\substack{all \\ data}} (y_{ij} - \bar{y}_{i.})^2 \tag{7.23}$$

where $\bar{y}_{i.}$ is the mean of the three replicates for run i. Summing over all the data means that we first sum (7.23) over j to get the error sum of squares for any one run i, and then we sum over i to combine the information from all $I = 24$ runs. This has $J - 1 = 3 - 1 = 2$ degrees of freedom for each run, so the combined number of degrees of freedom is $I(J - 1) = 24(3 - 1) = 48$.

The sum of squares and degrees of freedom for runs can be obtained by subtracting the error terms from the total terms. Alternatively, we can write:

$$\text{Sum of squares for runs} = \text{SS(run)} = \sum_{\substack{all \\ data}} (\bar{y}_{i.} - \bar{y})^2 \tag{7.24}$$

and this has $(I - 1) = (24 - 1) = 23$ degrees of freedom. Summing over all the data means getting the sum of the squared deviations of the run means $\bar{y}_{i.}$ from the overall mean, $\bar{y}$ and then multiplying the answer by 3, since each $\bar{y}_{i.}$ is the average of 3 replicates. As before, the ratios of the sums of squares to the degrees of freedom give the mean squares; all the elements of the analysis of variance are collected into Table 7.23 and the data summaries are given in Table 7.24.

Table 7.23 *The ANOVA decompositions*

Source of variation	*Degrees of freedom*	*Sums of squares*	*Mean squares*	*Expected mean squares*
Between runs	$I-1$	$\sum_{\substack{all\\data}} (\bar{y}_{i.} - \bar{y})^2$	$MS(run) = \frac{SS(run)}{I-1}$	$\sigma_W^2 + J\sigma_R^2$
Error	$I(J-1)$	$\sum_{\substack{all\\data}} (y_{ij} - \bar{y}_{i.})^2$	$MS(error) = \frac{SS(error)}{I(J-1)}$	σ_W^2
Total	$IJ-1$	$\sum_{\substack{all\\data}} (y_{ij} - \bar{y})^2$		

Underneath the ANOVA table Minitab gives a table of variance components and corresponding standard deviations. These are calculated from the mean squares for runs and error. To understand the rationale underlying the calculations we must refer to the column of expected mean squares in Table 7.23.

Given the model (7.20), it can be shown that the expectations of (*i.e.*, the long-run averages for) the mean squares are those shown in the last column of Table 7.23. Thus, MS(error) estimates σ_W^2, the repeatability variance, while MS(run) estimates a combination of σ_W^2 and σ_R^2, *i.e.*, the ANOVA table gives a direct estimate of σ_W^2, but to obtain an estimate of σ_R^2 we need to subtract out the repeatability element from the MS(run), as implied by the last column of the table:

$$\hat{\sigma}_W^2 = MS(error) = 27.2$$

$$\hat{\sigma}_R^2 = \frac{(MS(run) - MS(error))}{J} = \frac{(179.9 - 27.2)}{3} = 50.9 \qquad (7.25)$$

Table 7.24 *ANOVA analysis*[a] *and variance component estimates for the control chart data*

Analysis of Variance for Potency

Source	DF	SS	MS	F	P
Run	23	4137.5341	179.8928	6.621	0.000
Error	48	1304.1748	27.1703		
Total	71	5441.7089			

Variance Components

Source	Var Comp.	% of Total	StDev
Run	50.907	65.20	7.135
Error	27.170	34.80	5.213
Total	78.078		8.836

[a]The *F*-test will not be considered here as what is of interest is to estimate the variance components. *F*-tests are discussed in the context of Example 2, below.

This is how the estimated variance components shown in Table 7.24 were calculated.

The total standard deviation of 8.84, shown in Table 7.24, is obtained from the sum of the two variance components. Since

$$\sigma_{total}^2 = \sigma_R^2 + \sigma_W^2$$

it follows that

$$\sigma_{total} = \sqrt{\sigma_R^2 + \sigma_W^2} = \sqrt{50.907 + 27.170} = 8.84$$

This is an estimate of the magnitude of the chance variability affecting single observations. It may be used, therefore, to assess their fitness for purpose. As noted above, it is a measure of the intra-laboratory intermediate precision of the assay.

The relative magnitudes of σ_R and σ_W are also of interest; for example, in a method development study, it might be pointless to try to reduce σ_W if σ_R were the dominant component of the total variability.

Model Validation. The model underlying our ANOVA, Equation (7.20), assumes independently and identically distributed Normal variables at each level of the hierarchy. These assumptions will now be investigated graphically.

Beginning at the bottom of the hierarchy, we ask if the chance variation within each run is Normally distributed. To assess this we calculate the residuals, *i.e.*, the differences between the three replicate test results within each run and the corresponding run means. We then combine all twenty four sets of three into a single set of 72 residuals. Figure 7.15 shows a Normal probability plot for the 72 residuals. The plotted points are close to a straight line, which suggests Normality. The *p*-value given at the bottom right-hand corner of the Figure is large, which supports the conclusion that the data may be considered Normal. Figure 7.16 is a plot of the residuals against the run number, which represents the time order in which the measurements were made. No systematic patterns are evident, consistent with the assumption of independence.

As an extra check, Figure 7.17 shows the residuals plotted against their corresponding run means. As we saw in both Chapter 1 and Chapter 6, it is often found that the variability of analytical results increases with the magnitude of the measurement. There is no such tendency here. Note the slightly odd patterns in the last two graphs. These arise because each set of three within-run residuals sums to zero – note, for example, the second and third last sets in Figure 7.17, in which one large value is balanced by two low values. Patterns like this are particularly strong when there are

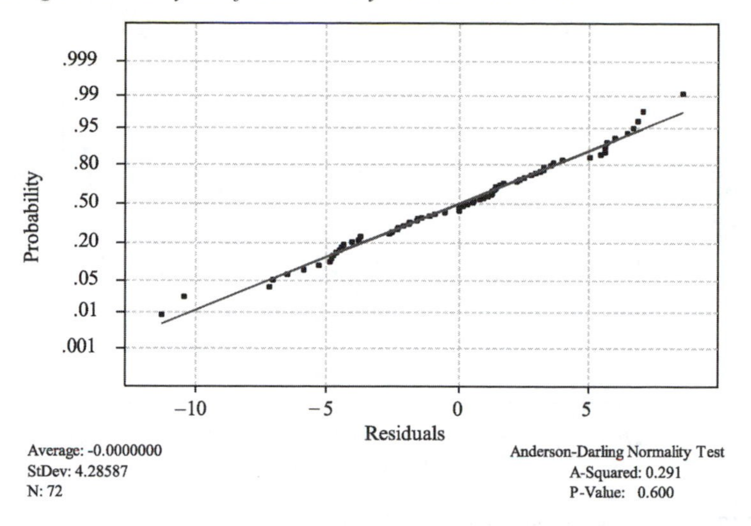

Figure 7.15 *A Normal plot for the control chart residuals*

only two replicates (as in the bilirubin data that are discussed in the next Example) and mirror images about zero will be seen in the residual plots.

Moving now to the top of the hierarchy, Figure 7.18 shows a Normal probability plot for the potency run-means. This graph suggests that the run-to-run variation is also Normally distributed. Figure 7.19 is an individuals control chart for the potency means. This graph suggests that the analytical system is in statistical control from run to run: there are no

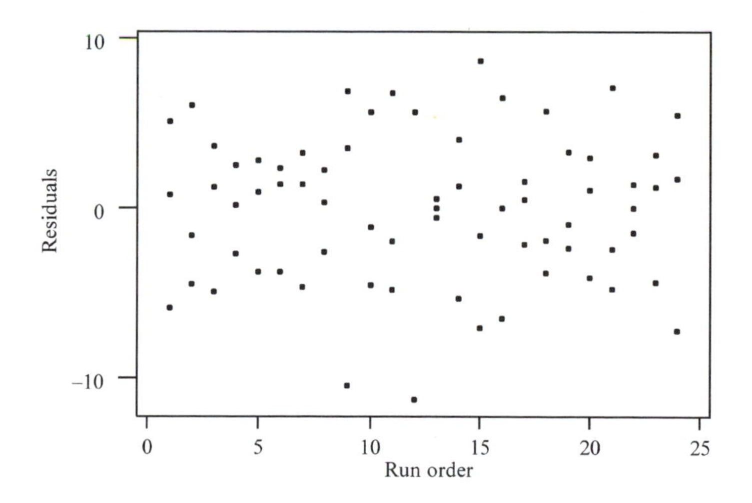

Figure 7.16 *The control chart residuals plotted against run order*

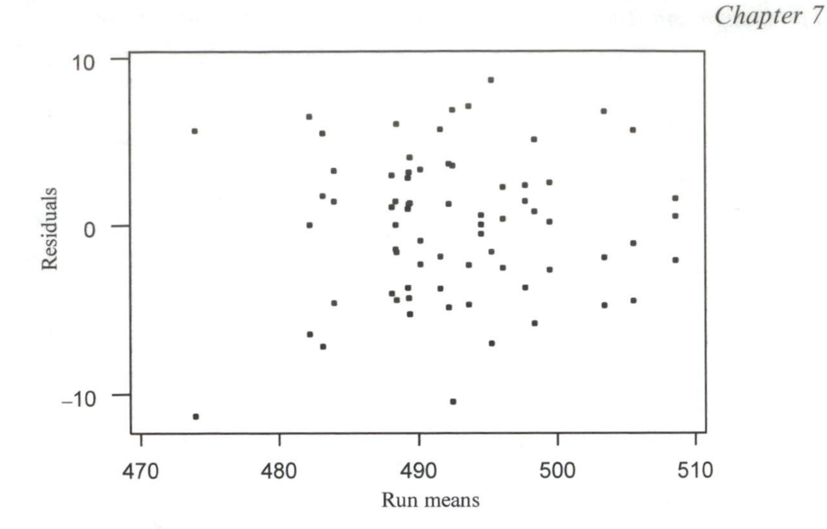

Figure 7.17 *The residuals plotted against their run means*

systematic patterns evident in the plot of the potency means against their time order.

The simple graphical analyses carried out above suggest that the model assumptions are realistic for the control chart data of Table 7.22.

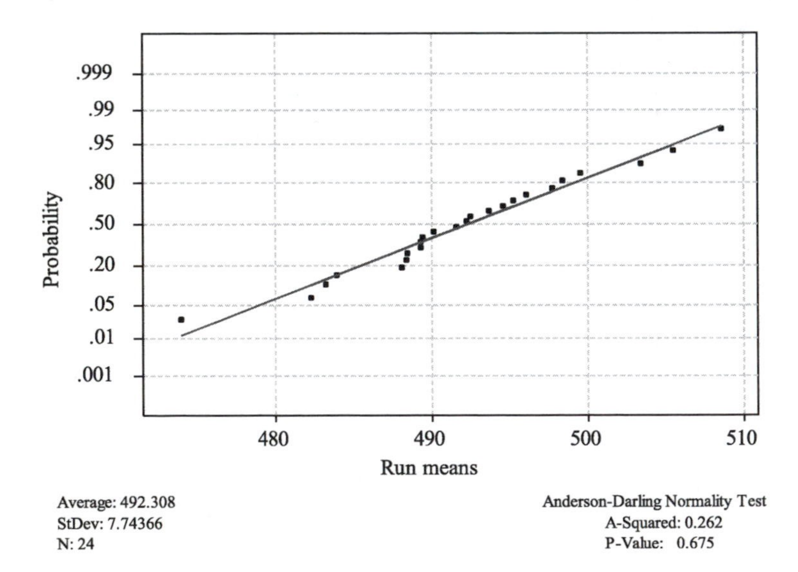

Figure 7.18 *Normal probability plot of the mean potencies for each run*

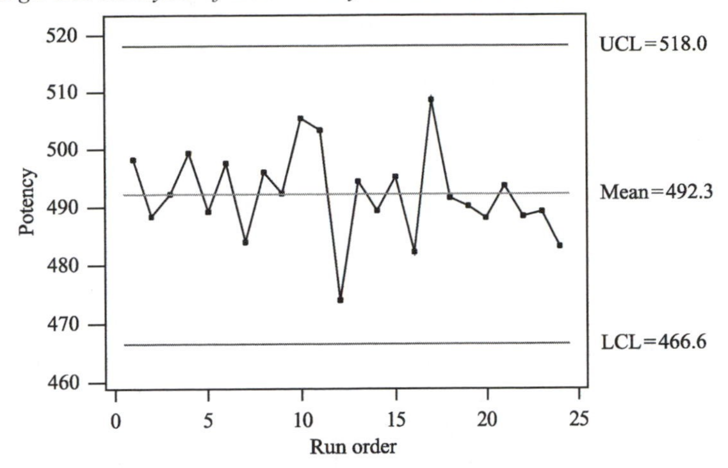

Figure 7.19 *An individuals chart for the run means*

Exercise

7.7 Exercise 2.12 Table 2.15, p 84, gave data on the routine measurement of zinc in potable water samples by flame atomic absorption spectrometry to ensure compliance with the EU drinking water directive. Every time a batch of test samples is measured the batch includes two replicates of a QC control sample spiked with 200 μg L^{-1} of zinc. For Exercise 2.12 a range chart and an individuals chart were drawn and suggested that the analytical system was stable over the data collection period.

For the zinc data, replicate the analyses that were carried out for the potency data in Example 1, including the residual analyses. Using the ANOVA summaries, calculate the variance components by hand, even if your software does the calculations for you – it is worth doing once by hand! From these obtain the corresponding standard deviations.

7.5.2 Example 2: Three Variance Components

The NCCLS document EP-5 'Evaluation of Precision Performance of Clinical Chemistry Devices; Approved Guideline'[1] proposes a standard approach to the estimation of intra-laboratory components of analytical variability. The guidelines allow for the analysis of three levels of variation, *viz.*, between-days, between-runs-within-days and within-run or repeatability variation. This design is a simple generalization of the two level hierarchical design discussed above. It is shown schematically in Figure 7.20.

A typical implementation of this design will involve making measurements on each of 20 days with two runs per day and two replicates per run. Table 7.25 shows data from a study carried out using

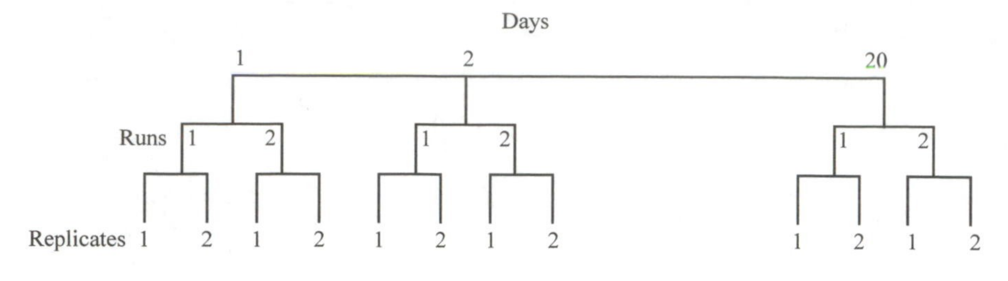

Figure 7.20 *A schematic representation of the NCCLS design*

this design. The analyte measured was the bilirubin concentration of a test kit; units are mg dL^{-1}. One of the purposes of carrying out studies like this is to assess whether the laboratory is able to achieve precision results comparable to those claimed by the equipment manufacturer. The EP-5 guidelines describe and illustrate a statistical significance test of whether the repeatability and total standard deviations arising in the study are consistent with the manufacturer's claims. We will address the same questions, but will answer them by calculating confidence intervals for the long-run standard deviations, using the method discussed in Chapter 3. First, the use of ANOVA to extract estimates of the relevant quantities from the data of Table 7.25 will be illustrated.

Table 7.25 *Measurements of bilirubin conc. (mg dL^{-1}) using the EP-5 design*

Day	*Run 1*		*Run 2*	
	Rep 1	*Rep 2*	*Rep 1*	*Rep 2*
1	2.44	2.43	2.42	2.42
2	2.39	2.41	2.38	2.38
3	2.39	2.39	2.39	2.38
4	2.37	2.35	2.36	2.37
5	2.40	2.39	2.40	2.39
6	2.38	2.40	2.40	2.40
7	2.37	2.38	2.42	2.41
8	2.42	2.41	2.37	2.37
9	2.39	2.38	2.42	2.41
10	2.43	2.43	2.37	2.36
11	2.38	2.39	2.38	2.38
12	2.40	2.40	2.42	2.40
13	2.43	2.44	2.45	2.44
14	2.42	2.41	2.42	2.43
15	2.37	2.38	2.42	2.41
16	2.41	2.42	2.42	2.40
17	2.40	2.39	2.41	2.40
18	2.45	2.47	2.43	2.44
19	2.33	2.33	2.36	2.36
20	2.43	2.44	2.34	2.35

Analysis of Variance. The model that underlies the ANOVA analysis of the data generated by the hierarchical design is that each observation, y_{ijk}, may be described as the sum of four components:

$$y_{ijk} = TV + D_i + R_{ij} + W_{ijk} \tag{7.26}$$

TV is the true value of the standard material being measured; it is regarded as a fixed quantity. The bias component for day i is D_i:, which is considered to vary at random around a mean of zero with standard deviation σ_D. The within-day run bias R_{ij}, where i labels the day and j the run, is also considered to vary randomly around a mean of zero; it has a standard deviation σ_R. Finally, the within-run-within-day repeatability error is W_{ijk}, where i labels the day, j labels the run and k the individual test results. This is assumed to vary at random around zero with standard deviation σ_W. All these random components are assumed Normally distributed and independent of each other. This model implicitly assumes that the magnitude of run-to-run variability is the same within all days and that the magnitude of the repeatability variation is the same for all runs. If the model is appropriate, the variance for individual measurements is:

$$Var(y_{ijk}) = \sigma_D^2 + \sigma_R^2 + \sigma_W^2 \tag{7.27}$$

ANOVA is used to estimate these three variance components.

Table 7.26 shows a nested or hierarchical ANOVA analysis of the bilirubin data. The total sum of squares and degrees of freedom are decomposed into components associated with days, runs within days, and repeatability/error. For each run, the sum of squared deviations of the two replicates around the run mean is a measure of the repeatability of the analytical system. The information from all runs is combined into a

Table 7.26 *ANOVA table and variance component estimates for the bilirubin study*

Analysis of Variance for Bilirubin

Source	DF	SS	MS	F	P
Day	19	0.0417200	0.0021958	2.07	0.057
Run(Day)	20	0.0212000	0.0010600	17.67	0.000
Error	40	0.0024000	0.0000600		
Total	79	0.0653200			

Source	Variance component	Error term	Expected Mean Square for Each Term
1 Day	0.00028	2	(3) + 2(2) + 4(1)
2 Run(Day)	0.00050	3	(3) + 2(2)
3 Error	0.00006		(3)

single sum of squares. Each pair of replicates provides $2 - 1 = 1$ degree of freedom in calculating the within-run or error sum of squares. The third rows of Tables 7.26 and 7.27 show that SS(error) has $20 \times 2 \times (2-1) = 40$ degrees of freedom, when the pairs of replicates from the duplicate runs on each of twenty days are taken into account. The corresponding mean square provides an estimate of σ_W^2, the repeatability variance.

Similarly, the variation between the mean results for the two runs within each day may be measured by a sum of squares, and these are combined to give a sum of squares associated with run-to-run variation. Note that once we average the two within-run replicates, the structure becomes a two-level hierarchy, exactly as we had for the control chart data of Example 1. As in the case of control charts above, SS(run) contains a component due to repeatability variation, also; the corresponding mean square will be adjusted later to give an estimate of the run variance, σ_R^2. There are $J = 2$ runs within each of $I = 20$ days, so the degrees of freedom for runs equals $I(J - 1) = 20(2 - 1) = 20$.

The variation between days is measured by the variation between the $I = 20$ daily means in the usual way, and the sum of squares has $I - 1 = 20 - 1 = 19$ degrees of freedom. Again, the day-to-day sum of squares reflects not only day-to-day variation, but also includes components due to run-to-run and repeatability variation. These will need to be extracted to give a pure estimate of the variance component, σ_D^2, describing to day-to-day variation alone.

Table 7.27 shows how Table 7.26 was calculated. Where one of the subscripts of y_{ijk} has been replaced by a dot, it means that the corresponding element has been averaged. Thus, $\bar{y}_{ij.}$ is the average of the two replicates in run j of day I, $\bar{y}_{i..}$ is the average of the four results for day I and $\bar{y}$ is the mean of all the data. The last column shows the expectations of the mean squares. These indicate how to estimate the variance components using the corresponding mean squares. Thus, the variance component estimates shown in Table 7.26 were calculated as follows:*

$$\hat{\sigma}_W^2 = MS(error) = 0.00006$$

$$\hat{\sigma}_R^2 = \frac{(MS(run) - MS(error))}{K} = \frac{(0.001060 - 0.000060)}{2} = 0.00050 \qquad (7.28)$$

$$\hat{\sigma}_D^2 = \frac{(MS(day) - MS(run))}{JK} = \frac{(0.002196 - 0.001060)}{4} = 0.00028$$

These estimates suggest that the repeatability variance, $\hat{\sigma}_W^2 = 0.00006$, is markedly smaller than the run-to-run variance, $\hat{\sigma}_R^2 = 0.00050$.

*Sometimes, due to chance variation, a result calculated in this manner will be negative. The variance component is set to zero if this happens.

Table 7.27 *ANOVA formulas*

Source of variation	*Degrees of freedom*	*Sums of squares*	*Mean squares*	*Expected mean squares*
Between days	$I-1$	$SS(day) = \sum_{\substack{all \\ data}} (\bar{y}_{i..} - \bar{y})^2$	$MS(day) = \frac{SS(day)}{I-1}$	$\sigma_W^2 + K\sigma_R^2 + JK\sigma_D^2$
Between runs within days	$I(J-1)$	$SS(run) = \sum_{\substack{all \\ data}} (\bar{y}_{ij.} - \bar{y}_{i..})^2$	$MS(run) = \frac{SS(run)}{I(J-1)}$	$\sigma_W^2 + K\sigma_R^2$
Within runs (error)	$IJ(K-1)$	$SS(error) = \sum_{\substack{all \\ data}} (y_{ijk} - \bar{y}_{ij.})^2$	$MS(error) = \frac{SS(error)}{IJ(K-1)}$	σ_W^2
Total	$IJK-1$	$SS(total) = \sum_{\substack{all \\ data}} (y_{ijk} - \bar{y})^2$		

The F-tests shown in the table compare the mean squares in each row to that in the row below it: thus, $F = 0.00106/0.00006 = 17.67$ is the ratio of MS(run) to MS(error). The null hypothesis is that σ_R^2 is zero; if this is true then the expectation of MS(run) is the same as that for MS(error) – this is the basis for the F-test. The p-value of $p = 0.000$ (corresponding to $F = 17.67$) indicates that the F-ratio is highly statistically significant. This means that there is strong statistical evidence for the existence of run-to-run variation over and above the repeatability variation. The ratio of MS(day) to MS(run) ($F = 2.07$, $p = 0.057$) is only marginally statistically significant. There is, therefore, only weak statistical evidence for a day-to-day component in the total variation.

Taking square roots of the variance components gives estimates of the standard deviations associated with the three sources of variability affecting individual measurements.

$$\hat{\sigma}_W = \sqrt{0.00006} = 0.008$$

$$\hat{\sigma}_R = \sqrt{0.00050} = 0.022$$

$$\hat{\sigma}_D = \sqrt{0.00028} = 0.017$$

The total measurement error that affects the precision of an individual test result is the combination of these three components. To obtain a standard deviation that characterizes this error, the variance components are added and the square root of the sum gives the required result:

$$\hat{\sigma}^2_{total} = \hat{\sigma}^2_D + \hat{\sigma}^2_R + \hat{\sigma}^2_W$$

and so

$$\hat{\sigma}_{total} = \sqrt{\hat{\sigma}^2_D + \hat{\sigma}^2_R + \hat{\sigma}^2_W}$$

$$= \sqrt{0.00028 + 0.00050 + 0.00006} = 0.029$$

Performance Assessment. It was suggested above that one reason for estimating variance components is to assess the performance of instruments against the claims made by their manufacturers. Suppose that for the bilirubin data the manufacturer had claimed that the repeatability standard deviation should be 0.007 and the total standard deviation should be 0.025. At first sight it would appear that our data fail to meet these standards, since both $\hat{\sigma}_W = 0.008$ and $\hat{\sigma}_{total} = 0.029$ are bigger than the claimed values. However, no account has been taken of the chance variation affecting these estimates – the long-run or true standard deviations may well be consistent with the manufacturer's claims.

To assess the effect of chance variation on the sample estimates, two essentially equivalent approaches are possible: significance tests or confidence intervals. The EP-5 guidelines carry out significance tests using the manufacturer's claims as the null hypothesis values for the standard deviations; the reader is referred to the guidelines for further details. Here, we will calculate confidence intervals for the long-run standard deviations; if the confidence intervals contain the claimed values, then there is no basis for rejecting these claims.

In Chapter 3 we saw how to calculate a confidence interval for a population or process variance σ^2 based on a sample variance s^2, calculated from n results. This was given by:

$$\frac{\nu s^2}{\chi^2_U} \leqslant \hat{\sigma}^2 \leqslant \frac{\nu s^2}{\chi^2_L}$$

where χ^2_L and χ^2_U are the values that leave $\alpha/2$ in each tail of the relevant chi-square distribution (see Table A5) and $\alpha = 0.05$ for a 95% confidence interval. The degrees of freedom of s^2 were $\nu = n - 1$ in the examples discussed in Chapter 3. The square roots of the variance bounds give confidence bounds for the standard deviation.

This method can be applied directly to the calculation of confidence intervals for the repeatability component of variance and the corresponding standard deviation. Thus, for the bilirubin example above, the MS(error) (another name for s^2) was 0.00006 and this had 40 degrees of freedom. A confidence interval for the repeatability variance component is given by:

$$\frac{IJ(K-1)MS(error)}{\chi_U^2} \leqslant \hat{\sigma}_W^2 \leqslant \frac{IJ(K-1)MS(error)}{\chi_L^2}$$

that is

$$\frac{40(0.00006)}{59.34} \leqslant \hat{\sigma}_W^2 \leqslant \frac{40(0.00006)}{24.43}$$

which gives

$$0.00004 \leqslant \hat{\sigma}_W^2 \leqslant 0.00010$$

The square roots of these bounds give confidence bounds of 0.006 and 0.010 for the repeatability standard deviation. Since these bounds include the manufacturer's claim of 0.007 we cannot reject this claim.

In order to carry out similar calculations for the other components of variance or for the total variance for an individual test result, we must first determine the (approximate) degrees of freedom associated with the estimator being used. These are given by a formula known as Satterthwaite's approximation. For any linear combination of k mean squares:

$$c_1\text{MS}_1 + c_2\text{MS}_2 + \ldots + c_k\text{MS}_k \tag{7.29}$$

where the cs are constants, the approximate degrees of freedom (df) are given by:

$$\text{df} = \frac{[c_1\text{MS}_1 + c_2\text{MS}_2 + \ldots + c_k\text{MS}_k]^2}{\frac{(c_1\text{MS}_1)^2}{v_1} + \frac{(c_2\text{MS}_2)^2}{v_2} + \ldots \frac{(c_k MS_k)^2}{v_k}} \tag{7.30}$$

where v_i are the degrees of freedom of MS_i. The resulting value will usually not be an integer and is rounded to give the approximate degrees of freedom. Note that this same approximation was used in Chapter 4 for the approximate two-sample t-test used when the standard deviations for the two groups are not the same.

Our estimate of the total variance was:

$$\hat{\sigma}_{total}^2 = \hat{\sigma}_D^2 + \hat{\sigma}_R^2 + \hat{\sigma}_W^2$$

Replacing each term on the right-hand-side by the combination of mean squares from which it was calculated (see Equations (7.28)) gives:

$$\hat{\sigma}_{total}^2 = 1/4\,[\text{MS(day)} - \text{MS(run)}] + 1/2\,[\text{MS(run)} - \text{MS(error)}] + \text{MS(error)}$$

or

$$\hat{\sigma}^2_{total} = {}^1\!/_4\,\text{MS(day)} + {}^1\!/_4\,\text{MS(run)} + {}^1\!/_2\,\text{MSE}$$

Accordingly, the degrees of freedom for $\hat{\sigma}^2_{total}$ are given by:

$$\begin{aligned} df_{total} &= \frac{[{}^1\!/_4\,\text{MS (day)} + {}^1\!/_4\,\text{MS(run)} + {}^1\!/_2\,\text{MS(error)}]^2}{\frac{({}^1\!/_4\,\text{MS(day)})^2}{19} + \frac{({}^1\!/_4\,\text{MS(run)})^2}{20} + \frac{({}^1\!/_2\,\text{MS(error)})^2}{40}} \\ &= \frac{[{}^1\!/_4(0.002196) + {}^1\!/_4(0.001060) + {}^1\!/_2(0.000060)]^2}{\frac{({}^1\!/_4(0.002196))^2}{19} + \frac{({}^1\!/_4(0.001060))^2}{20} + \frac{({}^1\!/_2(0.000060))^2}{40}} \\ &= 9.2 \end{aligned}$$

So the total variance has 9 degrees of freedom, approximately.

This result enables us to get confidence intervals for the long-run total variance and total standard deviation values. A 95% confidence interval for the total variance is given by:

$$\frac{9(0.00084)}{19.02} \leqslant \hat{\sigma}^2_{total} \leqslant \frac{9(0.00084)}{2.70}$$

$$0.00040 \leqslant \hat{\sigma}^2_{total} \leqslant 0.00280$$

The square roots of these bounds give 0.020 and 0.053 as confidence bounds for the total standard deviation. This interval contains the manufacturer's claim of 0.025 for the total standard deviation and so, although the observed sample estimate of 0.029 exceeds the claimed value, the deviation from the claimed value is consistent with just chance variation. Note that the EP-5 guidelines use one-sided tests; one-sided confidence intervals are equivalent to these.

Exercises

7.8 For the control chart data of Table 7.22, p 344, calculate 95% confidence intervals for the repeatability, runs and total standard deviations. When calculating degrees of freedom, always round the final result downwards. The chi-square multipliers given in Table 7.28 will be useful for this and the next exercise.

7.9 Obtain the same standard deviation estimates for the zinc data of Exercise 7.7 (see Table 2.15, p 84).

7.10 Isoflavones are a class of chemical compound found in a variety of plants. Soybeans contain relatively high levels. Isoflavones are of interest for various reasons, including their ability to inhibit certain cancers. Klump *et al.*[6] describe

Table 7.28 *Chi-square multipliers for confidence intervals*

df	*0.025*	*0.975*
16	6.91	28.85
24	12.40	39.36
30	16.79	46.98
32	18.29	49.48
35	20.57	53.20
48	30.75	69.02

an inter-laboratory collaborative trial in which 10 laboratories each determined isoflavone levels in a soy isolate sample, in duplicate on each of two days. Replicates are nested within days, which in turn are nested within laboratories. Thus, the data form a three-level hierarchy with the same structure as that of Example 2, above.

The data reproduced in Table 7.29 are the glycitin (one form of Isoflavone) results for a soy isolate, reported as aglycon equivalents in $\mu g\ g^{-1}$.

Carry out a nested ANOVA and compute the relevant standard deviations to assess the magnitudes of repeatability, day-to-day and laboratory-to-laboratory variation. Note that collaborative trials will be discussed in the next chapter, after which the importance of such measures should be clearer.

Note that in assessing the statistical assumptions underlying your analysis, the residuals from the three-level hierarchy relate to the distribution of within-run errors. If the replicates are averaged, then the structure becomes a two-level hierarchy and the analysis proceeds as for the control chart data of Example 1. The residuals from fitting the two-level model relate to the day-to-day variation within laboratories. A Normal plot of the laboratory means assesses the assumption of Normality of laboratory-to-laboratory variation.

Table 7.29 *Duplicate results for isoflavone inter-laboratory study* (Reprinted from the Journal of AOAC INTERNATIONAL, 2001, Vol. 84, pp 1865. Copyright, 2001, by AOAC INTERNATIONAL.)

Lab	*Day 1*	*Day 2*	*Lab*	*Day 1*	*Day 2*
1	199	187	6	199	205
	189	173		202	206
2	207	221	7	210	205
	238	225		217	210
3	197	203	8	216	219
	206	206		206	217
4	219	215	9	235	206
	231	222		237	208
5	227	221	10	234	237
	233	218		236	236

If, in carrying out the residual analysis, you discover outliers, exclude all the data from the corresponding laboratory.

7.6 CONCLUSION

In this chapter we saw how analysis of variance provides a general framework which can be used for the analysis of multi-factor experimental studies, where the factors may have several levels. Such studies are concerned with the comparison of the mean responses obtained under different configurations of an analytical system. The use of analysis of variance in such situations is sometimes described as 'fixed effects' ANOVA. It is fixed in the sense that the experimental conditions (say the levels of temperature or choice of mobile phase) could be set or fixed at the same setting in a repeat of the study.

The chapter introduced, also, the application of ANOVA to a very different mode of data collection: this involves observations made in such a way that if the study were to be replicated, different conditions would apply when the subsequent data were collected. Typical examples of such data are those collected for control charting, where several replicate measurements are made on a control material at a series of time points (often labelled 'runs'). If the data collection exercise were extended in time, the chance influences would change continually, so that, unlike fixing two temperatures or mobile phases, no two runs would be the same and it would not be possible to replicate the data collection under exactly the same set of conditions. What is of interest in this second situation is to estimate the magnitudes of the various components of the chance variation affecting the system under study. When ANOVA is used to do this, it is often referred to as 'random effects' ANOVA.

The data structure described by Figure 7.14 is, in fact, the same as that which underlies the one-way ANOVA analysis of Section 7.2, where replicate measurements were made within each of four laboratories. The difference between the analysis of Section 7.5 and that of Section 7.2 centres on the questions of interest in the two cases, and the underlying statistical models for the data are correspondingly different.

For the laboratory comparison study, interest centred on the four laboratory means – were they the same or different and, if different, which laboratories were different and how big were the differences. Thus, the F-test shown in Table 7.2 is a test of the equality of the four laboratory means. For the control chart data of Section 7.5 the runs are not of interest in themselves – they are only of interest to the extent that they reflect the chance run-to-run variation in the analytical system, the magnitude of which is one focus of the analysis; the other is concerned with the standard deviation of the within-run or repeatability variation. The *F*-test in Table 7.24 is a test of whether σ_R equals zero – this is clearly rejected since the *p*-value is less than 0.001. The contrast between the two examples emphasizes that different analyses can be applied to essentially the same data structures. It is the question that is to be answered that provides the key to determining which mode of analysis is appropriate.

7.7 REVIEW EXERCISES

7.11 Thompson and Wood[7] report the data shown in Table 7.30 as resulting from a homogeneity test of the material used in a proficiency testing scheme. Twelve containers were randomly sampled from the total set of containers to be used in the scheme. The material from each container was homogenized and two test portions were selected. The 24 test portions were analyzed in a randomized order under repeatability conditions. The data represent the copper content of soybean flour (μg g^{-1}).

The material is considered sufficiently homogeneous if either of the following conditions holds:

- The *F*-test from a one-way analysis of variance is not statistically significant;
- If the between sample standard deviation estimate is less than 30% of the target value for the laboratory-to-laboratory standard deviation. The target value for this test was 1.1 μg g^{-1}. Do the data support the conclusion that the bulk material is sufficiently homogeneous for the proficiency test?

Table 7.30 *Homogeneity test results*
(Reprinted from the Journal of AOAC INTERNATIONAL, 1993, Vol. 76, pp 926. Copyright, 1993, by AOAC INTERNATIONAL.)

Sample	*Copper*		*Sample*	*Copper*	
1	10.5	10.4	7	9.8	10.4
2	9.6	9.5	8	9.8	10.2
3	10.4	9.9	9	10.8	10.7
4	9.5	9.9	10	10.2	10.0
5	10.0	9.7	11	9.8	9.5
6	9.6	10.1	12	10.2	10.0

7.12 The data shown Table 7.31 below are percentage recoveries for a second drug residue, narasin, from the study by Matabudul *et al.* described in Example 2 of Section 7.3, p 330. These results were treated as pure replicates in Exercise

Table 7.31 *Drug residue recovery results (%)*
(Reprinted from the *Analyst*, 2002, Vol. 127, p 760. Copyright, 2002, Royal Society of Chemistry.)

	Spike level ng g^{-1}					
Matrix	*10*		*20*		*40*	
	101	73	95	102	98	97
Poultry	79	86	88	99	81	88
liver	91	104	87	93	110	82
Sheep	87	98	92	100	98	109
liver	89	99	108	97	79	91
	79	91	104	89	89	90
	77	104	76	89	91	101
Egg	92	98	98	88	88	98
	84	74	95	101	94	98

7.3, p 333, even though they were obtained on six different days. The first column in each cell of the table represents days 1–3 and the second days 4–6.

Treat day as a blocking factor, *i.e.*, consider the data as representing a 3 × 3 factorial study replicated on each of six days and carry out an ANOVA analysis (see Section 7.4 above). Before carrying out the ANOVA using software, write down the degrees of freedom that you expect to appear in each row of the ANOVA table – compare your values with those produced by the software. Carry out appropriate residual analyses. Interpret the results of your analysis.

7.13 The data shown in Table 7.32 come from a study of the factors that influence the performance of a formic acid biosensor, which is used for air monitoring. For more information and a reference to the original study see Exercise 5.8, p 229; more factors are given there and the labelling here is

Table 7.32 *Biosensor development study*
(Reprinted from the *Analyst*, 2001, Vol. 126, pp 2008. Copyright, 2001, Royal Society of Chemistry.)

A	*B*	*F*	*Results*			
−1	−1	−1	377	518	443	384
1	−1	−1	445	476	455	548
−1	1	−1	589	540	570	391
1	1	−1	534	579	542	590
−1	−1	1	528	425	491	497
1	−1	1	512	485	521	465
−1	1	1	441	544	590	553
1	1	1	697	570	662	610

consistent with that in Exercise 5.8. Several response variables were studied; a response labelled C2 is reported in the table below. Large responses are desirable. For the factors listed below, the first level is coded −1 and the second level as +1 in the data table:

A: Concentration of Medola's blue (mg g^{-1}) 0.3, 1.5
B: Concentration of the enzyme formate dehydrogenase (mg g^{-1}) 0.36, 0.792
F: Electrochemical potential (V) 0, 0.2

The response C2 is the current (nA) at steady state when the biosensor was exposed to a formic acid vapour concentration of 2.9 mg m^{-3}. There are four replicates at each of the 2^3 design points. Analyze the data as a three-way ANOVA model, carry out a residual analysis and set out your conclusions as to what are the important features of the study results. These data were the basis for Exercise 5.14, p 245, in which the standard 'effects analysis' of a 2^3 design was required. Compare the results of the two analyses.

7.14 Example 1 of Section 7.4 and Exercise 7.7 were concerned with the extraction of variance components (and their corresponding standard deviations) from control chart data. Carry out the same set of analyses for the total oxidized nitrogen (TON) data of Table 3.12, p 128. Note that runs 10, 18 and 21 were identified as outliers and should be excluded.

7.15 Exercise 7.10, p 358, was concerned with the analysis of data from a three-level hierarchical study of isoflavones in a soy isolate. The paper also contained the genistein test results shown in Table 7.33 as aglycon equivalents in μg g^{-1} for the analysis of a miso test material. Carry out the same analyses as were recommended in Exercise 7.10.

Table 7.33 *Isoflavone inter-laboratory study results*
(Reprinted from the Journal of AOAC INTERNATIONAL, 2001, Vol. 84, pp 1865. Copyright, 2001, by AOAC INTERNATIONAL.)

Lab	*Day 1*	*Day 2*	*Lab*	*Day 1*	*Day 2*
1	191	201	6	202	200
	197	241		199	205
2	220	234	7	201	196
	224	229		202	198
3	186	193	8	212	212
	185	194		211	212
4	194	193	9	212	209
	197	195		212	209
5	235	220	10	232	231
	221	222		228	224

7.8 REFERENCES

1. *Evaluation of Precision Performance of Clinical Chemistry Devices; Approved Guideline*, NCCLS-EP5-A, National Committee for Clinical Laboratory Standards, Wayne, PA, 1999.
2. M. Danaher, Personal communication.
3. D.K. Matabudul, I.D. Lumley and J.S. Points, *Analyst*, 2002, **127**, 760.
4. G.E.P. Box, W.G. Hunter and J.S. Hunter, *Statistics for Experimenters*, Wiley, New York, 1978.
5. D.C. Montgomery, *Design and Analysis of Experiments*, 3rd edn, Wiley, New York, 1991.
6. S.P. Klump, M.C. Allred, J.L. MacDonald and J.M. Ballam, *J. AOAC Int.*, 2001, **84**, 1865.
7. M. Thompson and R. Wood, *J. AOAC Int.*, 1993, **76**, 926.

CHAPTER 8

Assessing Measurement Quality

8.1 INTRODUCTION

This chapter is concerned with providing quantitative measures of the quality of test results. Two approaches are discussed and then brought together. The first approach is concerned with estimating the reproducibility of analytical measurements. We saw in Chapter 1 that reproducibility is a measure of the magnitude of the difference that might be expected between two measurements made on the same material, one in each of two laboratories. This concept is important for commercial and regulatory purposes in that it explicitly recognizes that differences are to be expected between measurements generated in different laboratories and it represents an approach to quantifying the magnitude of expected differences. In order to calculate the reproducibility, estimates of the sizes of the within and between-laboratory variation must be obtained; the different components of the total variation are measured by appropriately defined standard deviations. Estimates of these standard deviations may be obtained from an inter-laboratory collaborative trial. This is a long-established approach to providing measures of the fitness for purpose of analytical results.

In recent years a different approach to assessing measurement quality has been gaining momentum. This is concerned with evaluating the 'uncertainty' of measurements, where that word is given a specific technical meaning: 'a parameter associated with the result of a measurement, that characterizes the dispersion of the values that could reasonably be attributed to the measurand' – the parameter is usually a standard deviation. The method essentially involves breaking down the measurement procedure into a series of steps, assessing the chance variability associated with each step, in terms of a standard deviation, and then combining all the standard deviations into a single overall measure of the quality of the reported result. This approach is based on a guideline document produced by the ISO[1] in 1993: 'Guide to the expression of uncertainty in measurements' and is often referred to as the GUM approach. Eurachem[2] has produced a version of this guide, which is more suited to the needs of analytical chemists. Estimating uncertainty is the subject matter of Section 8.3.

Section 8.4 discusses how the two approaches might be integrated. The final section of the chapter reviews some aspects of the interplay between the statistical methods and the data collection designs discussed throughout the book.

8.2 INTER-LABORATORY COLLABORATIVE TRIALS

Inter-laboratory collaborative trials (CTs) take essentially the same form as the proficiency tests (PTs) introduced in Chapter 2: a common material is analyzed by a number of laboratories and the results are subjected to a statistical analysis. The aims, however, are different. Proficiency testing is concerned with providing laboratories with an assessment of their ability to perform particular types of analysis by allowing them to compare their results with those generated in peer laboratories or against an established assigned value, where such is available. Collaborative trials are concerned with establishing the performance characteristics of new analytical methods or of older methods adapted to new purposes.

In this section the statistical methods required for the analysis of CTs are introduced. Fuller accounts will be found in the AOAC 'Guidelines for collaborative study procedures to validate characteristics of a method of analysis'[3] and in the ISO set of documents 'Accuracy (trueness and precision) of measurement methods and results ISO-5725, parts 1–6.[4] Full discussions of the chemistry and logistical aspects of such trials will be found in these documents; they are not discussed here.

8.2.1 Estimating the Reproducibility Standard Deviation

Schaffler *et al.*[5] carried out a study to evaluate an ion chromatographic method for the analysis of sugars in beet and cane molasses. The data shown in Table 8.1 represent duplicate measurements of sucrose in beet molasses made by ten laboratories using high-performance anion-exchange with pulsed amperometric detection (HPAE–PAD).

Note that the data structure here is exactly the same as that encountered in Section 7.5 for the control chart data: replicates are nested under laboratories here, they were nested under runs earlier. Accordingly, the statistical model underlying our analysis will be the

Table 8.1 *Sucrose measurements for ten laboratories*
(Reprinted from the Journal of AOAC INTERNATIONAL, 1997, Vol. 80, pp 603. Copyright, 1997, by AOAC INTERNATIONAL.)

1	*2*	*3*	*4*	*5*	*6*	*7*	*8*	*9*	*10*
50.74	50.64	50.67	49.18	50.19	50.79	51.87	49.57	49.56	51.89
50.50	50.02	50.19	49.51	49.98	50.72	51.67	50.20	49.61	52.31

same as before with 'laboratory' replacing 'run'; we will use σ_L to denote the standard deviation that measures between-laboratory variation. Note, however, that since both measurements are made in a single analytical run in each laboratory, the standard deviation for laboratories, σ_L, includes the run bias component of total analytical variability. A more elaborate study design, involving several replicates in each of several runs, would be required to separate the two components (see Exercise 8.12 for an example of such a design). The definition of reproducibility limit, involving as it does the difference between two single measurements in each of two laboratories, implicitly confounds the run bias with the laboratory bias and, hence, the standard interlaboratory study design reflects this.

Table 8.2 shows the analysis of variance, the estimated variance components and their corresponding standard deviations.

The repeatability (σ_r^2) and laboratory (σ_L^2) variance components were estimated as follows:

$$\hat{\sigma}_r^2 = s_r^2 = MS(error) = 0.0723$$

$$\hat{\sigma}_L^2 = s_L^2 = \tfrac{1}{2}[MS(lab) - MS(error)] = \tfrac{1}{2}(1.5571 - 0.0723) = 0.7424 \qquad (8.1)$$

Table 8.2 *ANOVA analysis of sucrose data*

Analysis of Variance for Sucrose

Source	DF	SS	MS	F	P
Labs	9	14.0138	1.5571	21.535	0.000
Error	10	0.7230	0.0723		
Total	19	14.7369			

Variance Components

Source	Var Comp.	% of Total	StDev
Labs	0.742	91.12	0.862
Error	0.072	8.88	0.269
Total	0.815		0.903

Together these give an estimate of the reproducibility variance (σ_R^2):

$$\hat{\sigma}_R^2 = s_R^2 = \hat{\sigma}_r^2 + \hat{\sigma}_L^2 = 0.0723 + 0.7424 = 0.815 \tag{8.2}$$

The corresponding standard deviations are the square roots of these quantities: $\hat{\sigma}_r = 0.269$, $\hat{\sigma}_L = 0.862$, and $\hat{\sigma}_R = 0.903$; $\hat{\sigma}_R$ is labelled 'total' in Table 8.2. Note that the method for calculating the estimate of σ_L^2 will occasionally result in a negative value, which is obviously incorrect. In such cases the estimate is set to zero. This may occur if the repeatability precision is poor and chance variation results in a very large MSE.

The AOAC Official Methods of Analysis[6] data analysis guidelines give the following formulae for calculating $s_r = \hat{\sigma}_r$ and $s_R = \hat{\sigma}_R$ directly, without explicitly calculating an ANOVA table. Note that the guidelines use s_r and s_R, only; the $\hat{\sigma}$ notation is retained in parallel here to make the correspondence with the models discussed in the last chapter. The repeatability standard deviation is given by:

$$s_r = \sqrt{\frac{\sum_{i=1}^{L} d_i^2}{2L}} = \sqrt{\frac{1.4461}{20}} = \sqrt{0.0723} = 0.269 \tag{8.3}$$

where d_i is the difference between the two replicate results in laboratory i and L is the number of laboratories. The reproducibility standard deviation is calculated by:

$$s_R = \sqrt{\tfrac{1}{2}(s_d^2 + s_r^2)} = \sqrt{\tfrac{1}{2}(1.55709 + 0.0723)} = 0.903 \tag{8.4}$$

where:

$$s_d^2 = \frac{\sum_{i=1}^{L} (T_i - \bar{T})^2}{2(L-1)} \tag{8.5}$$

in which T_i is the sum of the two replicate results from laboratory i, and $\bar{T}$ is the mean of the T_i values across all laboratories.

The two methods for calculating the standard deviations are equivalent and give the same results, as shown above. Note, however, that Equations (8.3)–(8.5) are applicable only where there are two replicates. If more than two replicates are measured in each laboratory, these formulae will need to be replaced by others, which will effectively involve doing the ANOVA calculations.

While the guidelines permit the inclusion of more than two replicates in the study design, they do not recommend it. The ANOVA degrees of freedom give insight into why this is so. If there are J replicates in each laboratory, they contribute J − 1 degrees of freedom to estimating the repeatability standard deviation. Thus, if there are 10 laboratories each with 2 replicates, there will be 10 degrees of freedom for assessing repeatability variation. Increasing the number of replicates to 3 gives 20 degrees of freedom for assessing repeatability variation, but there are still only 10 pieces of information available to assess laboratory-to-laboratory variation. In our example, the laboratory-to-laboratory standard deviation is approximately three times the repeatability standard deviation and so it dominates the calculation of the reproducibility standard deviation; this will often be the case. Accordingly, it makes more sense to increase the number of laboratories or the number of levels (*e.g.*, concentrations) of the test material, rather than the number of within-laboratory replicates, where this is feasible.

Exercise

8.1 Kalra[7] carried out an inter-laboratory collaborative trial of methods for the determination of the pH of soils. Fifty-three laboratories were involved in the study and they each measured 10 blind duplicate samples of different soils using several methods. The data presented in Table 8.3 are the pH measurements of Malbis soil samples for the first 20 laboratories, using a single method.

Analyze the data using ANOVA and estimate the repeatability and reproducibility standard deviations. Use Equations (8.3)–(8.5) to obtain estimates, also, and confirm that the two methods give the same results.

Table 8.3 *Duplicate pH measurements of a soil by 20 laboratories* (Reprinted from the Journal of AOAC INTERNATIONAL, 1995, Vol. 78, pp 310. Copyright, 1995, by AOAC INTERNATIONAL).

Lab	*pH*		*Lab*	*pH*	
1	5.90	5.95	11	4.92	5.46
2	5.48	5.69	12	5.40	5.49
3	5.50	5.50	13	6.29	6.54
4	5.45	5.52	14	5.65	5.73
5	5.37	5.59	15	5.45	7.10
6	5.18	5.57	16	5.55	5.82
7	5.46	5.53	17	5.42	5.33
8	5.64	5.84	18	5.46	5.57
9	5.30	5.20	19	4.95	5.18
10	5.86	6.14	20	5.51	5.89

8.2.2 Data Scrutiny

The analysis of variance and variance estimation calculations carried out for the sucrose data implicitly assume that the data are well-behaved.

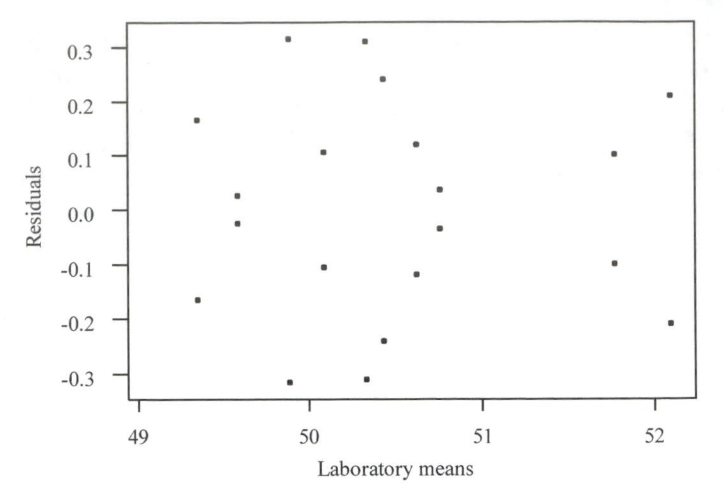

Figure 8.1 *Residuals versus laboratory means*

Whatever the possibility of controlling data quality within a single laboratory, it would be optimistic in the extreme for the organizer of a collaborative trial to assume, without checking, that no problems have occurred when a large number of laboratories are involved in a study. Accordingly, the guidelines recommend a series of steps designed to discover outlying observations or outlying laboratories.* Before introducing the formal statistical significance testing procedures recommended in the guidelines, we will carry out residual analysis, which has been the preferred approach to data scrutiny throughout the book.

Residual Analysis. The assumptions underlying the definitions of the repeatability and reproducibility critical limits are that the within-laboratory repeatability variation and the between laboratory variation are approximately Normal. Do the data support these assumptions? There is also an implicit assumption of constant within-laboratory variability involved in combining the within-laboratory standard deviations into a single value for repeatability. Residual analysis can shed light on all three assumptions.

Figure 8.1 shows the residuals plotted against the mean results for the ten laboratories. Apart from the usual oddities associated with having just two replicates, the plot shows no tendency for the variability to change systematically with the mean level of the measurements. The Normal plot of the residuals Figure 8.2, suggests

*Instead of identifying and excluding outliers, an alternative approach is to use 'robust' statistical methods that are not sensitive to the presence of outliers.[8,9]

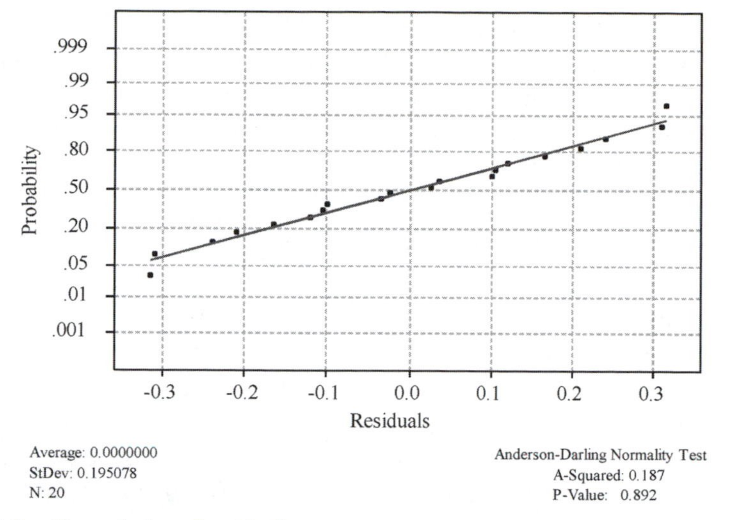

Figure 8.2 *Normal plot of residuals*

that the assumption that within-laboratory repeatability variation is Normal is reasonable.

The Normal plot of the ten laboratory means, Figure 8.3, shows two relatively large values. Normal plots based on small numbers of observations tend to be somewhat unstable, so we would want to interpret the plot with caution. The test-statistic is not statistically

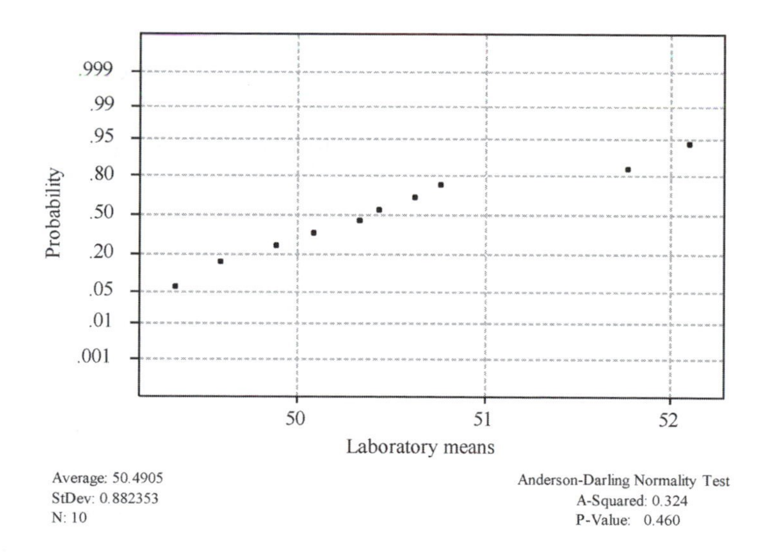

Figure 8.3 *Normal plot of laboratory means*

significant ($p = 0.46$), but again the test being based on a small sample will not be powerful. Eight of the 10 values are in a fairly good straight line, so we might reasonably expect that with more data the majority of laboratories would display Normal laboratory-to-laboratory variation. This, of course, is speculative.

Outlier Tests. The harmonized guidelines recommend a series of statistical tests to identify outlying observations or outlying laboratories. These are then excluded from the dataset before the within-laboratory and between-laboratory standard deviations are calculated. The steps involve using Cochran's test to identify laboratories with relatively large within-laboratory variability; this could be the result of poor repeatability performance or due to the presence of one or more outliers. This is followed by the application of Grubbs' test to identify laboratories whose mean values are very high or very low, on the assumption that the distribution of laboratory means is Normal. These tests are introduced below.

Cochran's test. Cochran's test asks if the largest of a set of sample variances (or equivalently, sample standard deviations) is statistically significantly larger than the rest of the set. The null hypothesis is that the data are samples from a set of Normal distributions, each with the same standard deviation. If the test-statistic:

$$100\frac{s_{\max}^2}{\sum_{i=1}^{L} s_i^2}\% \tag{8.6}$$

exceeds the critical value (Table A10 in the Appendix), the null hypothesis is rejected and we conclude that the repeatability standard deviation for the laboratory corresponding to the largest sample variance is out of line with the others. Note that the test is inherently one-sided.

The AOAC guidelines recommend a significance level of 0.025, *i.e.*, the critical value leaves an area of 0.025 in the right-hand tail of the sampling distribution. For the sucrose data the test statistic is:

$$100\frac{s_{\max}^2}{\sum_{i=1}^{L} s_i^2} = 100\frac{0.19845}{0.72305} = 27.4\%$$

Table A10 gives the critical value as 65.5%. Accordingly, the null hypothesis is not rejected. The test result suggests that the observed variances are consistent with a common underlying value.

Exercise

8.2 Apply Cochran's test to the pH data of Exercise 8.1. If you exclude one of the laboratories on the basis of the test, redo the test for the remaining laboratories to determine whether or not the next largest variance is inconsistent with the others. What do you conclude?

Note that applying a sequence of tests to a dataset carries the same implications for increased probabilities of type 1 error that were discussed in the context of multiple comparisons in Chapter 7. The approach described above is, however, that which is recommended in the guidelines.

Grubbs' Test. Grubbs' test is designed to identify outliers in data supposedly sampled from a Normal distribution. Figure 8.4 illustrates the simplest case, a single outlier.

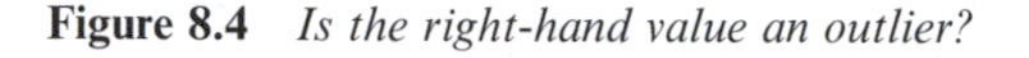

Figure 8.4 *Is the right-hand value an outlier?*

If the standard deviation, s, of the full set of test results is calculated, it will obviously be inflated by the extreme value on the right-hand side of Figure 8.4. Deleting the high value and re-calculating the standard deviation gives a smaller value, s_H, where H labels the high result. If the outlier had been on the left-hand side of the main body of data points then the standard deviation calculated after deleting the extreme value would be labelled s_L. The percentage decrease in the standard deviation on deletion of a single outlier is given by either:

$$100\left(1-\frac{S_H}{S}\right)\% \text{ or } 100\left(1-\frac{S_L}{S}\right)\% \tag{8.7}$$

The larger of these quantities is the test statistic for the presence of a single outlier. If this exceeds the critical value (Table A11 in the Appendix, again the guidelines recommend a significance level of 0.025) the null hypothesis that the extreme value comes from the same Normal distribution as the rest of the data is rejected.

Applying this test to the 10 laboratory means for the sucrose data results in a test-statistic of 18.6% since

$$100\left(1-\frac{0.718}{0.882}\right)\% = 18.6\% \text{ and } 100\left(1-\frac{0.833}{0.882}\right)\% = 5.6\%$$

The critical value for 10 laboratories is 42.8%. The sucrose test-statistic of 18.6% is very much less than this, so we do not reject the hypothesis that either the largest or smallest value is an outlier.

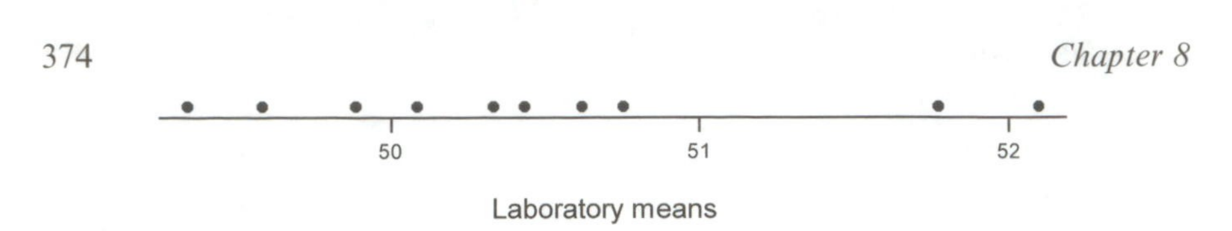

Figure 8.5 *Dotplot of the laboratory means*

Figure 8.5 is a dotplot of the mean sucrose results for the ten laboratories. This includes two relatively large values that are obviously possible outliers. It is now clear why removing the largest value does not have a major impact on the standard deviation – the second largest value is also relatively extreme, which means that the standard deviation does not reduce dramatically on removal of the largest value. What happens if we remove both large values?

The test-statistic is:

$$100\left(1 - \frac{0.498}{0.882}\right)\% = 43.5\%$$

Table A11 gives critical values for the removal of the two highest, two lowest or the largest and smallest values in a dataset. It is clear from the dotplot that removing the two largest values will have the greatest impact on the standard deviation, so we focus on that here. The critical value for ten laboratories is 56.4%. Accordingly, we fail to reject the hypothesis that the two large values come from the same distribution as the remainder of the data.

Even though we have failed to identify the two large values as outliers, they must be considered suspect. The power of the statistical test (its ability to reject the null hypothesis when the extreme values are indeed outliers, *i.e.*, not from the same distribution as the rest of the data) will be low for such a small dataset and we should not rely uncritically on the statistical test. Note how large a reduction in the standard deviation (56.4%) is required for statistical significance to be established.

Inter-laboratory collaborative trials typically involve several parameters being measured at several levels. Accordingly, there is more information available to assess the performance of the two laboratories that produced the large results. The trial organizer will naturally want to make decisions in the light of the overall set of results, so the outcomes of individual significance tests are only a part of the evidence that will be evaluated in the course of analyzing the study data.

Exercise

8.3 Calculate the mean pH result for each of the 20 laboratories involved in the collaborative trial presented in Exercise 8.1. Draw a dotplot of the data.

Apply Grubbs' tests for a single outlier (high (H) or low (L)) and for two outliers (HH, LL or HL). What do you conclude?

8.4 Re-do the ANOVA of the pH data of Exercise 8.1, but this time carry out a residual analysis. Do the residual plots give you more, less or the same information as the statistical tests you carried out in Exercises 8.2 and 8.3?

8.5 Normal plots based on small numbers of observations do not tend to give very straight lines, even when data are sampled from a Normal distribution. Accordingly, we should be slow to interpret apparent departures from a straight line as indicating non-Normality. One visual method of deciding whether the departure from linearity should be considered serious is to compare the plot to those produced by data simulated from a Normal distribution.

Simulate 19 sets of 10 random Normal variates with the same mean and standard deviation as those of Figure 8.3. Draw Normal plots for all sets. If Figure 8.3 is less like a straight line than all the others, then the hypothesis of Normality is cast in doubt, otherwise not.

8.2.3 Measuring the Trueness of a Method

The vast majority of collaborative trials are carried out to determine the precision characteristics of a measurement method, as described above. If the materials measured in the trial have an accepted reference value (this will be referred to as the true value, TV) then, in addition, the method bias (lack of 'trueness'), if any exists, may be estimated from the study results. The ISO Standard 5725-4 indicates that the reference materials could be:

- Certified reference materials;
- Materials manufactured for the purpose of the experiment with known properties;
- Materials whose properties have been established by measurements using an alternative measurement method whose bias is known to be negligible.

In the rest of the chapter the material will be referred to as a certified reference material (CRM), for convenience, but any of the above materials might be used.

If n replicate measurements are made in a single run in each of p laboratories, then the standard error of the mean of all the data is:

$$SE(\bar{y}) = \sqrt{\frac{(\sigma_L^2 + \sigma_r^2/n)}{p}} \tag{8.8}$$

where σ_r^2 is the within-run repeatability variance and σ_L^2 is the sum of the laboratory and run bias variances which, as we saw earlier, are confounded with each other by the study design. Note that increasing p, the number of laboratories, rather than n, the number of within-laboratory replicates, has a much greater impact in reducing this standard error. The method bias is estimated by the difference between the overall mean and the accepted reference value, $\bar{y} - TV$. If the various components of the variation are Normal and the reference value is considered a known constant, or determined with a precision that is negligible compared to the CT variation, then an approximate 95% confidence interval for the true method bias is given by:

$$(\bar{y} - TV) \pm 2\sqrt{(\hat{\sigma}_L^2 + \hat{\sigma}_r^2/n)/p} \qquad (8.9)$$

The method bias estimated from a collaborative trial may be used to adjust routine results obtained in any one laboratory.

Exercise

8.6 Fontaine and Eudaimon[10] carried out an inter-laboratory collaborative study of methods for the determination of the lysine, methionine and threonine in trade products or concentrated amino acid pre-mixes. Seventeen laboratories were involved in the study. The data presented in Table 8.4 are the lysine measurements for one material.

- Analyze the data using ANOVA and estimate the repeatability and reproducibility standard deviations.

Table 8.4 *Inter-laboratory study results for lysine*
(Reprinted from the Journal of AOAC INTERNATIONAL, 2000, Vol. 83, pp 771. Copyright, 2000 by AOAC INTERNATIONAL.)

Lab	*Lysine*		*Lab*	*Lysine*	
1	46.23	46.65	10	45.57	45.50
2	46.45	46.10	11	45.85	45.83
3	46.04	46.97	12	45.94	46.62
4	46.23	46.05	13	46.16	46.58
5	45.14	45.88	14	45.57	46.04
6	45.95	46.31	15	45.77	45.79
7	48.58	47.78	16	44.90	44.84
8	44.49	44.72	17	43.50	43.74
9	45.83	45.26			

- Apply Cochran's test to the laboratory variances and Grubbs' tests to the laboratory means. Draw a dotplot of the means. If any data are called into question by these tests, exclude the corresponding laboratories, re-do the ANOVA and re-calculate the repeatability and reproducibility standard deviations.

- Plot the residuals against the fitted values and draw Normal plots for both the residuals and laboratory means.

8.3 MEASUREMENT UNCERTAINTY

The competition associated with the growth in international trade during the second half of the twentieth century inevitably led to a focus on the quality of goods and services. This, in turn, implied a requirement that components, raw materials and finished goods (including food and drugs) should meet, or exceed, quality and regulatory standards. As a consequence, the quality of measurements, on the basis of which commercial and regulatory decisions are made, came under unprecedented scrutiny. To break down barriers to international trade, it became obvious that there was a need for measurements made in one country to be accepted by customers in other countries. This led to a system of multi-lateral agreements whereby laboratories might be accredited by national accreditation bodies as being competent to carry out particular types of measurement and this competency would be accepted by signatories to the multi-lateral agreement. The system involves the laboratory being audited regularly by the accreditation body in respect of both the laboratory's quality management system and its technical competence in relation to the calibration or testing methods for which it is accredited. The accreditation bodies themselves are audited by peer groups involved in the multi-lateral agreement. In this context of quality assurance and control, the ISO produced two key documents, 'Guide to the expression of uncertainty in measurement', 1993, [1]commonly referred to as GUM, and ISO/IEC 17025:1999, 'General requirements for the competence of calibration and testing laboratories',[11] which requires accredited laboratories to produce estimates of the uncertainties of the measurements they produce. The procedures set out in GUM for estimating measurement uncertainties are the focus of the following sections.

The GUM Approach. GUM defines uncertainty of measurement as 'a parameter associated with the result of a measurement, that characterizes the dispersion of values that could reasonably be attributed to the measurand'. The measurand is the particular quantity that is being measured, *e.g.*, the concentration of the analyte. The parameter is typically either a standard deviation, in which case it is referred to as a 'standard uncertainty', or the half-width of an interval, when it is referred to as an 'expanded uncertainty'. The interval is formed by multiplying the standard uncertainty by a 'coverage factor (k)'. The understanding is that the interval 'result $\pm$ k standard uncertainty' will contain a large fraction of the values that might reasonably be attributed to the measurand. The factor k is typically 2 (sometimes 3) and the interval is analogous to a confidence interval.

Guidance on the GUM approach to the estimation of measurement uncertainties in analytical chemistry laboratories has been provided by the Eurachem document 'Quantifying uncertainty in analytical measurements' (1993, revised 2000).[2] The approach is 'bottom-up': it examines in detail every aspect of the measurement process and assesses the contribution of each element to the overall likely variation in test results. Specifically the process involves:

- Describing the measurement procedure in detail
- Writing down the expression that gives the final test result
- Identifying all the sources of variability affecting each element in that expression
- Quantifying the magnitudes of each of these sources in terms of a standard deviation
- Combining these into a single standard deviation that describes the variability of the measured quantity
- Reporting the result as $y \pm ku_c(y)$ in which y is the test result, *i.e.*, the best estimate of the quantity being measured, k is the coverage factor, and $u_c(y)$ is the estimated combined standard uncertainty.

Working through these steps requires a comprehensive understanding of the chemistry of the method being studied. Detailed case studies are beginning to appear in the literature in recent years; the revised Eurachem Guide, in particular, is a rich source of information on how to carry out an uncertainty assessment. Here, only a short account is given. The purpose is to provide a simple introduction to the two main statistical issues that need to be addressed. The first of these is concerned with the statistical distributions that may be used to described measurement variability, and with obtaining the standard deviations of these distributions. The second relates to the rules for combining the standard deviations for the various contributing sources of chance variation into a single standard deviation that will represent the combined standard uncertainty of the measurand. Rather than discuss these questions in the abstract, we will consider a relatively simple example of an analytical procedure and discuss the statistical ideas in that context. The example is illustrative, rather than definitive, and details which may be important in some circumstances (such as buoyancy in weighing) are ignored for simplicity of presentation.

8.3.1 Example: Preparing a Stock Solution

Preparation of stock solutions that are used in the calibration of a GC system for the analysis of distilled spirits involves weighing approximately 1 g of a compound into a tared 100 mL volumetric flask, adding approximately 90 mL of high purity ethanol, shaking the flask, and making the volume up to the mark with ethanol. The concentration of the solution is given by:

$$C = \frac{mP}{V} \text{ g mL}^{-1} \tag{8.10}$$

where m is the mass of the compound placed in the flask, P is the purity of the compound and V is the volume of the flask. The uncertainty associated with each of these parameters, m, P, V, must be quantified in order to calculate a combined standard uncertainty for C. Figure 8.6 provides a 'route map' to help readers navigate their way through the technical details. There are three broad headings, corresponding to the three inputs, which will be treated in turn. Within these headings there

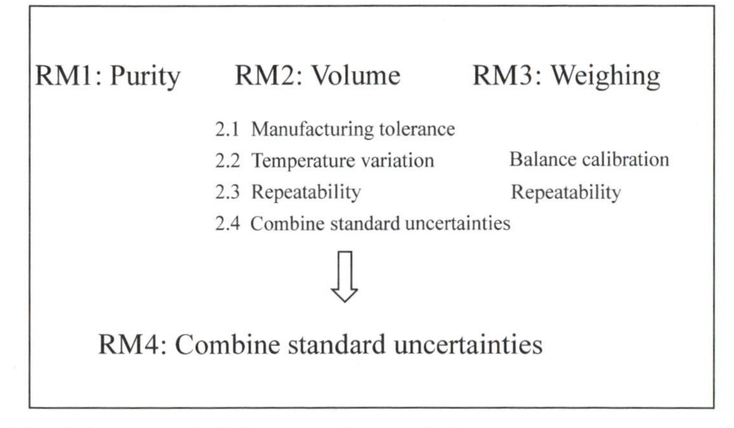

Figure 8.6 *A 'route map' of the uncertainty analysis*

are sub-headings to be considered. Finally, the estimates for the separate standard uncertainties must be combined into a single combined standard uncertainty for concentration.

RM1: Purity

The compound is sold with a specification that its purity is not less than 99%. This is generally taken to mean that the purity is equally likely to be anywhere in the interval 99.0–100%. For the purposes of the calculation, the purity is expressed as a fraction, *i.e.*, between 0.9 and 1.0, rather than as a percentage. The Uniform or Rectangular distribution, illustrated in Figure 8.7, is a model for chance variation, where results are considered equally likely on an interval. As for any probability distribution, the total area under the curve is one. For a Uniform distribution, the standard deviation is given by the half-width divided by the square root of three. Thus, the standard uncertainty for the compound purity is:

$$u(P) = SD(P) = 0.005/\sqrt{3} = 0.0029$$

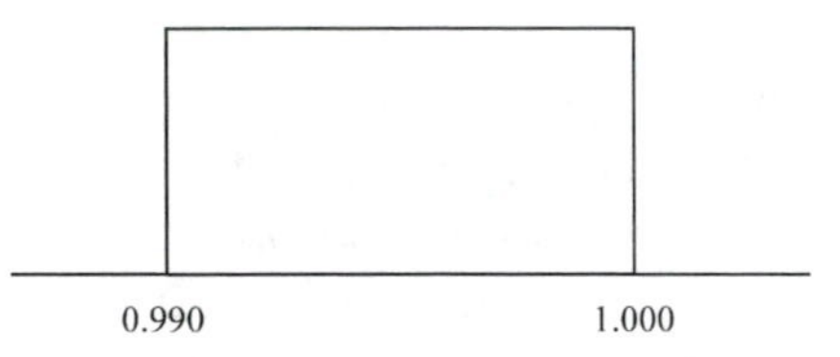

Figure 8.7 *A Uniform distribution for purity, 0.995 ± 0.005*

RM2: Volume

The manufacturer's tolerance for the 100 mL class A volumetric flask is 100 ± 0.08 mL, at 20°C. The flask is not filled under very tightly controlled temperature conditions and so the volume of the liquid will be affected by temperature fluctuations in the laboratory. Also, the filling-to-the-mark operation will be subject to some chance variation. Accordingly, these three sources of variability need to be taken into account in assessing the uncertainty of the liquid volume.

RM2.1: Manufacturing Variability. In discrete manufacturing quality control assessments of process capability, it is generally assumed that variation between individual manufactured units follows a Normal distribution. However, since the manufacturer has not provided any information on the status of the tolerance, ± 0.08 mL, (for example, by

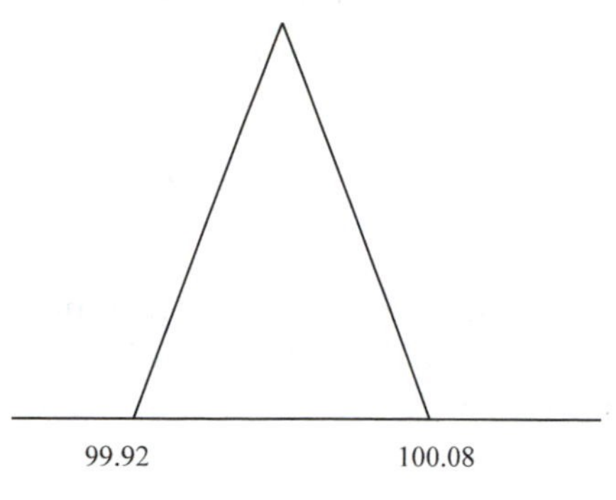

Figure 8.8 *A Triangular distribution*

saying that it contains 95 or 99% of units produced), we are forced to make some assumption about the nature of the chance variability. We might, for example, assume a Uniform distribution, as we did for purity. Instead, a Triangular distribution, as illustrated in Figure 8.8, is assumed here.

This distribution is such that the individual units (flask volumes) are more likely to be close to the centre of the tolerance interval, 100 mL, than they are to its end-points – a more plausible assumption than uniformity in this case. As for the Uniform distribution, the standard deviation of a Triangular distribution is related in a simple way to the half-width: the standard deviation is the half-width divided by the square root of six. Thus, the standard uncertainty associated with the flask volume manufacturing tolerance is:

$$u(\text{manufacturing}) = SD(\text{manufacturing}) = \frac{0.08}{\sqrt{6}} = 0.033\,\text{mL}$$

RM2.2: Temperature Variation. The laboratory temperature is controlled to vary by no more than 4°C from 20°C. We will assume that when concentration calculations are made, the volumes are not corrected for temperature variations away from 20°C, the temperature at which the flask was calibrated. Accordingly, allowance must be made for this variation when calculating uncertainties. The temperature variation will be assumed to follow a Uniform distribution with end-points at 16°C and 24°C.

The coefficient of volume expansion for ethanol is considerably greater than that for the glass, so the volume expansion of the glass will be ignored in the calculation. The coefficient of volume expansion for ethanol is $\alpha = 10.5 \times 10^{-4}\,^{\circ}\text{C}^{-1}$, which means that the variation in the nominal volume of $V = 100\,\text{mL}$, due to temperature variation of $\Delta T = \pm 4\,^{\circ}\text{C}$, is

$$V \alpha \Delta T$$

or

$$(100\,\text{mL}) \times (10.5 \times 10^{-4}\,{}^{\circ}\text{C}^{-1}) \times (\pm\,4\,{}^{\circ}\text{C})\,\text{mL} = \pm\,0.420\,\text{mL}$$

Since the temperature variation is assumed Uniformly distributed, with a half-width of 4 °C, so also the change in volume due to temperature variation, will be Uniformly distributed with half-width 0.420 mL. The standard deviation of this distribution is the half-width divided by the square root of 3, so the contribution to the standard uncertainty in volume, due to temperature variation is:

$$u(\text{temperature}) = \frac{0.420}{\sqrt{3}} = 0.242\,\text{mL}$$

RM2.3: Repeatability. The effect of chance variation introduced into the final result by the analyst's filling to the mark was estimated empirically by the standard deviation of 20 fill-and-weigh operations. This gives the standard uncertainty directly:

$$u(\text{repeatability}) = SD(\text{repeatability}) = 0.025\,\text{mL}$$

The repeatability variation would be expected to follow a Normal distribution. However, this does not affect our uncertainty estimation, as it did for the Uniform and Triangular distributions, as the standard deviation is obtained directly.

RM2.4: Combining the Standard Uncertainty Contributions for Volume. Each of the three elements that together determine the uncertainty associated with the flask fill-volume has now been determined. The effects of the three factors are considered to be additive. This implies that to calculate the combined standard uncertainty, the squares of the three standard deviations are summed and the square root is taken. Thus, the total variance is:

$$\hat{\sigma}^2_{total} = \hat{\sigma}^2(\text{manufacturing}) + \hat{\sigma}^2(\text{temperature}) + \hat{\sigma}^2(\text{repeatability}) \qquad (8.11)$$

which gives the combined standard uncertainty (standard deviation) as:

$$u(\text{volume}) = \sqrt{u^2(\text{manufacturing}) + u^2(\text{temperature}) + u^2(\text{repeatability})} \qquad (8.12)$$

$$= \sqrt{(0.033)^2 + (0.242)^2 + (0.025)^2}$$

$$= 0.246\,\text{mL}$$

Note the dominance of the temperature component in determining the combined uncertainty of the volume. Thus, if we wish to improve the quality of measurements (in terms of reducing the standard uncertainty) attention should be focused on temperature control, rather than analyst training to improve repeatability.

RM3: Weighing

A full analysis of the uncertainties involved in weighing involves a number of factors, as listed in the Eurachem Guide[2] (p.117). In the following simplified treatment, only two factors are considered. These are the uncertainty given on the balance calibration certificate and the repeatability standard deviation for check weighings.

The calibration certificate for the balance quotes a measurement uncertainty of 0.52 mg, using a coverage factor of 2. This means that there is a standard uncertainty of 0.26 mg associated with balance calibration. This is counted twice to allow for both the taring and weighing operations. Twenty check weighings of a 1 g check weight gave a standard deviation of 1.2×10^{-4} g. Accordingly, the uncertainty associated with the weighing operation is given by:

$$u(m) = \sqrt{2u^2(\text{calibration}) + u^2(\text{repeatability})}$$

$$= \sqrt{2(0.26)^2 + (0.12)^2} = 0.387\,\text{mg}$$

RM4: The Combined Uncertainty for Concentration

For a particular stock solution the mass of the compound was found to be 1.018 g, the nominal purity was 0.995 and the nominal flask volume was 100 mL. This gives the concentration of the solution as:

$$C = \frac{mP}{V} = \frac{1.0180 \times 0.995}{100} = 10.129\,\text{g}\,\text{L}^{-1}$$

To obtain the combined standard uncertainty for concentration, we need to combine the three standard uncertainties associated with the three variables in the equation. Since the variables are not additive, we do not sum the corresponding variances, as was done when combining the elements that determined the uncertainty of the fill volume. When several independent variables are either multiplied or divided by each other, as here, the squares of the coefficients of variation are additive (this is an approximate relationship). Thus,

$$\frac{SD^2(C)}{C^2} = \frac{SD^2(m)}{m^2} + \frac{SD^2(P)}{P^2} + \frac{SD^2(V)}{V^2} \tag{8.13}$$

gives the combined standard uncertainty as:

$$\begin{aligned} u_c(C) &= C\sqrt{\frac{u^2(m)}{m^2} + \frac{u^2(P)}{P^2} + \frac{u^2(V)}{V^2}} \\ &= 10.129\sqrt{\left(\frac{0.387}{1.0180}\right)^2 + \left(\frac{0.0029}{0.995}\right)^2 + \left(\frac{0.246}{100}\right)^2} \\ &= 3.85\,\mathrm{g\,L^{-1}} \end{aligned} \tag{8.14}$$

The expanded uncertainty $U(C)$ is found by multiplying the combined standard uncertainty by a coverage factor of 2.

8.3.2 Discussion

This example illustrates the two main statistical ideas involved in carrying out an uncertainty assessment. These are the nature of the frequency or probability distributions that may need to be deployed, and the rules for combining the component standard uncertainties (expressed as standard deviations) into a final combined value.

The information used in bottom-up uncertainty assessments may be obtained empirically, as was, for example, the repeatability standard deviation for the fill-to-the-mark operation in determining the uncertainty associated with the volume in our example. This is referred to as Type A by GUM. Type B information comes in the form of manufacturers' tolerances, calibration certificates, or just plain 'professional judgements'. Where no other information is currently available, it may be necessary to make a judgement as to the likely variation that is to be expected for a particular system component. In such cases the Uniform or Triangular distributions will typically be used to express the 'degree of belief' in the likely parameter values that may be encountered. As we have seen, these distributions are completely specified by their end-points, so it is a simple matter to calculate the relevant standard deviation (uncertainty) once the distribution is specified.

The rules for combining the standard uncertainties are simple to implement. Where the individual independent sources of chance variation are combined linearly, *i.e.*, by addition or subtraction, together with possible multiplication by constants, then the square of the combined standard deviation is the sum of the squares of the individual standard deviations, multiplied, where appropriate, by the constants squared. Symbolically, if the quantity to be evaluated is:

$$T = aX + bY - cZ$$

where a, b and c are constants, and X, Y and Z are random variables, *i.e.*, system parameters that are subject to chance variation, then the standard deviation of T is:

$$SD(T) = \sqrt{a^2SD^2(X) + b^2SD^2(Y) + c^2SD^2(Z)} \tag{8.15}$$

This formula was applied in combining the three separate components involved in determining the uncertainty in volume.

Where the elements are combined in a multiplicative fashion, as in the final calculation of concentration, then it is the squares of the coefficients of variation that are additive (as indicated before, this is an approximate result). Thus, if the resultant, T, is given by:

$$T = \frac{aXY}{Z}$$

then, the standard deviation of T is obtained from the components by:

$$CV^2(T) = CV^2(X) + CV^2(Y) + CV^2(Z)$$

and

$$SD(T) = T\sqrt{\frac{SD^2(X)}{X^2} + \frac{SD^2(Y)}{Y^2} + \frac{SD^2(Z)}{Z^2}} \tag{8.16}$$

where the letters have the same meaning as before. Note that the constant, a, drops out of Equation (8.16).

One statistical issue which may arise in practice has not been discussed, as it did not arise in our example. This is the possibility that there may be correlations between the inputs, X, Y, Z, to the equation that determines the test result. The simple combination rules, stated above, apply only when the variation in each of the inputs is independent of the variation in the others. When this is not the case the formulae have to be extended to take account of the non-independence. Further information may be found in the references cited above.[1,2]

The GUM approach provides a very detailed analysis of the analytical system, in terms of the factors contributing to overall measurement uncertainty. This will often be the starting point for process improvement – once the sources of the larger uncertainties are identified, it is a natural next step to attempt to reduce their influence. Also, the GUM analysis need only be carried out once, provided there is evidence of continuing performance to the standards implied by the stated uncertainty; this can be provided by the Internal and External Quality Control measures discussed in Chapter 2.

Exercises

8.7 Suppose that the procedure described above required the use of water, rather than ethanol, as the solvent and that the range of temperature variation in the laboratory was $\pm 3°C$ rather than $\pm 4°C$. Suppose also that the manufacturer's volume tolerance was ± 0.10 mL instead of ± 0.08 mL and that the repeatability standard deviation remains unchanged. Re-calculate the standard uncertainty for the liquid volume. The coefficient of volume expansion for water is $2.1 \times 10^{-4}\ °C^{-1}$.

8.8 If laboratory temperature were measured and the liquid volume calculation adjusted accordingly, then the uncertainty component associated with laboratory temperature variation could be eliminated (assume, for simplicity, that the uncertainty associated with the temperature adjustment is negligible). Redo the calculations for both the example in the text and for the revised assumptions of Exercise 8.7 and assess the impact exclusion of the temperature variation effect would have on the volume uncertainty.

8.4 AN INTEGRATED APPROACH TO MEASUREMENT UNCERTAINTY

Two approaches to the assessment of measurement quality have now been described. The first, the collaborative trial approach, is 'top-down' in that it deals simultaneously with the totality of procedures involved in making a measurement by studying the outcomes from the analysis of the same material in many different laboratories. The second, the GUM approach, is 'bottom-up': it involves a detailed examination of all aspects of the measurement SOP individually, usually within a single laboratory, with a view to combining the contributing sources of variability into a single figure, the combined standard uncertainty, that describes the quality of measurements produced by the analytical system.

The collaborative trial is a well-established approach to providing measures of the fitness-for-purpose of analytical measurements. It is widely accepted and well-understood within the analytical community. The GUM approach, on the other hand, is more recent, has its origins in physical metrology, and is both less well-understood and less accepted currently by chemists. See, for example, the trenchant criticisms made by Horwitz.[12] As long ago as 1995, the Analytical Methods Committee of the Royal Society of Chemistry[13] suggested that collaborative trial data might be used to produce estimates of measurement uncertainty, thus bringing the two approaches together.

This suggestion is attractive in that, where collaborative trial results are available, it can save the laboratory from having to carry out the detailed studies required for the GUM approach. It should also mean that different laboratories, carrying out the same analyses, would report the same measurement uncertainties for their results. Examination of the results of the IMEP study discussed in Chapter 1 shows that different laboratories can have radically different opinions as to the quality of the results they produce, when measuring the same analyte. The laboratories involved in the IMEP study did not all use the same analytical methods, but the variations in reported uncertainties, shown in Figure 1.2 of Chapter 1, do not reflect only method differences. The collaborative trial approach also has the advantage that it is entirely empirical; it does not require a detailed mathematical description of the analytical system and is, therefore, less open to the possibility of incomplete model specification with consequential under-estimation of the total uncertainty affecting the results. Of course, any laboratory intending to use collaborative trial data for assessing the uncertainty of its own results must be able to perform at a level consistent with the conditions obtaining during the collaborative trial. Specifically, this implies appropriate validation studies to assess laboratory bias and precision and on-going IQC and participation in proficiency testing to guarantee continued performance at this standard.

In very many cases collaborative trial measures of variability will not be sufficient in themselves as estimates of the uncertainties associated with routine measurements. They will need to be augmented to provide a realistic measure of the uncertainty of the results produced by a laboratory. This will be necessary, for example, when the collaborative trial shows the method to be biased and when factors that affect routine measurements have not been included in the protocol operating during the collaborative trial. Environmental test samples provide an example of the second situation. Environmental test samples may arrive at the laboratory wet, inhomogeneous and coarsely divided. For a collaborative trial the corresponding test material may already have been dried, finely powdered and homogenized.[14] The operations involved in bringing the material routinely received by the laboratory to the same state as the collaborative trial material, including sub-sampling, need to be assessed for their contributions to overall uncertainty. These contributions must then be combined with the reproducibility standard deviation from the collaborative trial, to arrive at a better assessment of the overall uncertainty.

The reproducibility measures from collaborative trials will also need to be augmented to allow for method bias adjustment of routine test results. We saw earlier that collaborative trials can provide a measure of method bias, where the material involved in the trial is a certified reference material (CRM). The bias is given by:

$$\hat{\delta} = \bar{y} - TV \tag{8.17}$$

where $\bar{y}$ is the average result for all the laboratories and TV is the accepted reference value or certified value for the material. If the reported results are routinely adjusted for the method bias then allowance has to be made for the uncertainty associated with the bias adjustment. This is given by:

$$u(\hat{\delta}) = \sqrt{(\hat{\sigma}_L^2 + \hat{\sigma}_r^2/n)/p + \hat{\sigma}^2(TV)} \tag{8.18}$$

where n within-run replicates are obtain in each of p laboratories and $\hat{\sigma}(TV)$ is the uncertainty associated with the CRM.

It is clear, then, that, in order to use collaborative trial data for estimating measurement uncertainty, a GUM-like description of the analytical system is required. Its purpose is to identify all aspects of the SOP that contribute to the variability of test results, but which are not fully represented in the collaborative study design. At the time of writing, a draft ISO Standard[14] (ISO/DTS 21748: Guide to the Use of Repeatability, Reproducibility and Trueness Estimates in Measurement Uncertainty Estimation) is in circulation. This provides guidance on the integrated approach to the assessment of measurement uncertainty. The AMC paper[13] referred to earlier provides an excellent discussion of the issues involved. It also contains a lucid account of measurement uncertainty in general and, as the title of the paper suggests, its implications for analytical science.

8.5 CONCLUDING REMARKS

A short review of some of the statistical ideas discussed in various places in the book, in the light of the general measurement model[4,13,15] which underlies the material in this chapter, may help some readers to obtain a clearer overall perspective of how the different methods are inter-related and complementary.

The measurement model specifies that a single test result may be viewed as composed of a number of elements:

$$\begin{aligned}\text{result} = \text{true value} + \text{method bias} + \text{laboratory bias} + \text{run bias}\\ + \text{repeatability error}\end{aligned}$$

in which the different terms were described by Thompson as forming a 'ladder of errors'. For an extended discussion covering a wide range of analytical practices, refer to Thompson.[15] Note the correspondence between this model and the hierarchical ANOVA models discussed in Chapter 7.

Random or Fixed Terms? The status of the different elements of this model will vary depending on the viewpoint of the user Different terms in the model may be regarded as fixed constants in some circumstances and as random variables in other circumstances. If, for example, a multinational company is concerned about differences between results produced in two or more *particular* laboratories, then the laboratory biases are considered fixed effects and the concern will be to estimate differences between the results produced in these particular laboratories. Such comparisons were carried out in Chapter 4 for two laboratories and in Section 7.2 for four different laboratories from one multinational company.

In Chapter 1 and in Section 8.2 of this chapter, on the other hand, the reproducibility was taken to be concerned with the difference between two measurements, one from each of any two laboratories. In the absence of the type of comparative study referred to above, differences between any two laboratories will be unpredictable. From a statistical perspective this means that the laboratory bias component of the model is considered a random variable. It will usually be assumed to be approximately Normally distributed with zero mean and a standard deviation that describes the laboratory-to-laboratory variation in results. The estimation of this standard deviation is of interest when characterizing the performance of an analytical method. Thus, in the analysis of collaborative trials the laboratories are regarded as a random sample from the population of all laboratories that might use the method under study. In practice, for most trials the laboratories are not randomly selected; the organizers will use other laboratories with which they are in contact or which have similar technical interests. The extent that such non-random selection departs from the ideal and results in the study not being representative is, of course, an issue that needs to be considered at the design stage.

Comparing Laboratories. For comparisons between the mean results from two laboratories using the same method, the method bias (if any) will cancel and the differences should reflect the differences between the laboratory biases. If the two laboratories use different methods or variations of the same method, as was the case for the example of

Section 7.2, then the method and the laboratory biases are confounded with each other, so 'difference between laboratories' includes both factors. In either case, the data collected to assess the presence of relative bias should include all the variance components lower down the ladder of errors. In particular, this should include run-to-run variation so that a valid statistical test can be carried out or, equivalently, so that the confidence interval used to measure the relative bias will fully reflect the variation in the analytical system. In practice, this means that data should be obtained from different runs. Recall that the replicate observations for the comparison of four laboratories in Section 7.2 were obtained on each of six different days, in each laboratory. This ensured that run-to-run variation was represented in the standard error used in the statistical tests and confidence interval estimation.

In Chapter 4, a simulation study was used to demonstrate how a very high error rate was obtained for a laboratory comparison, where all the data came from a single analytical run in each laboratory. Fifty-eight percent of tests rejected the hypothesis of no relative bias when this was true by design. The model shows the cause of the problem. When all the data come from single runs, only repeatability variation is embodied in the standard error of the mean difference. Effectively, the run biases become confounded with the laboratory and method biases, and, even where these latter biases are zero, the difference between the two run biases is unlikely to be zero. Since the run biases will sometimes be relatively large, the test gives spurious results. The more within-run replicates that are obtained, the smaller the calculated standard error will become, and the more likely this is to happen. Although the objective of the study was to test for a laboratory difference, in fact, the t-test really tests for a laboratory plus run bias difference, because of the inadequacy of the design. As was pointed out in Chapter 4, the data are not fit-for-purpose.

Assessing Method Bias. The same hierarchical principle applies when it comes to measuring method bias. We saw in Section 8.2 that the trueness of a method can be evaluated from the results of a collaborative trial, where the material analyzed is a CRM. The standard error used in testing for bias is based on all the components in the model that are below method in the hierarchy, *i.e.*, laboratory, run bias and repeatability error. It is given by $SE(\bar{y}) = \sqrt{(\sigma_L^2 + \sigma_R^2 + \sigma_r^2/n)/p}$, where σ_L refers to the laboratory bias, only, and σ_R explicitly represents the run bias standard deviation. The sum of σ_L^2 and σ_R^2 was represented by σ_L^2 in both Sections 8.2 and 8.4 in conformity with the usage in the international collaborative trial guidance documents; they are separated here as σ_R^2 will be used later without reference to σ_L^2.

If we attempt to assess the method bias by measuring a CRM n times in each of several runs, k say, within a single laboratory, then the calculated standard error of the mean result will estimate $\sqrt{(\sigma_R^2 + \sigma_w^2/n)/k}$, where σ_R^2 is, again, the run bias variance and σ_w^2 is the within-run repeatability variance. If the laboratory performs to the same standard as the laboratories involved in the relevant collaborative trial, then it would be expected that $\sigma_w = \sigma_r$. Note that the standard error of the mean of the laboratory results does not include a laboratory standard deviation, since only one laboratory is involved in the study. When used as the denominator of a test of the hypothesis that the mean of the study results equals the CRM assigned value, apart from chance variation, this standard error provides a test of the hypothesis that the method bias *and the laboratory bias* are both zero. Since laboratory variation is not contained in the standard error, the laboratory term is implicitly confounded with the next higher term in the measurement model, which is the method term. Thus, what is intended as a test of the method bias really addresses the question 'does the method give unbiased results *in this laboratory*?' This is, of course, a valid and important question from the perspective of the individual laboratory, but it does not address the more general question regarding the trueness of the results produced by this method, *in general*.

In Chapter 3 we discussed the use of a confidence interval based on control chart data for estimating the potency of a batch of tablets used as a control material in monitoring the stability of an analytical system. It should be clear from the discussion above that the average of the plotted data values essentially estimates the sum of the true value, the method bias and the laboratory bias. The interpretation of the result as a measure of the batch potency (true value) implies that it is reasonable to assume that method and laboratory biases are either absent or they cancel each other. This implies that there is no 'local bias'. The role of within-laboratory method validation studies is to provide assurance that an assumption of no 'local bias' is valid.

Even if there is a 'local bias', the output of the analytical system may be quite acceptable for monitoring a production system. Provided the biases remain stable, changes in the output of the production system will be reflected in the analytical results and will signal action, as required. The assumption here is that the production system is varying around a level sufficiently far from any specification limit such that the analytical bias is not in danger of producing misleading results. What is required of the analytical system is that it should signal changes in what is currently considered a satisfactory process.

Within-laboratory Method Comparison. Laboratories will often wish to assess new methods with a view to replacing older ones. A standard approach would be to measure one or more materials several times by both methods and to compare the average results to determine if a relative bias exists. If all measurements are made in single analytical runs for each method, as might be done, for example, if tablets were being measured by two different instrumental methods, then the study design is exactly the same as that discussed above for laboratory comparisons. The comparison involves the difference between the sums of the method, laboratory and run biases for each method (note that the laboratory biases could be different for the two methods) and not simply the difference between the method biases. To assess whether or not the two methods give comparable results *in this laboratory*, measurements should be in different runs. Again, the hierarchy principle operates: to make comparisons between terms in the model (here the 'local bias' for the two methods) ensure that data are generated in such a way that all terms lower down the ladder of errors are embodied in the standard error that will be used for assessing the study results. Of course, if the run bias is negligible in relation to the repeatability errors, then it can be eliminated from the model and the problem will not arise.

Averaging out Laboratory Biases. In Chapter 2 we considered how the precision of routine results could be improved by moving from a replication scheme that involved within-run replicates to one where replicates appeared in different analytical runs. This same principle is extended by moving up the ladder of errors and requiring different measurements to be made in different laboratories. For example, suppose a consignment of ore is to be assayed to determine the percentage of the analyte of commercial interest or the percentage of an important impurity present. If a sample is taken and split between several laboratories then the laboratory biases will tend to average out, whereas if the same number of measurements are made in a single laboratory then all measurements contain that laboratory's bias component.

Averaging over Samples. The question of sampling was touched on in Chapter 3, but not discussed in any detail. This was because the main concern was to provide an account of the use of statistical methods within the QC laboratory and sampling mainly happens outside the laboratory. It may be helpful, however, to view the role of sampling in the context of our measurement model. The true value that appears in the measurement model is that of the laboratory test portion. If this has been sampled from a non-homogeneous material (*e.g.*, the bulk consignment referred to above) then the model can be extended by writing the laboratory true value as $TV_{lab\text{-}samp}$, and the consignment true value as TV_{bulk}. The true values of laboratory samples will vary randomly with a standard deviation $\sigma_{sampling}$ and assume the same status as the other elements in the extended measurement model. The relative magnitudes of the different variance components will then suggest how to improve the estimation of the measurand (TV_{bulk}). Since the lower terms (repeatability and run-bias) are likely to be negligible in relation to sampling variability, generally there will be little point in replication at a level below the laboratory term. In practice, the most effective strategy is likely to be to take several samples and to measure them at most twice in each of several laboratories; the duplication being required for consistency checks rather than for improving precision.

8.6 REVIEW EXERCISES

8.9 Kalra[7] carried out an inter-laboratory collaborative study of methods for the determination of the pH of soils. Fifty-three laboratories were involved in the study and they each measured 10 blind duplicate samples of different soils using several methods. The data presented in Table 8.5 are the pH measurements of saline soil samples for the first 20 laboratories, using a single method. The method used here is an alternative to that used in Exercise 8.1.

Analyze the data using ANOVA and estimate the repeatability and reproducibility standard deviations. Use Equations (8.3)–(8.5) to obtain estimates, also, and confirm that the two methods give the same results. Carry out a residual analysis.

Table 8.5 *Duplicate pH measurements of a soil by 20 laboratories* (Reprinted from the Journal of AOAC INTERNATIONAL, 1995, Vol. 78, pp 310. Copyright, 1995, by AOAC INTERNATIONAL.)

Lab	*pH*		*Lab*	*pH*	
1	9.08	9.10	11	8.61	9.24
2	9.35	9.35	12	9.44	9.43
3	9.35	9.35	13	9.14	9.19
4	9.53	9.46	14	9.41	9.42
5	8.76	8.74	15	9.47	9.39
6	9.09	9.05	16	9.32	9.37
7	9.26	9.24	17	9.29	9.14
8	9.29	9.27	18	9.28	9.32
9	7.00	9.71	19	9.36	9.25
10	9.09	9.03	20	8.93	8.93

8.10 Apply Cochran's test to the pH data of Exercise 8.9. If you exclude any of the laboratories on the basis of the test, re-do the test for the remaining laboratories to determine whether or not the next largest variance is inconsistent with the others. What do you conclude?

8.11 Calculate the mean pH result for each of the 20 laboratories involved in the collaborative trial presented in Exercise 8.9. Draw a dotplot of the data. Apply Grubbs' tests for a single outlier (high (H) or low (L)) and for two outliers (HH, LL or HL). What do you conclude?

8.12 In Exercise 7.10, p 358, replicate measurements of isoflavones in a soy isolate were presented in the form of a three-level hierarchy, in which replicates were nested within days, which in turn were nested within laboratories.

- For each of the 20 pairs of duplicates, calculate the sample variance. Draw a dotplot of the variances. Carry out Cochran's test. If you find a variance significantly larger than the rest, exclude all the data from that laboratory. Carry out Cochran's test on the remaining variances.

- For the remaining laboratories, average the within-day replicates. Calculate the within-laboratory variances (*i.e.*, based on the daily means). Carry out Cochran's test on the variances. What do you conclude?

- Compare your results with the conclusions drawn on the basis of residual analysis for Exercise 7.10.

8.13 Exercise 7.15, p 363, was based on the study referred to above, but the data analyzed were for genistein in a miso test material. Carry out the same analyses as for Exercise 8.12 and compare your results to the conclusions you arrived at as a result of residual analysis in Exercise 7.15.

8.14 Fontaine and Eudaimon[10] carried out an inter-laboratory collaborative study of methods for the determination of the lysine, methionine and threonine in trade products or concentrated amino acid pre-mixes. Seventeen laboratories were involved in the study. Lysine measurements were analyzed in Exercise 8.6. The data presented in Table 8.6 are the methionine measurements for a different material.

Analyze the data and estimate the repeatability and reproducibility standard deviations.

8.15 The 26 means from which the final individuals chart of the case study in Chapter 2, p 55, was drawn are shown in Table 8.7. Recall that the laboratory manager was suspicious about two of the results, which were lower than the rest. Carry out a Grubbs' test to determine if the smallest two values can be regarded as outliers.

Table 8.6 *Inter-laboratory study results for methionine*
(Reprinted from the Journal of AOAC INTERNATIONAL, 2000, Vol. 83, pp 771. Copyright, 2000 by AOAC INTERNATIONAL.)

Lab	*Methionine*		*Lab*	*Methionine*	
1	26.95	26.79	10	26.96	26.65
2	26.88	26.63	11	26.75	26.74
3	26.45	26.78	12	26.76	26.92
4	26.73	26.58	13	26.15	26.22
5	27.07	26.85	14	25.72	25.48
6	26.19	26.25	15	25.91	26.34
7	26.28	27.05	16	25.63	26.00
8	26.96	26.84	17	16.35	16.38
9	26.67	26.95			

Table 8.7 *Mean potency results for control chart data*
(Reprinted from the *Analyst*, 1999, Vol. 124 pp 433. Copyright, 1999, Royal Society of Chemistry.)

498.38	492.46	508.58
488.47	505.50	491.57
492.22	503.41	490.17
499.48	473.93	488.14
463.68	494.54	493.65
489.31	489.45	488.42
497.71	495.24	489.33
483.96	482.25	483.20
496.05	463.40	

Table 8.8 *The effects from the* 2^{6-1} *fractional factorial study*

35.94	−1.56	−26.56	6.19
20.19	−5.69	6.19	20.69
7.31	16.06	−0.06	8.06
−4.69	−4.81	−15.56	−8.94
14.94	−11.56	−1.94	1.06
14.94	−4.44	1.31	−13.69
16.56	6.06	7.56	−15.31
0.94	−16.69	−1.56	

8.16 Exercise 5.8, p 229, involved the analysis of a 2^{6-1} fractional factorial design. The 31 effects are shown in Table 8.8. Draw a dotplot and confirm that A = 35.94 and CE = −26.56 appear to be outliers. Carry out a Grubbs' test to determine whether they are statistically significantly different from the rest.

8.17 Do a web search for the Eurachem website, download the latest guide on estimating uncertainties, and study the examples. Apply the methods to some of your own systems.

8.7 REFERENCES

1. *Guide to the Expression of Uncertainty in Measurement*, International Organization for Standardization, Geneva, Switzerland, 1993.
2. *Quantifying Uncertainty in Analytical Measurement*, Eurachem, London, 1995.
3. AOAC International, *J. AOAC Int.*, 1995, **78**, 143A.
4. *Accuracy (Trueness and Precision) of Measurement Methods and Results*, Parts 1–6, ISO 5725, International Organization for Standardization, Geneva, 1994.
5. K. Schaffler, C.M.J. Day-Lewis, M. Clarke and J. Jekot, *J. AOAC Int.*, 1997, **80**, 603.
6. *Official Methods of Analysis of AOAC International*, W. Horwitz (ed), AOAC International, Arlington, VA, 2000.
7. Y. Kalra, *J. AOAC Int.*, 1995, **78**, 310.
8. Analytical Methods Committee, *Analyst*, 1989, **114**, 1693.
9. Analytical Methods Committee, *Analyst*, 1989, **114**, 1699.
10. J. Fontaine and M. Eudaimon, *J. AOAC Int.*, 2000, **83**, 771.
11. *General Requirements for the Competence of Calibration and Testing Laboratories*, ISO/IEC 17025, International Organization for Standardization, Geneva, 1999.
12. *W*. Horwitz, *J. AOAC Int.*, 1998, **81**, 785.
13. Analytical Methods Committee, *Analyst*, 1995, **120**, 2303.
14. *Guide to the Use of Repeatability, Reproducibility and Trueness Estimates in Measurement Uncertainty Estimation*, ISO/DTS 21748, International Organization for Standardization, Geneva, 2002.
15. M. Thompson, *Analyst*, 2000, **125**, 2020.

Solutions to Exercises

CHAPTER 1

1.1. The required values are ± 1.645 and ± 2.575, but two decimal places are commonly used.

1.2. The fraction further than 1 unit from the true value is 0.21; the fraction further than 1.5 units is 0.06.

1.4. The standard deviations are: oxolinic acid 2.978, flumequine 3.580.

1.5. The relative standard deviation for oxolinic acid is 3.1% and for flumequine 4.5%.

1.7. Estimated repeatability limit for oxolinic acid is 8.25 and for flumequine 9.92.

1.8. For the casein data the combined standard deviation is 0.01356 and the estimated repeatability limit is 0.0376.

1.9. For the 10 laboratories the pooled standard deviation is 0.04295, which gives an estimated repeatability limit of 0.119 and an estimated repeatability critical difference (for two replicates) of 0.084. When the large standard deviation is excluded the pooled standard deviation is 0.03333, which gives an estimated repeatability limit of 0.092 and an estimated repeatability critical difference (for two replicates) of 0.065.

1.10. The probability that a single measurement will be either less than 84% or greater than 86% is 0.046. To reduce this to less than 0.001 use $n = 3$ replicates.

1.11. Because the analytical range is so narrow, it would be surprising if a trend in standard deviation were apparent. For the UV data the combined standard deviation is 0.3029, the estimated repeatability limit is 0.840 and the estimated repeatability critical difference is 0.594. For the NIR data the combined standard deviation is 0.1994, the estimated repeatability limit is 0.553 and the estimated repeatability critical difference is 0.391.

1.12. The combined standard deviation for all the data is 0.0338 and the estimated repeatability limit is 0.094. When run 20 is deleted the combined standard deviation becomes 0.0307 and the estimated repeatability limit 0.085.

1.13. When $n = 1$ the probability that the result is less than 100 ng g^{-1} is 0.145. This reduces to 0.017 when $n = 4$ and to 0.0007 when $n = 9$.

1.14. The combined standard deviation for the duplicate results is 0.340 and the repeatability limit estimate is 0.942.

CHAPTER 2

2.1. Fusels data. Set A has centre line 1289, UCL = 1371 and LCL = 1207. All points in control. Set B has centre line 1340, UCL = 1390 and LCL = 1289. Points 7 and 21 are outside the action limit. Set C has centre line 1262, UCL = 1319 and LCL = 1205. Points 12 and 13 are between the warning and action limit.

2.2. Glucose data. The centre line is 98.44, the UCL = 106.2, the LCL = 90.62. The overall standard deviation of the data is 2.4514; this gives UCL = 105.8 and LCL = 91.09. Points 17 and 18 are now between the warning and lower control limits.

2.3. In-process assay data. The average range is the centre line and equals 2.872, UCL = 7.393: point 18 is a big outlier. When the three replicates corresponding to run 18 are excluded the centre line changes to 2.240 and the UCL = 5.767. All points are in control.

2.4. Sucrose data. The centre line is $\bar{R} = 0.045$, UCL $= 0.1470$. There are two out of control points, 5 and 33 and a run of 9 points from run 20 onwards. When 5 and 33 are excluded, $\bar{R}$ becomes 0.0395 and UCL = 0.1291.

2.5. Glucose data. The average range is 0.0396 and UCL = 0.1294. When point 20 is excluded $\bar{R} = 0.0373$ and UCL = 0.1218.

2.6. Case study data. Note that when you delete points there are two ways of dealing with the gaps: you can estimate the standard deviation by taking account of the gaps or you can simply close up the gap and use all the remaining data. The latter is simpler and is what was done in the case study and for the summaries here. After exclusion of the two low values CL = 492.3, UCL = 518.0 and LCL = 466.0.

2.7. Pharmaceutical laboratory control material data. The control charts show the system to be in control. The range charts has a centre line at 0.4912 and UCL = 1.605. The individuals charts has centre line 94.18, UCL = 96.10 and LCL = 92.26. The replication scheme comparison table for the standard error of a mean based on n replicates is given below.

n	*1*	*2*	*3*	*4*	*5*
1-run	0.73	0.66	0.64	0.63	0.62
n-run	0.73	0.52	0.42	0.37	0.33

2.8. Case study data after exclusion of runs 5 and 20, as well as 6, 11 and 28. The replication scheme comparison table for the standard error of a mean based on n replicates is given below.

n	*1*	*2*	*3*	*4*	*5*
1-run	9.62	8.85	8.58	8.44	8.36
n-run	9.62	6.80	5.55	4.81	4.30

2.9. UV/NIR data. The production system does appear to be in control as neither chart shows instability. The centre line for the UV chart is 84.63, UCL = 86.42, LCL = 82.84. For the NIR chart, CL = 84.66, UCL = 85.93, LCL = 83.38.

2.10. Proficiency test data. Apart from the two relatively low values, all the other suspended solids results are within the range $-1 < z < +1$. This is unexpected. We saw that for a standard Normal distribution approximately 68% of values should lie in this range. These results indicate that the standard deviation being achieved by the laboratory is smaller than that used by the scheme organizers to standardize the results. The standard deviation of the z values is 0.45, including the two relatively low values, and 0.30 when they are excluded; a value of 1.0 would be expected if the laboratory standard deviation coincided with the one used to calculate the z scores. This indicates that the laboratory's performance is highly consistent and exceeds expectations for this assay. The runs of points either below the centre line of $z = 0$, at the beginning of the sequence, or above the centre line towards the end of the data, suggest small but consistent biases with respect to the assigned values being used to standardize the results.

Similar comments apply to the BOD values from point 9 onwards. The COD values are also highly consistent, with a standard deviation of 0.31. The mean value of 0.13 suggests a small upward bias.

2.11. The fructose data are in control. The centre line for the range chart is 0.0356 and the UCL is 0.1168. For the individuals chart the centre line is 99.23, UCL = 105.90 and LCL = 92.57.

2.12. The zinc data are in control by the laboratory's signalling rules. The centre line of the range chart is 2.367 and UCL = 7.733. For the individuals chart the centre line is 195.6, UCL = 210.6 and LCL = 180.7.

CHAPTER 3

3.1. Glucose data. Mean = 98.443 SD = 2.451, t-statistic = −4.49. Since the two-tailed critical value for a t-distribution with 49 degrees of freedom, for a significance level of 0.05 is 2.01, this result is highly statistically significant. The importance, or otherwise, of such a bias is not, of course, a statistical issue.

Note that critical values not available in Table A3 in Appendix A can be calculated in most good statistics or spreadsheet packages. However, exact values will only rarely be required: our result is obviously highly significant, since the critical values for 40, 50 and 60 degrees of freedom are all around 2.

3.2. Total oxidized nitrogen data. The critical value for a two-tailed test, based on 34 degrees of freedom, is 2.03. The calculated t-statistic is 1.86, so there is no evidence of bias here.

3.3. We use as our decision criterion the result of a one tailed t-test of the hypothesis that the average impurity level is 3% or less. If we use a significance level of 0.05, then the one-tailed critical value is 2.02, since the SD has 5 degrees of freedom. The test statistic is $t = 2.07$, so we reject the hypothesis and conclude that the impurity level is higher than 3%.

3.4. We need to specify the risks of type I and II errors we are willing to take in choosing the sample sizes. Suppose we say $\alpha = \beta = 0.05$. Differences of 0.5, 0.4, 0.25 and 0.1 correspond to D values of 2, 1.6, 1.0 and 0.4, respectively. These give sample sizes of 6, 8, 16 and 84.

3.5. No specifications for the decision criteria have been given in the exercise. This is realistic! In practice, someone has to make judgements on what would be appropriate values. Suppose we take $\alpha = \beta = 0.05$, as before. Now what difference from 3% would be important? If we take a value of 0.5%, then $D = 1$ gives a sample size of 13. If we take 1% then $D = 2$ gives a sample size of 5.

3.6. The 95% confidence interval is given by the mean plus or minus the t-multiplier times the standard error of the mean, where this is the standard deviation for individual results divided by the square root of 50. The multiplier is 2.01. The answer is 97.746 to 99.140%. Note that this does not include 100%. This means that the set of possible long-run recoveries does not include 100%, so the system is biased downwards. This is consistent with the t-test (as it must be).

3.7. The sample size is 35 so the calculated s value has 34 degrees of freedom; the corresponding t-multiplier is 2.03. The confidence interval is 5.049 to 5.083. Since this includes 5.05, there is no evidence of bias. The results of the two analyses are consistent, as they must be, since they are only different ways of processing the same numbers.

3.8. Note that with the exclusion of the two low values the mean of the remaining 24 values rises by 2.21 units to 492.31. The 95% confidence interval for the long-run mean becomes 489.04 to 495.58. The width of the interval has also reduced by approximately two units from 8.71 to 6.54.

3.9. The mean of the 23 remaining results is 240.856 and the SD is 3.338. The 95% t-multiplier for 22 degrees of freedom is 2.07. These summaries give a 95% confidence interval of 239.42 to 242.30.

3.10. For UV the mean is 84.631, SD = 0.606. The t-multiplier for 24 degrees of freedom is 2.06. Accordingly the confidence interval is 84.38 to 84.88. For the NIR data the mean is 84.656, the SD = 0.416, so the 95% confidence interval is 84.49 to 84.83.

3.11. The mean of the 50 fructose recovery results is 99.229, the SD = 2.082. The t-distribution critical value for 49 degrees of freedom is 2.01. Accordingly the t-statistic of -2.62 is statistically significant. The 95% confidence interval is 98.64 to 99.82%, which does not include 100%. Both analyses point to the fructose method being biased downwards, though the practical significance of such a bias is another matter. The statistical analysis simply indicates that the observed bias is unlikely to be an artifact of the chance variation.

3.12. The standard deviation of the 10 results is 0.4406. The chi-square factors will be the same as those in the text example, so the confidence interval for the system variance is 0.092 to 0.647 and that for the intermediate precision standard deviation is 0.30 to 0.80.

3.13. For the oxolinic acid data the standard deviation is 2.9782. The chi-square factors for 95% confidence, with 5 degrees of freedom are 0.83 and 12.83. These give confidence limits for the standard deviation of 1.86 and 7.31. The corresponding factors for 99% confidence are 0.41 and 16.75. The 99% confidence interval is 1.63 to 10.40.

The factors will be the same for the flumequine data, as the number of observations is the same. The standard deviation is 3.5802 and the 95% confidence limits are 2.24 and 8.79. The 99% limits are 1.96 and 12.50.

3.14. The casein data gave a combined standard deviation of 0.01356, based on 9 degrees of freedom. The chi-square factors for a 95% confidence interval are 2.70 and 19.02. Upper and lower confidence limits are found for σ^2 and then inserted into the formulae for the two precision measures. The repeatability limits are 0.0259 and 0.0686. The repeatability critical difference ($n = 5$) limits are 0.0116 and 0.0307.

3.15. The 95% confidence limits for the variance of the oxolinic acid data are 3.4566 and 53.4317 (if you square the rounded standard deviations in the solution to Exercise 3.13, you will get a slightly different answer). Inserting these into the formulae for repeatability limit and repeatability critical difference ($n = 6$) gives confidence limits for the quantities. The results are: repeatability limit 5.15 and 20.26; repeatability critical difference 2.10 and 8.27.

The corresponding results for the flumequine data are: repeatability limit 6.20 and 24.36; repeatability critical difference 2.53 and 9.94.

3.16.–3.18 The Normal plots support the Normality assumption in all cases.

3.19. Minitab gave the following *p*-values for the Anderson–Darling Normality test: 0.000 (29 means), 0.037 (26 means), and 0.675 (24 means).

3.20. The following Minitab output gives the required t-tests and confidence intervals:

Test of mu = 10.8 vs mu not = 10.8

Variable	N	Mean	StDev	SE Mean
A	10	10.7160	0.1059	0.0335
B	10	10.6690	0.1149	0.0363

Variable	95.0% CI	T	P
A	(10.6403, 10.7917)	−2.51	0.033
B	(10.5868, 10.7512)	−3.60	0.006

The tests are statistically significant and the intervals do not include the certified value of 10.8%. Thus, both laboratories are biased downwards. Whether the biases are of any importance is a separate question.

For Laboratory A the 95% confidence limits for the standard deviation are 0.073 to 0.193, while the 99% limits are 0.065 to 0.242. For Laboratory B the 95% confidence limits for the standard deviation are 0.079 to 0.210, while the 99% limits are 0.071 to 0.262.

3.21. The following table contains the sample sizes implied by the various specifications.

Diff. = 0.1		Power	
		0.95	0.99
Signif.	0.05	16	21
Level	0.01	22	28
Diff. = 0.2		**Power**	
		0.95	0.99
Signif.	0.05	6	7
Level	0.01	8	10

3.22. The Normal plots do not reject the assumption of Normality: for the UV means the Anderson–Darling test gives a p-value of 0.25; the corresponding value for the NIR means is 0.34.

3.23. The Normal plots for the proficiency test data are reasonably straight in all cases and none of the Anderson–Darling tests are statistically significant. The following Minitab output give the tests and confidence intervals:

Test of mu = 0 vs mu not = 0

Variable	N	Mean	StDev	SE Mean
BOD	34	−0.0724	0.3576	0.0613
COD	43	0.1256	0.3116	0.0475
Susp-Solids	41	−0.0490	0.2969	0.0464

Variable	95.0% CI	T	P
BOD	(−0.1971, 0.0524)	−1.18	0.247
COD	(0.0297, 0.2215)	2.64	0.012
Susp-Solids	(−0.1427, 0.0447)	−1.06	0.297

The t-tests and confidence intervals suggest that the means for suspended solids and BOD are consistent with a long-run value of zero, *i.e.*, that the analytical systems are unbiased in analyzing these parameters. Recall, however, that the run charts for these two datasets suggested that the data contain clusters of runs above and below the zero line. If this is the case then it suggests a shifting mean level. The t-test compares the data to a single mean (zero) and assesses whether the observed mean (−0.072 for BOD) can be considered just randomly different from this value. It takes no account of the time order in the data. Accordingly, runs of data above and below the centre line of zero will balance each other. This is an example of where a simple graphical analysis may tell us more than a sophisticated mathematical test. The COD value is statistically significant ($p = 0.012$) and gives a confidence interval of 0.03 to 0.22, supporting the suggestion of a small upwards bias indicated in Exercise 2.10.

3.24. The Normal plot is very straight – the Anderson–Darling p-value is 0.86. The t-test of the hypothesis that the long-run mean is 200 is highly statistically significant ($t = -4.27$). The 95% confidence interval for the long-run mean is 193.52 to 197.72. If we subtract 200 from these values we get a confidence interval for the bias.

CHAPTER 4

4.1. The combined $s^2 = 0.0685$, which gives a standard error of 0.151 for the difference between the two sample means. This gives a t-statistic of 4.17, which is highly statistically significant, since the critical value for a test based on 10 degrees of freedom is 2.23 for a significance level of 0.05. On average, the analysts get different results.

4.2. The combined standard deviation is 0.174. The t-statistic is 3.45; this has an associated p-value of 0.003, which is good evidence of a relative bias between the laboratories. Whether a difference of 0.269 is of any practical consequence is a separate question. The role of the statistical test is to determine whether or not such a difference is likely to be the result of the chance variation affecting the two systems. Here the answer is no.

4.3. The average difference was 0.63, the standard error estimate was 0.151, the critical value was 2.23. Hence the 95% confidence interval is 0.63 ± 0.34. Since this interval does not contain zero, it means that the corresponding t-test will reject the hypothesis of a zero difference – this was the result we obtained in Exercise 4.1.

4.4. The average difference was 0.269, the combined standard deviation was 0.174, the estimated standard error of the difference between the sample means was 0.0778, the critical value was 2.10 (18 degrees of freedom). Hence, the 95% confidence interval is 0.269 ± 0.163. Since this interval does not contain zero, it means that the corresponding t-test will reject the hypothesis of a zero difference – this was the result we obtained in Exercise 4.2. We conclude that a relative bias of this magnitude does exist.

4.5. The F-statistic for the control chart data is 1.590. The F-critical value for 5,5 degrees of freedom, using a two-tailed significance level of 0.05 is 7.1. Accordingly, we do not reject the hypothesis of equal long-run standard deviations.

For the coconut data of Exercise 4.2 the F-statistic is 1.35, which is not statistically significant since the critical value for 9 and 9 degrees of freedom is 4.03, for a two-tailed significance level of 0.05.

4.6. The standard deviations are 0.374 and 1.723 for Laboratories A and B, respectively. These give an F-ratio of 21.18, which is highly statistically significant when compared to the critical value of

4.03, given above.

Minitab gives the following output for a *t*-test on the data – note that equal standard deviations are not assumed by this test.

Two-sample T for A vs B

	N	Mean	StDev	SE Mean
A	10	79.292	0.374	0.12
B	10	77.73	1.72	0.54

Difference = mu A − mu B
Estimate for difference: 1.557
95% CI for difference: (0.296, 2.818)
T-Test of difference = 0 (vs not =): T-Value = 2.79 P-Value = 0.021 DF = 9

4.7. While the Normal plot of the residuals does not give a particularly nice looking straight line, the Anderson–Darling significance test for Normality gives a *p*-value of 0.52 which is very far from being statistically significant.

4.8. The Minitab analysis of the reduced dataset is shown below.

Paired T for Lab-1 – Lab-2

	N	Mean	StDev	SE Mean
Lab-1	5	89.460	0.340	0.152
Lab-2	5	90.010	0.297	0.133
Difference	5	−0.5500	0.1173	0.0524

95% CI for mean difference: (−0.6956, −0.4044)
T-Test of mean difference = 0 (vs not = 0): T-Value = −10.49 P-Value = 0.000

Note that the *t*-statistic has increased markedly from 6.58 to 10.49 (the sign just depends on which way the differences are calculated). The reason for this is that although deleting the large difference reduces the average difference from 0.637 to 0.550, it has a very marked effect on the standard deviation of differences; this changes from 0.237 to 0.117.

4.9. The mean difference is 0.786 and the standard deviation of differences is 2.564. When we test the hypothesis that the long-run mean is zero, we obtain a test statistic of $t = 2.17$, which has an associated *p*-value of 0.03. This is marginally statistically significant, indicating a recovery difference. The 95% confidence interval for the long-run recovery difference is 0.06 to 1.51%.

4.10. For the laboratory comparison data, Minitab gives the following output.

```
Two-sample T for Lab-1 vs Lab-2

          N    Mean     StDev   SE Mean
Lab-1     5    89.460   0.340   0.15
Lab-2     5    90.010   0.297   0.13

Difference = mu Lab-1 − mu Lab-2
Estimate for difference: −0.550
95% CI for difference: (−1.028, −0.072)
T-Test of difference = 0 (vs not = ): T-Value = −2.72
P-Value = 0.030 DF = 7
```

While the t-value is still statistically significant, there has been a very marked reduction in the t-value, from 10.5 to 2.7. This arises because the batch-to-batch variation now becomes part of the variation embodied in the pooled s^2 used in the denominator of the test. This is inappropriate.

The corresponding analysis for the glucose and fructose data is shown below.

```
Two-sample T for F-recovery vs G-recovery

            N    Mean    StDev   SE Mean
F-recove    50   99.23   2.08    0.29
G-recove    50   98.44   2.45    0.35

Difference = mu F-recovery − mu G-recovery
Estimate for difference: 0.786
95% CI for difference: (−0.117, 1.688)
T-Test of difference = 0 (vs not =): T-Value = 1.73  P-Value = 0.087 DF = 98
Both use Pooled StDev = 2.27
```

The run-to-run variability, which is incorporated in the pooled standard deviation, has so inflated the standard error of differences that the mean difference is no longer statistically significantly different from zero. The paired analysis is more sensitive and detects the presence of a difference in recovery rates; treating the two sets of results as independent (they are not!) loses sensitivity and means we are no longer able to detect the long-run difference.

4.11. The individuals chart does not show instability, when judged against the two rules used by the laboratory, *viz*., the 3-sigma rule and a runs rule that requires nine points in a row on one side of the centre line as a signal of a shift.

The plot of the differences against the means shows no systematic patterns and the Normal plot is reasonably straight (Anderson–Darling p-value $= 0.10$). The assumptions required for our t-test and confidence interval appear to be acceptable here.

4.12. Before tackling this problem we need to choose values for α and β which will define our attitude to the risks of both type I and II error; say we choose $\alpha = \beta = 0.05$. The sample sizes are then given by Table A7 as 8 and 27.

4.13. For a standard deviation of 0.20 the sample sizes are 27 and 8, for relative biases of 0.2 and 0.4, respectively. When the standard deviation is 0.15, Table A7 does not give exact results. For a relative bias of 0.2 $D = 1.3$ gives 17 and $D = 1.4$ gives 15. A sample size of 16 or 17 would be appropriate. Similarly for a relative bias of 0.4 the D value of 2.67 lies between two tabulated values. If we had to rely on the sample size table then it would be better to be more conservative and use 6. However, Minitab gives a sample size of 5 for the set of specifications we are considering here.

4.14. Here we are dealing with paired data, so Table A4 is the one we use. For a power level of 0.90 $D = 0.67$ falls between $D = 0.65$ which corresponds to a sample size of 27 and $D = 0.70$ with sample size 24. For power 0.95 the corresponding values are 33 and 29. We could either interpolate or be conservative and use the larger sample size.

4.15. A difference of 2% corresponds to $D = 0.8$, which gives sample sizes of 23 and 32 for significance levels of $\alpha = 0.05$ and $\alpha = 0.01$, respectively.

A difference of 3% corresponds to $D = 1.2$, which gives sample sizes of 12 and 16 for significance levels of $\alpha = 0.05$ and $\alpha = 0.01$, respectively.

4.17. The standard deviations of the before and after data are 2.060 and 2.323; these give an F-ratio of 1.27. Since the critical value for a test with significance level 0.05 is 3.3 (12 and 12 degrees of freedom), the result is not statistically significant – the evidence does not suggest a change in response variability. The t-test is highly statistically significant, suggesting a systematic shift: the t-statistic is 4.09, while the critical value using a significance level of 0.05 (24 degrees of freedom) is 2.06. The 95% confidence interval is 1.75 to 5.30. The Normal plot of the residuals is reasonably straight with an Anderson–Darling p-value of 0.29, suggesting at least approximate Normality.

4.18. In Exercise 1.11 we found the standard deviation for the UV data to be 0.3029 and that for the NIR data to be 0.1994. In each case the estimate has 25 degrees of freedom. The F-ratio is 2.31 and since the critical value for a two-sided test using a significance level of 0.05 is 2.23, this result rejects the null hypothesis of equal long-run standard deviations (or variances).

4.19. The three charts suggest that the statistical assumptions are acceptable. The t-test gives a t-statistic of $t = 0.90$ which has an associated p-value of 0.374. There is no evidence of a recovery difference between glucose and sucrose.

4.20. The standard deviations for glucose, sucrose and fructose (with their degrees of freedom) are 0.0307 (43), 0.0328 (42) and 0.0306 (44), respectively. The F-statistics (with their critical values) are S:G 1.14 (1.84), G:F 1.003 (1.82) and S:F 1.15 (1.83). Accordingly none of the comparisons is statistically significant – there is no evidence of differences between the repeatability standard deviations for the three sugars.

CHAPTER 5

5.1. The effects are: Injection Volume 98.23, Flow Rate – 19.23, and Interaction 8.77. The MSE (Mean Square Error) = 14.8, which gives a standard error for effects of 2.22. All effects are clearly highly statistically significant. The residual plots are good, so the statistical assumptions underlying the analysis are likely to be valid.

Even inspecting the raw data makes it obvious that the dominant effect is Injection Volume – this would be expected in advance. A study like this would be carried out not to verify such an obvious result, but rather to determine if the effect of a change in Flow Rate depends on the Injection Volume level. Here it does:

the difference between the means for the two levels of Flow Rate are 28 at the low level of Injection Volume and 10.5 at the higher level. Is the difference at the higher level statistically significant? The standard error for comparing two means on the square of means is 3.14, since each mean is based on three test results. The *t*-statistic is $10.5/3.14 = 3.34$, which is statistically significant since the critical *t*-value is 2.31, based on eight degrees of freedom. The best operating conditions are at high Injection Volume and low Flow Rate.

5.2. The effects are: Injection Volume 97.6, Temperature – 3.125, and Interaction 2.05. The MSE = 20.9, which gives a standard error for effects of 2.29. Only Injection Volume is statistically significant. This means that changing Initial Temperature (over the range in the study) has no effect on the Peak Area. The residual plots are good, so the statistical assumptions underlying the analysis are likely to be valid.

5.3. The three effects are Pressure: 52, CO_2: 11.75 and Interaction: 11.25. The MSE is 98.1 which gives a standard error for effects of 4.95. Accordingly, all effects are statistically significant at the $\alpha = 0.05$ significance level (critical *t*-statistic is 2.18, 12 degrees of freedom). The best set of operating conditions is obviously at the high levels of both factors, where the mean response was 75.5. Note the strong interaction: the point estimate of the effect of CO_2 is 0.5 at low pressure and 23 at the higher level of pressure. The residual plots are good, so the statistical assumptions underlying the analysis are likely to be valid.

5.4. The A main effect is highly significant and it is clear that higher responses are generated at the higher level of A. The CE interaction is also statistically significant. None of the other effects are significant. Inspection of the four means shows that when both C and E are at their low levels the response is significantly lower than for the other three combinations (which are not significantly different from each other). The estimated standard error for comparing any pair of means is $S\hat{E} = \sqrt{2s^2/8} = 15.79$, where the combined s^2 is 997.9, based on 24 degrees of freedom. Using a *t*-multiplier of 2.06 gives the value of 32.54 as the amount by which two means must differ to be statistically significantly different.

A plot of the residuals *versus* the fitted values is satisfactory, as is the Normal probability plot of the residuals.

5.5. Both of the two-way interactions involving pressure are statistically significant. The MSE is 19.75. This means that when we wish to compare any two points on either of the interaction squares the smallest difference that will be statistically significant is $2.31\sqrt{\frac{2\times 19.75}{4}} = 7.26$. In both cases the mean corresponding to the high levels of both factors is clearly the best, in terms of recovery. Since all three factors must be considered together, we note that while the best response is at the high level of all three (62.5), the next best mean, corresponding to both pressure and CO_2 at their high levels and temperature at its low level (55.0), is not statistically significantly different from the best response. The smallest difference that will produce a statistically significant result for comparisons between means on the three-factor cube is 10.27. Note, though, that the means at the corners of the cube are the average of only two replicates. Unless there were reasons why the lower temperature might be preferable, then the high levels of all three factors give the best operating conditions. If lower temperature were desirable, then further studies might be carried out to determine if the temperature change is, in fact, important, given that we have decided to operate at the higher levels of pressure and CO_2. The residual plots are good.

5.8. Most of the points in the Normal plot of the contrasts form a good straight line. Two of the contrasts (A = 35.94 and CE = −26.56) are far away from the rest, though not especially out of line with the straight line. Since we have a measure of pure chance variation we can investigate these further. The standard error of the effects is 7.95 since $s = 22.49$ and since the s value has 4 degrees of freedom an effect will be expected to be larger than $2.78\times 7.95 = 22.1$ before it is considered statistically significantly different from zero. This criterion shows both A and CE as significant. None of the other contrasts are either out of line with the majority of the points in the Normal probability plot or exceed 22.1. At this stage we could ignore factors B, D, F and re-analyze the data as a replicated 2^3 study to see if our conclusions stand up and to examine the nature of the effects we have identified as important. However, this was done in Exercise 5.4, so we will not repeat the analysis here.

5.10. I labelled the capitals as −1 and the small letters as +1. This gave the following design generators:

$$\mathbf{D} = -\mathbf{AB} \quad \mathbf{E} = -\mathbf{AC} \quad \mathbf{F} = -\mathbf{BC} \quad \mathbf{G} = \mathbf{ABC}$$

The largest effects for each of the three responses are smaller than 5% of the average response for that preservative. Accordingly, the system can be considered robust.

5.11. The estimated contrasts are:

Acetonitrile	−1.730
Water	−0.575
GAA	0.235
Wavelength	−3.080
Column	2.030
Flow Rate	2.745
Acetonitrile*Flow Rate	2.690

The largest of these is only 0.6% of the nominal capsule potency – effects of this order will not usually be considered of any practical importance. Accordingly the system can be considered robust.

5.12. The Normal plots of the contrasts are reasonably good for three of the responses. The plot of the half-hour data is probably the least convincing straight line. For the half-hour data the contrast associated with temperature (1.39) is 5.4% of the average dissolution percentage of 25.8%. This exceeds the set criterion, so temperature should be flagged for special care, or further investigation. Temperature also produces the largest contrast at two hours, although the 5% criterion is not breached in this case.

5.13. The Normal probability plot of the contrasts shows two large values B = 89.5 and A = 50.6. B is out of line with the others and is immediately suggested as important. A, although large, is not clearly out of line with the other contrasts in the plot. However, the estimated standard error of an effect, based on the pure replicate standard deviation $s = 42.0$ is 14.85, which means that any effect bigger than $2.78 \times 14.84 = 41.3$ is statistically significant. Accordingly, both A and B main effects (making the reasonable assumption that 5-fis do not exist) are important; in each case a higher response is given at the higher level of the factor. The next largest effect was F and, although it is not statistically significant, it appears worthwhile to include it in a subsequent exploratory analysis as suggested in the next exercise.

5.14. The two main effects A and B are again highly statistically significant. F is on the margin of being statistically significant (the effect is 38.125, the estimated $SE = 19.8$, $t = 1.93$, p-value = 0.07) so it suggests that the higher level of F may well give somewhat better responses. None of the factors appear to interact. The residual plots are reasonable, suggesting that the statistical assumptions are acceptable.

5.15 The estimated contrasts are:

Ph	−0.043
Mobile phase	−0.443
Buffer conc	0.207
Flow	−0.022
Temperature	0.027
Column	0.238

The largest of these is less than 0.5% of the nominal capsule potency – effects of this order will not usually be considered of any practical importance. Accordingly the system can be considered robust.

CHAPTER 6

6.1. The test statistic to test the hypothesis that the true intercept is 100% is (101.159 – 100)/0.685 = 1.69. This is not statistically significant as the critical values for 5 degrees of freedom, using a significance level of 0.05, is 2.57. The 95% confidence interval for the true intercept is 101.159 ± 2.57(0.685), *i.e.*, 101.16 ± 1.76.

The interpretation of the confidence interval is that the data are consistent with true intercept values anywhere between 99.40 and 101.92%. Since the interval contains 100%, we cannot reject this null hypothesis value – the interpretations of the confidence interval and the significance test are the same.

6.2. The fitted regression line is given in the table below.

The regression equation is
Dissolution = 84.0− 1.51 Month

Predictor	Coef	SE Coef	T	P
Constant	83.987	1.375	61.09	0.000
Month	− 1.5053	0.1382	− 10.89	0.000

There are 6 observations, so the standard errors have 4 degrees of freedom. The *t*-multiplier for 95% confidence intervals is 2.78.

The 95% confidence interval for the slope is given by –1.505 ± 2.78×(0.138), which gives bounds of – 1.89 and – 1.12. This means that the dissolution rate decreases by between 1.89 and 1.12 percentage points per month from the beginning of the stability programme. Since the interval does not contain zero the slope is statistically different from zero – this is clear, also, from the *t*-value of –10.89 in the table. The confidence interval for the intercept is calculated in the same way. The *t*-value of 61.09 in the table tests the hypothesis that the true intercept is zero – not a sensible test here but is the default in the software. The *t*-statistic for testing that the true intercept is 100% is – 11.65, which is highly statistically significant.

6.3. The 95% confidence interval for mean batch potency at 6 months is obtained by replacing the value 15 by 6 in the calculations in the text for the 15 month interval. The result is 100.15 ± 1.22. The measured potency at 6 months was 98.4%.

6.5. The prediction interval for 24 months is 39.3 to 56.4%.

6.6. The ANOVA table for the stability dissolution data is shown below.

Analysis of Variance

Source	DF	SS	MS	F	P
Regression	1	475.84	475.84	118.66	0.000
Residual Error	4	16.04	4.01		
Total	5	491.88			

The r^2 value is the ratio of the SS Regression to the Total SS and is 0.97. Note that this is often expressed in percentage terms in the output of statistics packages.

6.7. The fitted regression line and summary statistics are given below.

The regression equation is
Peak-Area = 0.566 + 140 conc

Predictor	Coef	SE Coef	T	P
Constant	0.5665	0.4734	1.20	0.253
Conc	139.759	2.889	48.38	0.000

The inverse prediction of concentration at a peak area of 20 is 0.139.

6.8. The point predictor was 0.139. The 95% prediction limits for the concentration are 0.139 ± 0.017.

6.9. The fitted line and summary statistics are given below.

The regression equation is
Peak-Area = 0.566 + 140 conc

Predictor	Coef	SE Coef	T	P
Constant	0.5665	0.4734	1.20	0.253
Conc	139.759	2.889	48.38	0.000

The t-statistic for the test that the true intercept is zero is $t = 1.2$ which is small. The hypothesis of a zero intercept cannot be rejected.

6.10. The output for a zero intercept calibration line is shown below.

The regression equation is
Peak-Area = 143 conc

Predictor	Coef	SE Coef	T	P
Noconstant				
Conc	142.535	1.748	81.54	0.000

The point predictor corresponding to a peak area of 20 is 0.140. The 95% prediction limits for the concentration are 0.140 ± 0.017. This is virtually identical to the result obtained when the intercept was included in the regression fit.

6.11. The mean of the blank results is 2.658 and the standard deviation is 0.551. These give an LOD of 4.31, *i.e.* 0.1% of the sample weight.

6.12. The residual plots for the HPLC calibration data are reasonable. The p-value for the Anderson–Darling Normality test is 0.43.

6.13. A scatterplot of the data suggests a very strong linear relationship. However, when a linear regression line is fitted and the residuals are plotted against the fitted values a clear non-linear relationship is apparent. The data were presented by the AMC as a demonstration of the fact that the correlation coefficient (here 0.9998) is not a suitable statistic for assessing linearity.

6.14. Again a strong linear relationship is suggested by the scatterplot. However, the residuals *versus* fitted values plot is similar to

Figure 6.22 – the variability clearly increases with response magnitude (or, equivalently, with concentration).

6.15. The parameter estimates and their standard errors are given in the table below.

Parameter	*Estimate*	*Estimated standard error*
Intercept	13.02	7.87
Slope	4.74	0.20

The predictions and the corresponding error bounds are given below.

Y_o	*Point prediction*	*Prediction bounds*
100	18.4	± 10.1
600	123.9	± 31.5

Note that for these data the procedure which involves carrying out a weighted regression of the standard deviations in order to obtain the weights for fitting the calibration line, leads to slightly narrower bounds at the lower response level, ± 9.0 rather than ±10.1, and slightly wider bounds at the higher response level, ±35.8 rather than ± 31.5. These differences are much less than those between either of the WLS results and those for the OLS regression. A lot is gained by moving from an OLS fit to a weighted fit, but the difference between the two WLS fits is relatively small.

6.16. The residuals *versus* fitted values plot shows increasing variability with the magnitude of responses, as we saw for OLS in Exercise 6.14. In the weighted residuals plot this trend has disappeared entirely, suggesting that the weighted analysis has adjusted the data appropriately.

The table below gives the inverse predictions and the corresponding bounds. Note that these calculations are prone to rounding error so retain as many decimal places as possible until the final presentation of results.

Y_o	*Point prediction*	*Prediction bounds*
1000	13.6	±0.9
5000	68.6	±5.0

6.17. The fitted regression line parameters and standard errors are given below.

The regression equation is
Absorbance = 0.00260 + 0.00213 Conc. + .000000 conc-sq

Predictor	Coef	SE Coef	T	P
Constant	0.002600	0.002881	0.90	0.390
Conc.	0.00212790	0.00002945	72.26	0.000
conc-sq	0.00000033	0.00000006	5.15	0.001
S = 0.002278				

The *t*-test on the squared term is highly statistically significant, indicating that the non-linearity is unlikely to be an artifact of the chance variation. The residual plots are good. Note that the ratio of the coefficient to the standard error as shown in the last line does not equal the *t*-value. The *t*-value is correct, the problem is the number of leading zeros in the other entries. It is often useful to multiply the response by a large number before carrying out the analysis to ensure that the interesting digits will be visible in the data summaries (especially the ANOVA).

6.18. The Minitab polynomial regression analysis of the iron data, including a lack-of-fit test is shown below.

The regression equation is
Response = 26.0 + 21.3 Conc. + 0.00331 conc-sq

Predictor	Coef	SE Coef	T	P
Constant	26.00	28.81	0.90	0.390
Conc.	21.2790	0.2945	72.26	0.000
conc-sq	0.0033133	0.0006435	5.15	0.001

S = 22.78 R-Sq = 100.0%

Analysis of Variance

Source	DF	SS	MS	F	P
Regression	2	74298898	37149449	71616.13	0.000
Residual Error	9	4669	519		
Lack of Fit	3	869	290	0.46	0.722
Pure Error	6	3800	633		
Total	11	74303567			

The F-ratio for lack of fit is small (0.46) and the corresponding p-value of 0.722 indicates that there is no reason to doubt the adequacy of the polynomial model for the responses.

6.19. The scatterplot suggests a strong linear relationship. The regression line has an r^2 value of 0.98. The fitted line and prediction interval are shown below.

The regression equation is
Total-RS = 0.148 + 0.0151 Time

Predictor	Coef	SE Coef	T	P
Constant	0.148400	0.008546	17.36	0.000
Time	0.015133	0.001163	13.01	0.001

S = 0.01103 R-Sq = 98.3%

Analysis of Variance

Source	DF	SS	MS	F	P
Regression	1	0.020612	0.020612	169.32	0.001
Residual Error	3	0.000365	0.000122		
Total	4	0.020977			

Predicted Values for New Observations

New Obs	Fit	SE Fit	95.0% CI	95.0% PI
1	0.42080	0.01480	(0.37369, 0.46791)	(0.36204, 0.47956)

New Obs	Time
1	18.0

There are 3 degrees of freedom for error so the t-multiplier for the intervals is 3.18. The slope coefficient plus or minus 3.18 times its estimated standard error gives a confidence interval for the rate of growth of the related substances. The intercept coefficient plus or minus 3.18 times its estimated standard error gives a confidence interval for the amount of related substances present at the beginning of the stability study. The prediction interval does not contain 0.673, so either the growth rate is not linear, or it has changed between 12 and 18 months, or a problem arose either in the storage or measurement of the product.

6.20. The coefficients and their estimated standard errors for both regression lines are given in the table below.

	Weights	$\hat{\beta}_o$	$S\hat{E}(\hat{\beta}_o)$	$\hat{\beta}_1$	$S\hat{E}(\hat{\beta}_1)$
OLS	None	50.026	5.844	1403.65	8.77
WLS	1/(OLS-fitted values)	52.340	2.356	1399.21	7.35

The main difference between the two sets of results is the smaller estimated standard error for the intercept given by the WLS analysis; it is less than half that for the OLS fit. The inverse predictions are given below.

	OLS regression		*WLS regression*	
Y_o	$\hat{X}_o$	*OLS bounds*	$\hat{X}_o$	*WLS bounds*
100	0.0356	±0.0234	0.0341	±0.0077
1500	1.0330	±0.0235	1.0346	±0.0287

The WLS bounds are more than three times wider for the predicted %Si value corresponding to a response of 1500 than they are for the lower value of 100. Inspection of Equation (6.28) shows that there are two reasons for this. One is the distances of the two response values from the weighted mean of the Y values, which is 295.4: Equation (6.28) contains a term $(Y_o - \bar{Y}_w)^2$ that results in the higher response value having wider prediction bounds. The second reason is the term $1/w_o$; since $w_o = 1/Y_o$, the inverse is the response itself, *i.e.*, $1/w_0 = Y_o$, which is very much larger for the second prediction.

The two sets of OLS bounds, on the other hand, are virtually the same. This is because the unweighted mean of the Y values is 781.5, which is almost exactly half way between the two responses, 100 and 1500. It is inappropriate that the two intervals should be the same, since larger responses are subject to greater chance variation and this should be reflected in the error bounds for %Si.

The residuals fan outwards when plotted against the fitted values; for the weighted residuals this tendency has almost entirely disappeared.

CHAPTER 7

7.1. The ANOVA table is given below.

Analysis of Variance for TCB-recoveries

Source	DF	SS	MS	F	P
Variant	3	11874.5	3958.2	40.34	0.000
Error	12	1177.5	98.1		
Total	15	13052.0			

The LSD = 15.27 and HSD = 20.8, so by either criterion the mean of the results for variant C (75.5) is statistically significantly bigger than the next largest mean, which is variant A (52.5). The residual plots are good, so the statistical assumptions underlying the *F*-test and comparisons of means are likely to be valid.

7.2.

Analysis of Variance for TCB

Source	DF	SS	MS	F	P
PRESS	1	10816.0	10816.0	110.23	0.000
CO2	1	552.3	552.3	5.63	0.035
PRESS*CO2	1	506.3	506.3	5.16	0.042
Error	12	1177.5	98.1		
Total	15	13052.0			

Note the correspondences between this and the ANOVA table in Exercise 7.1. The total and error sums of squares and degrees of freedom are the same. The three degrees of freedom (and corresponding sum of squares) for variant are split between the three effects. The analysis of means will be identical as that for Exercise 7.1, so it need not be repeated here.

7.3. The ANOVA table is shown below.

Analysis of Variance for Narasin

Source	DF	SS	MS	F	P
Spike	2	280.78	140.39	1.66	0.201
Matrix	2	58.11	29.06	0.34	0.711
Spike*Matrix	4	136.78	34.19	0.41	0.804
Error	45	3799.17	84.43		
Total	53	4274.83			

None of the *F*-ratios is statistically significant, so there is no evidence of differences in recovery rates for the nine matrix/spike level combinations.

7.4. Since the effects analysis and the ANOVA are equivalent the results here should be the same as for Exercise 5.4: the main effect of A and the CE interaction are statistically significant. Refer to the solution for Exercise 5.4 for further details on the bio-sensor data.

7.5. The ANOVA table is shown below.

Analysis of Variance for ABZ-SO

Source	DF	SS	MS	F	P
PRESS	1	6162.3	6162.3	312.01	0.000
CO2	1	420.3	420.3	21.28	0.002
TEMP	1	12.3	12.3	0.62	0.454
PRESS*CO2	1	225.0	225.0	11.39	0.010
PRESS*TEMP	1	324.0	324.0	16.41	0.004
CO2*TEMP	1	1.0	1.0	0.05	0.828
PRESS*CO2*TEMP	1	56.3	56.3	2.85	0.130
Error	8	158.0	19.8		
Total	15	7359.0			

Note, for example, that in Exercise 5.5 the effect for the pressure–temperature interaction was 9.0, while its standard error was 2.222. The ratio squared gives 16.41, which is the corresponding *F*-statistic. All other aspects of the analysis are the same. Consult the solution to Exercise 5.5 for further comments.

7.6. The ANOVA table is shown below.

Analysis of Variance for %Recovery

Source	DF	SS	MS	F	P
Spike	2	329.33	164.67	2.19	0.125
Matrix	2	94.33	47.17	0.63	0.539
Spike*Matrix	4	533.33	133.33	1.77	0.153
Day	5	984.89	196.98	2.62	0.038
Error	40	3005.44	75.14		
Total	53	4947.33			

None of the experimental factors is statistically significant, so there is no evidence that percentage recovery is dependent on spike level or matrix type. The blocking factor is significant indicating variation from day to day (run to run) in recovery rates. The residual plots support the assumptions of constant variability and Normality.

7.7. The residual analyses for the zinc data support the model assumptions. The ANOVA gives MSE = 4.6167 and MS(run) = 62.2305. These give the following standard deviations: SD(run) = 5.414, SD(within-run) = 2.149, SD(total) = 5.824.

7.8. The following table shows the 95% confidence bounds for the three standard deviations of the control chart data of Table 7.22.

	Lower bound	*Upper bound*
Run	5.31	10.86
repeatability	4.35	6.51
total	7.17	11.53

7.9. The following table shows the 95% confidence bounds for the three standard deviations of the zinc control chart data of Exercise 7.7.

	Lower bound	*Upper bound*
Run	4.23	7.53
repeatability	1.72	2.87
total	4.68	7.70

7.10. The first three-level ANOVA shows a large pair of residuals for the data generated on day 1 in laboratory 2. When the data from laboratory 2 are excluded, the residual plots are satisfactory. The ANOVA and variance component estimates are shown below.

Analysis of Variance for Glycitin

Source	DF	SS	MS	F	P
Laborato	8	6745.7222	843.2153	5.790	0.008
Day_1	9	1310.7500	145.6389	6.355	0.000
Error	18	412.5000	22.9167		
Total	35	8468.9722			

Variance Components

Source	Var Comp.	% of Total	StDev
Laborato	174.394	67.42	13.206
Day_1	61.361	23.72	7.833
Error	22.917	8.86	4.787
Total	258.672		16.083

If we average the replicates and carry out an ANOVA of the two-level hierarchy corresponding to days nested within laboratories, then the pair of residuals corresponding to laboratory 9 is large. This means that the results from the two days are relatively further apart than those for other laboratories. If we exclude laboratory 9, also, and recomputed the ANOVA for the three-level hierarchy then we get the following results.

Analysis of Variance for Glycitin

Source	DF	SS	MS	F	P
Lab	7	6459.7188	922.8170	15.716	0.000
day-	8	469.7500	58.7188	2.300	0.074
Error	16	408.5000	25.5313		
Total	31	7337.9688			

Variance Components

Source	Var Comp.	% of Total	StDev
Lab	216.025	83.68	14.698
day-	16.594	6.43	4.074
Error	25.531	9.89	5.053
Total	258.150		16.067

The Normal plot of laboratory means, at the top of the hierarchy, is satisfactory. Note the very large impact that exclusion of laboratory 9 has had on the standard deviation for day-to-day variation.

7.11. The one-way analysis of variance of the Proficiency Test data gives MS(sample)= 0.23133, MSE=MS(analytical) = 0.06125, $F =$ 3.78 ($p = 0.015$). Thus, the sampling mean square is statistically significantly larger than the analytical mean square. The estimates of the corresponding standard deviations are SD(sampling) = 0.29 and SD(analytical) = 0.25. Since the SD(sampling) is less than 30% of the target value of 1.1 the material may be considered sufficiently homogeneous for use in the proficiency test.

7.12. The ANOVA table is shown below.

Analysis of Variance for narasin

Source	DF	SS	MS	F	P
Spike	2	280.78	140.39	1.63	0.209
Matrix	2	58.11	29.06	0.34	0.716
Spike*Matrix	4	136.78	34.19	0.40	0.810
Day	5	345.94	69.19	0.80	0.555
Error	40	3453.22	86.33		
Total	53	4274.83			

None of the experimental factors is statistically significant, so there is no evidence of varying recovery levels. Note also that the blocking factor (Day) is also not significant. This indicates that the day-to-day variation (if any) is small in comparison with the residual, or within-run, variation. The residual plots are good, so our statistical assumptions are likely to be valid.

7.13. These data are the same as those analyzed by effects analysis in Exercise 5.14. Since the effects analysis and the ANOVA are equivalent the results here should be the same: the main effects of A and B are strongly significant, while that for F is marginally significant. In all three cases the higher factor level gives larger, *i.e.*, better responses. None of the interactions are statistically significant. Refer to the solution for Exercise 5.14 for further details.

7.14. The residual plots for the TON data are good: the residuals *versus* fitted values plot shows no systematic trends and the Normal plot of residuals gives a very straight line. The ANOVA gives MS(run) = 0.005120 and MSE = 0.001045. The estimated standard deviations are SD(run) = 0.045 and SD(within-run) = 0.032; together these combine to give SD(total) = 0.056.

7.15. When the three-level model is fitted the replicates from day 2 in laboratory 1 are outliers. Exclude laboratory 1 and re-fit. Now the data from day 1 in laboratory 5 are outliers. Again, we exclude all the data from laboratory 5. The residuals plots are now acceptable. If we average the replicates, we can fit the two-level model with daily means nested under laboratories. The residuals from this model are also acceptable. Averaging the data from the two days gives eight laboratory means: these give an acceptable Normal plot. Accordingly, the statistical assumptions for the remaining four test results from eight laboratories are considered valid. The ANOVA, the variance components, and standard deviation estimates are shown below.

Analysis of Variance for Genistein

Source	DF	SS	MS	F	P
Laborato	7	5755.4688	822.2098	33.517	0.000
Day	8	196.2500	24.5313	4.876	0.003
Error	16	80.5000	5.0313		
Total	31	6032.2188			

Variance Components

Source	Var Comp.	% of Total	StDev
Laborato	199.420	93.10	14.122
Day	9.750	4.55	3.122
Error	5.031	2.35	2.243
Total	214.201		14.636

CHAPTER 8

8.1. The table below gives the ANOVA analysis of the Kalra data, including estimates of the repeatability standard deviation (error) and reproducibility standard deviation (total).

Analysis of Variance for pH

Source	DF	SS	MS	F	P
Lab	19	4.3736	0.2302	2.444	0.027
Error	20	1.8834	0.0942		
Total	39	6.2570			

Variance Components

Source	Var Comp.	% of Total	StDev
Lab	0.068	41.93	0.261
Error	0.094	58.07	0.307
Total	0.162		0.403

8.2. Laboratory 15 gives the highest variance (1.361). When compared to the sum of all the variances (1.883) this gives a ratio of 72.3%, which is statistically significant when compared to the critical value of 42.8%. When laboratory 15 is excluded, the next largest variance is that from laboratory 11 – it gives a value of 0.146. This represents 28.0% of the sum of the variances from the 19 laboratories. This ratio is not statistically significant when compared to the critical value of 44.3%.

8.3. The standard deviations, test statistics and critical values are shown below. None of the test-statistics are significant, so the tests do not identify any of the laboratories as outlying.

	SD	*Test statistic*	*Critical value*
All	0.339		
Delete 1 high value	0.289	14.8%	23.6%
Delete 1 low value	0.323	4.9%	
Delete 2 high values	0.240	29.4%	33.2%
Delete 2 low values	0.313	7.8%	
Delete 1 high + 1 low	0.270	20.5%	35.4%

8.4. The two results from laboratory 15 are obvious outliers in both the residuals *versus* fitted values plot and the Normal plot of residuals. When laboratory 15 is excluded the residual plots are good.

8.6. The ANOVA and calculated standard deviations for the lysine data are shown in the output below.

Analysis of Variance for Lysine

Source	DF	SS	MS	F	P
Laboratory	16	29.6053	1.8503	16.478	0.000
Error	17	1.9089	0.1123		
Total	33	31.5142			

Variance Components

Source	Var Comp.	% of Total	StDev
Laboratory	0.869	88.56	0.932
Error	0.112	11.44	0.335
Total	0.981		0.991

The total standard deviation is the square root of the sum of the error (repeatability) variance and laboratory variance. Hence, it is the reproducibility standard deviation. The residuals *versus* fitted values plot is reasonably good and the Normal plot of the residuals gives a very straight line.

Cochran's test for the maximum within-laboratory variance gives a test statistic of 22.6%, which is small compared to the critical value of 47.8%. The summary statistics for Grubbs' tests are given in the table below.

	SD	*Test statistic*	*Critical value*
All	0.962		
Delete H	0.776	19.3%	26.9%
Delete L	0.797	17.2%	
Delete H + L	0.561	41.7%	40.1%

The test for the highest and lowest means being outliers is statistically significant. Exclude these and re-calculate the ANOVA summaries. The results are shown below: note the large change in the laboratory SD, with little change to the repeatability SD.

Analysis of Variance for Lysine					
Source	DF	SS	MS	F	P
Laboratory	14	8.7998	0.6286	6.043	0.001
Error	15	1.5601	0.1040		
Total	29	10.3599			
Variance Components					
Source	Var Comp.	% of Total	StDev		
Laboratory	0.262	71.60	0.512		
Error	0.104	28.40	0.323		
Total	0.366		0.605		

8.7. The coefficient of volume expansion for water is only one fifth of that for ethanol – this has a big impact on the calculation of the standard uncertainty for volume, which becomes 0.060 instead of 0.246.

8.8. The temperature effect dominates the calculation of the standard uncertainty for volume, when ethanol is used. When it is excluded, the standard uncertainty decreases from 0.246 to 0.041. When water is used (Exercise 8.7) the reduction is much less marked, from 0.060 to 0.048.

8.9. The first ANOVA analysis of the Kalra data shows very large residuals for laboratory 9. When this is excluded and the ANOVA re-run, the residuals for laboratory 11 are very large. Excluding laboratory 11, also, results in a satisfactory fit, with good residual plots. The ANOVA table and standard deviation estimates are shown below.

Analysis of Variance for pH					
Source	DF	SS	MS	F	P
Lab	17	1.2807	0.0753	45.583	0.000
Error	18	0.0298	0.0017		
Total	35	1.3105			
Variance Components					
Source	Var Comp.	% of Total		StDev	
Lab	0.037	95.71		0.192	
Error	0.002	4.29		0.041	
Total	0.038			0.196	

8.10. The sum of the variances of the pH data is 3.90. Laboratory 9 has a variance of 3.67, which is 94.1% of the total. This is highly statistically significant when compared to the critical value of 42.8%. When laboratory 9 is deleted the sum becomes 0.2282 and the largest variance is 0.1985, which relates to laboratory 11. The ratio of 87.0% is highly significant; the critical value is 44.3%. The variance for laboratory 17 is the next largest (0.01125), but is only 37.8% of the total (0.02875) and is not statistically significant, since the critical value is 46.0%. These results agree with the decisions made on the basis of the residual analysis in Exercise 8.9.

8.11. Applying Grubbs' tests to the 20 laboratory means for the pH data gives the summary values in the table below.

	SD	*Test statistic*	*Critical value*
All 20 laboratories	0.275		
Delete 1 high value	0.272	1.16%	23.60%
Delete 1 low value	0.202	26.80%	
Delete 2 high values	0.272	1.42%	33.20%
Delete 2 low values	0.171	37.76%	
Delete 1 high + 1 low	0.196	28.94%	35.40%

The lowest laboratory mean (number 9) is an outlier, by the single outlier test. Deleting two low values also gives a statistically significant result, but this may be due to the very low value for laboratory 9 dominating the calculation. If this is excluded and the test for a single outlier is carried out on the 19 remaining means, the lowest value is for laboratory 5. Deleting this value reduces the standard deviation of the set by 15.5%, which is not statistically significant; the critical value is 24.6%.

8.12. The largest of the 20 replicate sample variances for the glycitin data is 480.5 (corresponding to day 1 in laboratory 2). This is 53.3% of the sum of the variances and so it is statistically significant when compared to the critical value of 42.8%. The residual analysis produced the same result in Exercise 7.10. When the data from laboratory 2 are excluded, the largest variance is 98; this is 23.3% of the sum of the variances which is 420.5. This is not statistically significant when compared to the critical value of 46%. When the replicates are averaged and the variances of the daily means (within laboratories) are calculated, the largest variance, corresponding to laboratory 9, is 420.5 which is 64.2% of the total. This falls just short of the critical value of 69.3%. Laboratory 9 was identified as having large residuals (for daily means) in Exercise 7.10. It was noted there that the exclusion of laboratory 9 has a very big impact on the calculated day-to-day standard deviation. The organizer of the trial has a difficult decision to make in such circumstances.

8.13. The largest of the 20 replicate sample variances for the genistein data is 800 (corresponding to day 2 in laboratory 1). This is 80.1% of the sum of the variances and so it is statistically significant when compared to the critical value of 42.8%. The residual analysis produced the same result in Exercise 7.15. When the data from laboratory 1 are excluded, the largest variance is 98 (corresponding to day 1 in laboratory 5); this is 54.3% of the sum of the variances which is 180.5. This is statistically significant when compared to the critical value of 46%. When the data from laboratory 5 are excluded the result of Cochran's test is not statistically significant. When the replicates are averaged and the variances of the daily means (within laboratories) are calculated, the largest variance, corresponding to laboratory 2, is 45.125 which is 46% of the total of 98.125. The critical value is 73.6% so this is not statistically significant. The residual analysis of Exercise 7.15 led to exactly the same conclusions.

8.14. The first ANOVA on the methionine data gives a residuals *versus* fitted values plot which shows one pair of residuals as apparently large. The plot also indicates that laboratory 17 gives a very low mean value relative to the other laboratories. The Cochran test-statistic is the ratio of 0.296 (laboratory 7) to 0.732, the sum of the variances for all laboratories. This equals 40.5% which is not statistically significant when compared to the critical value of 47.8%. Grubbs' test for a single outlier gives a reduction in the standard deviation of the 17 means of 83.4% when laboratory 17 is excluded. This is highly statistically significant, as the critical value is 26.9%.

The ANOVA for the 16 laboratories is given below.

Analysis of Variance for Methionine

Source	DF	SS	MS	F	P
Laboratory	15	5.1605	0.3440	7.524	0.000
Error	16	0.7316	0.0457		
Total	31	5.8922			

Variance Components

Source	Var Comp.	% of Total	StDev
Laboratory	0.149	76.54	0.386
Error	0.046	23.46	0.214
Total	0.195		0.441

8.15. The standard deviation for all the means is 10.78, while it is 7.74 for the reduced set of 24. This is a percentage reduction of 28.2%. Since the critical value is 27.1% the two low values may be considered as not belonging to the same distribution as the rest.

8.16. The standard deviation for all the effects is 13.27, while it is 10.87 for the reduced set of 29. This is a percentage reduction of 18.1%. Since the critical value is about 26.0% (for $n = 30$), we cannot reject the null hypothesis that all effects come from the same distribution.

APPENDIX

Statistical Tables

A1 Standard Normal Table
A2 Factors for constructing control charts
A3 Critical values for Student's t-distribution
A4 Sample size tables – one-sample t-tests
A5 Critical values for the Chi-square distribution
A6 Critical values for the F-distribution
A7 Sample size tables – two-sample t-tests
A8 Sample size tables for comparing standard deviations
A9 Critical values for the Studentized range distribution
A10 Critical values (2.5% one-tail) for Cochran's test
A11 Critical values for Grubbs' test (2.5% two-tail, 1.25% one-tail)

A1 *Standard Normal Table*

z	*.00*	*.01*	*.02*	*.03*	*.04*	*.05*	*.06*	*.07*	*.08*	*.09*
0.0	.5000	.5040	.5080	.5120	.5160	.5199	.5239	.5279	.5319	.5359
0.1	.5398	.5438	.5478	.5517	.5557	.5596	.5636	.5675	.5714	.5753
0.2	.5793	.5832	.5871	.5910	.5948	.5987	.6026	.6064	.6103	.6141
0.3	.6179	.6217	.6255	.6293	.6331	.6368	.6406	.6443	.6480	.6517
0.4	.6554	.6591	.6628	.6664	.6700	.6736	.6772	.6808	.6844	.6879
0.5	.6915	.6950	.6985	.7019	.7054	.7088	.7123	.7157	.7190	.7224
0.6	.7257	.7291	.7324	.7357	.7389	.7422	.7454	.7486	.7517	.7549
0.7	.7580	.7611	.7642	.7673	.7704	.7734	.7764	.7794	.7823	.7852
0.8	.7881	.7910	.7939	.7967	.7995	.8023	.8051	.8079	.8106	.8133
0.9	.8159	.8186	.8212	.8238	.8264	.8289	.8315	.8340	.8365	.8389
1.0	.8413	.8438	.8461	.8485	.8508	.8531	.8554	.8577	.8599	.8621
1.1	.8643	.8665	.8686	.8708	.8729	.8749	.8770	.8790	.8810	.8830
1.2	.8849	.8869	.8888	.8907	.8925	.8944	.8962	.8980	.8997	.9015
1.3	.9032	.9049	.9066	.9082	.9099	.9115	.9131	.9147	.9162	.9177
1.4	.9192	.9207	.9222	.9236	.9251	.9265	.9279	.9292	.9306	.9319
1.5	.9332	.9345	.9357	.9370	.9382	.9394	.9406	.9418	.9429	.9441
1.6	.9452	.9463	.9474	.9484	.9495	.9505	.9515	.9525	.9535	.9545
1.7	.9554	.9564	.9573	.9582	.9591	.9599	.9608	.9616	.9625	.9633
1.8	.9641	.9649	.9656	.9664	.9671	.9678	.9686	.9693	.9699	.9706
1.9	.9713	.9719	.9726	.9732	.9738	.9744	.9750	.9756	.9761	.9767
2.0	.9773	.9778	.9783	.9788	.9793	.9798	.9803	.9808	.9812	.9817
2.1	.9821	.9826	.9830	.9834	.9838	.9842	.9846	.9850	.9854	.9857
2.2	.9861	.9864	.9868	.9871	.9875	.9878	.9881	.9884	.9887	.9890
2.3	.9893	.9896	.9898	.9901	.9904	.9906	.9909	.9911	.9913	.9916
2.4	.9918	.9920	.9922	.9925	.9927	.9929	.9931	.9932	.9934	.9936
2.5	.9938	.9940	.9941	.9943	.9945	.9946	.9948	.9949	.9951	.9952
2.6	.9953	.9955	.9956	.9957	.9959	.9960	.9961	.9962	.9963	.9964
2.7	.9965	.9966	.9967	.9968	.9969	.9970	.9971	.9972	.9973	.9974
2.8	.9974	.9975	.9976	.9977	.9977	.9978	.9979	.9979	.9980	.9981
2.9	.9981	.9982	.9983	.9983	.9984	.9984	.9985	.9985	.9986	.9986
3.0	.9987	.9987	.9987	.9988	.9988	.9989	.9989	.9989	.9990	.9990
3.1	.9990	.9991	.9991	.9991	.9992	.9992	.9992	.9992	.9993	.9993
3.2	.9993	.9993	.9994	.9994	.9994	.9994	.9994	.9995	.9995	.9995
3.3	.9995	.9995	.9996	.9996	.9996	.9996	.9996	.9996	.9996	.9997
3.4	.9997	.9997	.9997	.9997	.9997	.9997	.9997	.9997	.9997	.9998

A2 *Factors for constructing control charts*

Sample size	A_2	D_3	D_4	d_n
2	1.88	0	3.27	1.128
3	1.02	0	2.57	1.693
4	0.73	0	2.28	2.059
5	0.58	0	2.11	2.326
6	0.48	0	2.00	2.534
7	0.42	0.08	1.92	2.704
8	0.37	0.14	1.86	2.847
9	0.34	0.18	1.82	2.970
10	0.31	0.22	1.78	3.078

X-bar Charts. A_2 is used together with the average range to calculate action limits for X-bar charts:

$$\text{Control limits} = \text{Centre line} \pm A_2 \times \bar{R}$$

Range Charts. D_3 and D_4 give 3-sigma limits for the range chart when multiplied by the average range $\bar{R}$.

Calculating Standard Deviation from Range. An estimate of the standard deviation may be obtained from a range, R, or average range, $\bar{R}$, based on samples of size n using the d_n factor in Table A2 as follows:

$$\hat{\sigma} = \frac{\bar{R}}{d_n}$$

where n is the sub-group size.

A3 *Critical values for Student's t-distribution*

deg. freedom \ *one tail* / *two tails*	*.050* / *.100*	*.025* / *.050*	*.010* / *.020*	*.005* / *.010*
1	6.31	12.71	31.82	63.66
2	2.92	4.30	6.96	9.92
3	2.35	3.18	4.54	5.84
4	2.13	2.78	3.75	4.60
5	2.02	2.57	3.36	4.03
6	1.94	2.45	3.14	3.71
7	1.89	2.36	3.00	3.50
8	1.86	2.31	2.90	3.36
9	1.83	2.26	2.82	3.25
10	1.81	2.23	2.76	3.17
11	1.80	2.20	2.72	3.11
12	1.78	2.18	2.68	3.05
13	1.77	2.16	2.65	3.01
14	1.76	2.14	2.62	2.98
15	1.75	2.13	2.60	2.95
16	1.75	2.12	2.58	2.92
17	1.74	2.11	2.57	2.90
18	1.73	2.10	2.55	2.88
19	1.73	2.09	2.54	2.86
20	1.72	2.09	2.53	2.85
21	1.72	2.08	2.52	2.83
22	1.72	2.07	2.51	2.82
23	1.71	2.07	2.50	2.81
24	1.71	2.06	2.49	2.80
25	1.71	2.06	2.49	2.79
26	1.71	2.06	2.48	2.78
27	1.70	2.05	2.47	2.77
28	1.70	2.05	2.47	2.76
29	1.70	2.05	2.47	2.76
30	1.70	2.04	2.46	2.75
40	1.68	2.02	2.42	2.70
50	1.68	2.01	2.40	2.68
60	1.67	2.00	2.39	2.66
100	1.66	1.98	2.36	2.63
200	1.65	1.97	2.35	2.60
∞	1.65	1.96	2.33	2.58

A4 *Sample size tables – one-sample t-tests*

	$\alpha = 0.025$ (one-tail) $\alpha = 0.05$ (two-tails)				$\alpha = 0.05$ (one-tail) $\alpha = 0.10$ (two-tails)			
D \ β	0.01	0.05	0.1	0.2	0.01	0.05	0.1	0.2
0.40	117	84	68	52	100	70	55	41
0.45	93	67	54	41	80	55	44	32
0.50	76	54	44	34	65	45	36	27
0.55	63	45	37	28	54	38	30	22
0.60	53	39	32	24	46	32	26	19
0.65	46	33	27	21	39	28	22	17
0.70	40	29	24	19	34	24	19	15
0.75	35	26	21	16	30	21	17	13
0.80	31	23	19	15	27	19	15	12
0.85	28	21	17	13	24	17	14	11
0.90	25	19	16	12	21	15	13	10
0.95	23	17	14	11	19	14	11	9
1.0	21	16	13	10	18	13	11	8
1.1	18	13	11	9	15	11	9	7
1.2	15	12	10	8	13	10	8	6
1.3	13	10	9	7	11	8	7	6
1.4	12	9	8	7	10	8	7	5
1.5	11	8	7	6	9	7	6	5
1.6	10	8	7	6	8	6	6	5
1.7	9	7	6	5	8	6	5	4
1.8	8	7	6	5	7	6	5	4
1.9	8	6	6	5	7	5	5	4
2.0	7	6	5	5	6	5	4	4
2.5	6	5	5	4	5	4	4	3

A4 *Sample size tables – one-sample t-tests*

D \ β	$\alpha = 0.005$ (one-tail) $\alpha = 0.01$ (two-tails)				$\alpha = 0.01$ (one-tail) $\alpha = 0.02$ (two-tails)			
	0.01	*0.05*	*0.1*	*0.2*	*0.01*	*0.05*	*0.1*	*0.2*
0.40	154	115	97	77	139	102	85	66
0.45	123	92	77	62	110	81	68	53
0.50	100	75	63	51	90	66	55	43
0.55	83	63	53	42	75	55	46	36
0.60	71	53	45	36	63	47	39	31
0.65	61	46	39	32	55	41	34	27
0.70	53	40	34	28	47	35	30	24
0.75	47	36	30	25	42	31	26	21
0.80	41	32	27	22	37	28	24	19
0.85	37	29	24	20	33	25	21	17
0.90	34	26	22	18	30	23	19	16
0.95	31	24	20	17	27	21	18	14
1.0	28	22	19	16	25	19	16	13
1.1	24	19	16	14	21	16	14	12
1.2	21	16	14	12	18	14	12	10
1.3	18	15	13	11	16	13	11	9
1.4	16	13	12	10	14	11	10	9
1.5	15	12	11	9	13	10	9	8
1.6	13	11	10	8	12	10	9	7
1.7	12	10	9	8	11	9	8	7
1.8	12	10	9	8	10	8	7	6
1.9	11	9	8	7	10	8	7	6
2.0	10	8	8	7	9	7	7	6
2.5	8	7	6	6	7	6	6	5

A5 *Critical values for the Chi-square distribution*

	Left-hand tail			*Right-hand tail*		
DF	*.005*	*.025*	*.050*	*.050*	*.025*	*.005*
1	0.00	0.00	0.00	3.84	5.02	7.88
2	0.01	0.05	0.10	5.99	7.38	10.60
3	0.07	0.22	0.35	7.81	9.35	12.84
4	0.21	0.48	0.71	9.49	11.14	14.86
5	0.41	0.83	1.15	11.07	12.83	16.75
6	0.68	1.24	1.64	12.59	14.45	18.55
7	0.99	1.69	2.17	14.07	16.01	20.28
8	1.34	2.18	2.73	15.51	17.53	21.95
9	1.73	2.70	3.33	16.92	19.02	23.59
10	2.16	3.25	3.94	18.31	20.48	25.19
11	2.60	3.82	4.57	19.68	21.92	26.76
12	3.07	4.40	5.23	21.03	23.34	28.30
13	3.57	5.01	5.89	22.36	24.74	29.82
14	4.07	5.63	6.57	23.68	26.12	31.32
15	4.60	6.26	7.26	25.00	27.49	32.80
16	5.14	6.91	7.96	26.30	28.85	34.27
17	5.70	7.56	8.67	27.59	30.19	35.72
18	6.26	8.23	9.39	28.87	31.53	37.16
19	6.84	8.91	10.12	30.14	32.85	38.58
20	7.43	9.59	10.85	31.41	34.17	40.00
21	8.03	10.28	11.59	32.67	35.48	41.40
22	8.64	10.98	12.34	33.92	36.78	42.80
23	9.26	11.69	13.09	35.17	38.08	44.18
24	9.89	12.40	13.85	36.42	39.36	45.56
25	10.52	13.12	14.61	37.65	40.65	46.93
26	11.16	13.84	15.38	38.89	41.92	48.29
27	11.81	14.57	16.15	40.11	43.19	49.65
28	12.46	15.31	16.93	41.34	44.46	50.99
28	12.46	15.31	16.93	41.34	44.46	50.99
30	13.79	16.79	18.49	43.77	46.98	53.67
40	20.71	24.43	26.51	55.76	59.34	66.77
50	27.99	32.36	34.76	67.50	71.42	79.49
60	35.53	40.48	43.19	79.08	83.30	91.95
100	67.33	74.22	77.93	124.34	129.56	140.17
120	83.85	91.57	95.70	146.57	152.21	163.65
200	152.24	162.73	168.28	233.99	241.06	255.26

A6 *Critical values for the F-distribution*

v_1 numerator and v_2 denominator degrees of freedom
5% critical values

v_2 \ v_1	1	2	3	4	5	6	7	8	10	12	24	∞
1	161.4	199.5	215.7	224.6	230.2	234.0	236.8	238.9	241.9	243.9	249.1	254.3
2	18.5	19.0	19.2	19.2	19.3	19.3	19.4	19.4	19.4	19.4	19.5	19.5
3	10.1	9.6	9.3	9.1	9.0	8.9	8.9	8.8	8.8	8.7	8.6	8.5
4	7.7	6.9	6.6	6.4	6.3	6.2	6.1	6.0	6.0	5.9	5.8	5.6
5	6.6	5.8	5.4	5.2	5.1	5.0	4.9	4.8	4.7	4.7	4.5	4.4
6	6.0	5.1	4.8	4.5	4.4	4.3	4.2	4.1	4.1	4.0	3.8	3.7
7	5.6	4.7	4.3	4.1	4.0	3.9	3.8	3.7	3.6	3.6	3.4	3.2
8	5.3	4.5	4.1	3.8	3.7	3.6	3.5	3.4	3.3	3.3	3.1	2.9
9	5.1	4.3	3.9	3.6	3.5	3.4	3.3	3.2	3.1	3.1	2.9	2.7
10	5.0	4.1	3.7	3.5	3.3	3.2	3.1	3.1	3.0	2.9	2.7	2.5
12	4.7	3.9	3.5	3.3	3.1	3.0	2.9	2.8	2.8	2.7	2.5	2.3
15	4.5	3.7	3.3	3.1	2.9	2.8	2.7	2.6	2.5	2.5	2.3	2.1
20	4.4	3.5	3.1	2.9	2.7	2.6	2.5	2.4	2.3	2.3	2.1	1.8
30	4.2	3.3	2.9	2.7	2.5	2.4	2.3	2.3	2.2	2.1	1.9	1.6
40	4.1	3.2	2.8	2.6	2.4	2.3	2.2	2.2	2.1	2.0	1.8	1.5
120	3.9	3.1	2.7	2.4	2.3	2.2	2.1	2.0	1.9	1.8	1.6	1.3
∞	3.8	3.0	2.6	2.4	2.2	2.1	2.0	1.9	1.8	1.8	1.5	1.0

A6 *Critical values for the F-distribution*

v_1 numerator and v_2 denominator degrees of freedom
2.5% critical values

v_2 \ v_1	1	2	3	4	5	6	7	8	10	12	24	∞
1	647.8	799.5	864.2	899.6	921.8	937.1	948.2	956.6	968.6	976.7	997.3	1018.3
2	38.5	39.0	39.2	39.2	39.3	39.3	39.4	39.4	39.4	39.4	39.5	39.5
3	17.4	16.0	15.4	15.1	14.9	14.7	14.6	14.5	14.4	14.3	14.1	13.9
4	12.2	10.6	10.0	9.6	9.4	9.2	9.1	9.0	8.8	8.8	8.5	8.3
5	10.0	8.4	7.8	7.4	7.1	7.0	6.9	6.8	6.6	6.5	6.3	6.0
6	8.8	7.3	6.6	6.2	6.0	5.8	5.7	5.6	5.5	5.4	5.1	4.8
7	8.1	6.5	5.9	5.5	5.3	5.1	5.0	4.9	4.8	4.7	4.4	4.1
8	7.6	6.1	5.4	5.1	4.8	4.7	4.5	4.4	4.3	4.2	3.9	3.7
9	7.2	5.7	5.1	4.7	4.5	4.3	4.2	4.1	4.0	3.9	3.6	3.3
10	6.9	5.5	4.8	4.5	4.2	4.1	3.9	3.9	3.7	3.6	3.4	3.1
12	6.6	5.1	4.5	4.1	3.9	3.7	3.6	3.5	3.4	3.3	3.0	2.7
15	6.2	4.8	4.2	3.8	3.6	3.4	3.3	3.2	3.1	3.0	2.7	2.4
20	5.9	4.5	3.9	3.5	3.3	3.1	3.0	2.9	2.8	2.7	2.4	2.1
30	5.6	4.2	3.6	3.2	3.0	2.9	2.7	2.7	2.5	2.4	2.1	1.8
40	5.4	4.1	3.5	3.1	2.9	2.7	2.6	2.5	2.4	2.3	2.0	1.6
120	5.2	3.8	3.2	2.9	2.7	2.5	2.4	2.3	2.2	2.1	1.8	1.3
∞	5.0	3.7	3.1	2.8	2.6	2.4	2.3	2.2	2.0	1.9	1.6	1.0

A7 *Sample size tables – two-sample t-tests*

	$\alpha = 0.025$ (one-tail) $\alpha = 0.05$ (two-tails)				$\alpha = 0.05$ (one-tail) $\alpha = 0.10$ (two-tails)			
D \ β	0.01	0.05	0.1	0.2	0.01	0.05	0.1	0.2
0.70	76	55	44	34	66	45	36	26
0.75	67	48	39	29	57	40	32	23
0.80	59	42	34	26	50	35	28	21
0.85	52	37	31	23	45	31	25	18
0.90	47	34	27	21	40	28	22	16
0.95	42	30	25	19	36	25	20	15
1.00	38	27	23	17	33	23	18	14
1.10	32	23	19	15	27	19	15	11
1.20	27	20	16	12	23	16	13	10
1.30	23	17	14	11	20	14	11	9
1.40	20	15	12	10	17	12	10	8
1.50	18	13	11	9	15	11	9	7
1.60	16	12	10	8	14	10	8	6
1.70	14	11	9	7	12	9	7	6
1.80	13	10	8	6	11	8	7	5
1.90	12	9	7	6	10	7	6	5
2.00	11	8	7	6	9	7	6	4
2.10	10	8	6	5	8	6	5	4
2.20	9	7	6	5	8	6	5	4
2.30	9	7	6	5	7	5	5	4
2.40	8	6	5	4	7	5	4	4
2.50	8	6	5	4	6	5	4	3
3.00	6	5	4	4	5	4	3	3

A7 *Sample size tables – two-sample t-tests*

	$\alpha=0.005$ (one-tail) $\alpha=0.01$ (two-tails)				$\alpha=0.01$ (one-tail) $\alpha=0.02$ (two-tails)			
D \ β	*0.01*	*0.05*	*0.1*	*0.2*	*0.01*	*0.05*	*0.1*	*0.2*
0.70	100	75	63	50	90	66	55	43
0.75	88	66	55	44	79	58	48	38
0.80	77	58	49	39	70	51	43	33
0.85	69	51	43	35	62	46	38	30
0.90	62	46	39	31	55	41	34	27
0.95	55	42	35	28	50	37	31	24
1.00	50	38	32	26	45	33	28	22
1.10	42	32	27	22	38	28	23	19
1.20	36	27	23	18	32	24	20	16
1.30	31	23	20	16	28	21	17	14
1.40	27	20	17	14	24	18	15	12
1.50	24	18	15	13	21	16	14	11
1.60	21	16	14	11	19	14	12	10
1.70	19	15	13	10	17	13	11	9
1.80	17	13	11	10	15	12	10	8
1.90	16	12	11	9	14	11	9	8
2.00	14	11	10	8	13	10	9	7
2.10	13	10	9	8	12	9	8	7
2.20	12	10	8	7	11	9	7	6
2.30	11	9	8	7	10	8	7	6
2.40	11	9	8	6	10	8	7	6
2.50	10	8	7	6	9	7	6	5
3.00	8	6	6	5	7	6	5	4

A8 *Sample size tables for comparing standard deviations*

one-tail α = 0.025 *two-tail α = 0.05*			*one-tail α = 0.05* *two-tail α = 0.10*		
β	*0.05*	*0.10*		*0.05*	*0.10*
R	*N*		*R*	*N*	
1.1	1433	1159	1.1	1194	945
1.2	393	319	1.2	328	260
1.3	191	155	1.3	160	127
1.4	117	95	1.4	98	78
1.5	82	66	1.5	68	54
1.6	61	50	1.6	51	41
1.7	49	40	1.7	41	33
1.8	40	33	1.8	34	27
1.9	34	28	1.9	29	23
2.0	30	24	2.0	25	20
2.1	26	22	2.1	22	18
2.2	23	19	2.2	20	16
2.3	21	18	2.3	18	15
2.4	19	16	2.4	17	14
2.5	18	15	2.5	15	13
2.6	17	14	2.6	14	12
2.7	16	13	2.7	13	11
2.8	15	12	2.8	13	10
2.9	14	12	2.9	12	10
3.0	13	11	3.0	11	9

A9 *Critical values for the Studentized range distribution*

(Reproduced from *The Analysis of Variance*, H. Scheffe, © 1966. This material is used by permission of John Wiley & Sons, Inc.)

$\alpha = 0.05$ DF	*Number of means* 2	3	4	5	6	7	8	9	10	11	12	13	14	15	16	17	18	19	20
1	18.0	27.0	32.8	37.1	40.4	43.1	45.4	47.4	49.1	50.6	52.0	53.2	54.3	55.4	56.3	57.2	58.0	58.8	59.6
2	6.08	8.33	9.80	10.9	11.7	12.4	13.0	13.5	14.0	14.4	14.7	15.1	15.4	15.7	15.9	16.1	16.4	16.6	16.8
3	4.50	5.91	6.82	7.50	8.04	8.48	8.85	9.18	9.46	9.72	9.95	10.2	10.3	10.5	10.7	10.8	11.0	11.1	11.2
4	3.93	5.04	5.76	6.29	6.71	7.05	7.35	7.60	7.83	8.03	8.21	8.37	8.52	8.66	8.79	8.91	9.03	9.13	9.23
5	3.64	4.60	5.22	5.67	6.03	6.33	6.58	6.80	6.99	7.17	7.32	7.47	7.60	7.72	7.83	7.93	8.03	8.12	8.21
6	3.46	4.34	4.90	5.30	5.63	5.90	6.12	6.32	6.49	6.65	6.79	6.92	7.03	7.14	7.24	7.34	7.43	7.51	7.59
7	3.34	4.16	4.68	5.06	5.36	5.61	5.82	6.00	6.16	6.30	6.43	6.55	6.66	6.76	6.85	6.94	7.02	7.10	7.17
8	3.26	4.04	4.53	4.89	5.17	5.40	5.60	5.77	5.92	6.05	6.18	6.29	6.39	6.48	6.57	6.65	6.73	6.80	6.87
9	3.20	3.95	4.41	4.76	5.02	5.24	5.43	5.59	5.74	5.87	5.98	6.09	6.19	6.28	6.36	6.44	6.51	6.58	6.64
10	3.15	3.88	4.33	4.65	4.91	5.12	5.30	5.46	5.60	5.72	5.83	5.93	6.03	6.11	6.19	6.27	6.34	6.40	6.47
11	3.11	3.82	4.26	4.57	4.82	5.03	5.20	5.35	5.49	5.61	5.71	5.81	5.90	5.98	6.06	6.13	6.20	6.27	6.33
12	3.08	3.77	4.20	4.51	4.75	4.95	5.12	5.27	5.39	5.51	5.61	5.71	5.80	5.88	5.95	6.02	6.09	6.15	6.21
13	3.06	3.73	4.15	4.45	4.69	4.88	5.05	5.19	5.32	5.43	5.53	5.63	5.71	5.79	5.86	5.93	5.99	6.05	6.11
14	3.03	3.70	4.11	4.41	4.64	4.83	4.99	5.13	5.25	5.36	5.46	5.55	5.64	5.71	5.79	5.85	5.91	5.97	6.03
15	3.01	3.67	4.08	4.37	4.59	4.78	4.94	5.08	5.20	5.31	5.40	5.49	5.57	5.65	5.72	5.78	5.85	5.90	5.96
16	3.00	3.65	4.05	4.33	4.56	4.74	4.90	5.03	5.15	5.26	5.35	5.44	5.52	5.59	5.66	5.73	5.79	5.84	5.90
17	2.98	3.63	4.02	4.30	4.52	4.70	4.86	4.99	5.11	5.21	5.31	5.39	5.47	5.54	5.61	5.67	5.73	5.79	5.84
18	2.97	3.61	4.00	4.28	4.49	4.67	4.82	4.96	5.07	5.17	5.27	5.35	5.43	5.50	5.57	5.63	5.69	5.74	5.79
19	2.96	3.59	3.98	4.25	4.47	4.65	4.79	4.92	5.04	5.14	5.23	5.31	5.39	5.46	5.53	5.59	5.65	5.70	5.75
20	2.95	3.58	3.96	4.23	4.45	4.62	4.77	4.90	5.01	5.11	5.20	5.28	5.36	5.43	5.49	5.55	5.61	5.66	5.71
24	2.92	3.53	3.90	4.17	4.37	4.54	4.68	4.81	4.92	5.01	5.10	5.18	5.25	5.32	5.38	5.44	5.49	5.55	5.59
30	2.89	3.49	3.85	4.10	4.30	4.46	4.60	4.72	4.82	4.92	5.00	5.08	5.15	5.21	5.27	5.33	5.38	5.43	5.47
40	2.86	3.44	3.79	4.04	4.23	4.39	4.52	4.63	4.73	4.82	4.90	4.98	5.04	5.11	5.16	5.22	5.27	5.31	5.36
60	2.83	3.40	3.74	3.98	4.16	4.31	4.44	4.55	4.65	4.73	4.81	4.88	4.94	5.00	5.06	5.11	5.15	5.20	5.24
120	2.80	3.36	3.68	3.92	4.10	4.24	4.36	4.47	4.56	4.64	4.71	4.78	4.84	4.90	4.95	5.00	5.04	5.09	5.13
∞	2.77	3.31	3.63	3.86	4.03	4.17	4.29	4.39	4.47	4.55	4.62	4.68	4.74	4.80	4.85	4.89	4.93	4.97	5.01

A9 *Critical values for the Studentized range distribution*

$\alpha = 0.10$	*Number of means*																		
DF	*2*	*3*	*4*	*5*	*6*	*7*	*8*	*9*	*10*	*11*	*12*	*13*	*14*	*15*	*16*	*17*	*18*	*19*	*20*
1	8.93	13.4	16.4	18.5	20.2	21.5	22.6	23.6	24.5	25.2	25.9	26.5	27.1	27.6	28.1	28.5	29.0	29.3	29.7
2	4.13	5.73	6.77	7.54	8.14	8.63	9.05	9.41	9.72	10.0	10.3	10.5	10.7	10.9	11.1	11.2	11.4	11.5	11.7
3	3.33	4.47	5.20	5.74	6.16	6.51	6.81	7.06	7.29	7.49	7.67	7.83	7.98	8.12	8.25	8.37	8.48	8.58	8.68
4	3.01	3.98	4.59	5.03	5.39	5.68	5.93	6.14	6.33	6.49	6.65	6.78	6.91	7.02	7.13	7.23	7.33	7.41	7.50
5	2.85	3.72	4.26	4.66	4.98	5.24	5.46	5.65	5.82	5.97	6.10	6.22	6.34	6.44	6.54	6.63	6.71	6.79	6.86
6	2.75	3.56	4.07	4.44	4.73	4.97	5.17	5.34	5.50	5.64	5.76	5.87	5.98	6.07	6.16	6.25	6.32	6.40	6.47
7	2.68	3.45	3.93	4.28	4.55	4.78	4.97	5.14	5.28	5.41	5.53	5.64	5.74	5.83	5.91	5.99	6.06	6.13	6.19
8	2.63	3.37	3.83	4.17	4.43	4.65	4.83	4.99	5.13	5.25	5.36	5.46	5.56	5.64	5.72	5.80	5.87	5.93	6.00
9	2.59	3.32	3.76	4.08	4.34	4.54	4.72	4.87	5.01	5.13	5.23	5.33	5.42	5.51	5.58	5.66	5.72	5.79	5.85
10	2.56	3.27	3.70	4.02	4.26	4.47	4.64	4.78	4.91	5.03	5.13	5.23	5.32	5.40	5.47	5.54	5.61	5.67	5.73
11	2.54	3.23	3.66	3.96	4.20	4.40	4.57	4.71	4.84	4.95	5.05	5.15	5.23	5.31	5.38	5.45	5.51	5.57	5.63
12	2.52	3.20	3.62	3.92	4.16	4.35	4.51	4.65	4.78	4.89	4.99	5.08	5.16	5.24	5.31	5.37	5.44	5.49	5.55
13	2.50	3.18	3.59	3.88	4.12	4.30	4.46	4.60	4.72	4.83	4.93	5.02	5.10	5.18	5.25	5.31	5.37	5.43	5.48
14	2.49	3.16	3.56	3.85	4.08	4.27	4.42	4.56	4.68	4.79	4.88	4.97	5.05	5.12	5.19	5.26	5.32	5.37	5.43
15	2.48	3.14	3.54	3.83	4.05	4.23	4.39	4.52	4.64	4.75	4.84	4.93	5.01	5.08	5.15	5.21	5.27	5.32	5.38
16	2.47	3.12	3.52	3.80	4.03	4.21	4.36	4.49	4.61	4.71	4.81	4.89	4.97	5.04	5.11	5.17	5.23	5.28	5.33
17	2.46	3.11	3.50	3.78	4.00	4.18	4.33	4.46	4.58	4.68	4.77	4.86	4.93	5.01	5.07	5.13	5.19	5.24	5.30
18	2.45	3.10	3.49	3.77	3.98	4.16	4.31	4.44	4.55	4.65	4.75	4.83	4.90	4.98	5.04	5.10	5.16	5.21	5.26
19	2.45	3.09	3.47	3.75	3.97	4.14	4.29	4.42	4.53	4.63	4.72	4.80	4.88	4.95	5.01	5.07	5.13	5.18	5.23
20	2.44	3.08	3.46	3.74	3.95	4.12	4.27	4.40	4.51	4.61	4.70	4.78	4.85	4.92	4.99	5.05	5.10	5.16	5.20
24	2.42	3.05	3.42	3.69	3.90	4.07	4.21	4.34	4.44	4.54	4.63	4.71	4.78	4.85	4.91	4.97	5.02	5.07	5.12
30	2.40	3.02	3.39	3.65	3.85	4.02	4.16	4.28	4.38	4.47	4.56	4.64	4.71	4.77	4.83	4.89	4.94	4.99	5.03
40	2.38	2.99	3.35	3.60	3.80	3.96	4.10	4.21	4.32	4.41	4.49	4.56	4.63	4.69	4.75	4.81	4.86	4.90	4.95
60	2.36	2.96	3.31	3.56	3.75	3.91	4.04	4.16	4.25	4.34	4.42	4.49	4.56	4.62	4.67	4.73	4.78	4.82	4.86
120	2.34	2.93	3.28	3.52	3.71	3.86	3.99	4.10	4.19	4.28	4.35	4.42	4.48	4.54	4.60	4.65	4.69	4.74	4.78
∞	2.33	2.90	3.24	3.48	3.66	3.81	3.93	4.04	4.13	4.21	4.28	4.35	4.41	4.47	4.52	4.57	4.61	4.65	4.69

A10 *Critical values (2.5% one-tail) for Cochran's test*
(Reprinted from the *Journal of AOAC INTERNATIONAL* 1995, Vol. 78, pp 143A. Copyright, 1995, by AOAC INTERNATIONAL.)

	Number of replicates per laboratory				
L labs	*2*	*3*	*4*	*5*	*6*
4	94.3	81.0	72.5	65.4	62.5
5	88.6	72.6	64.6	58.1	53.9
6	83.2	65.8	58.3	52.2	47.3
7	78.2	60.2	52.2	47.3	42.3
8	73.6	55.6	47.4	43.0	38.5
9	69.3	51.8	43.3	39.3	35.3
10	65.5	48.6	39.9	36.2	32.6
11	62.2	45.8	37.2	33.6	30.3
12	59.2	43.1	35.0	31.3	28.3
13	56.4	40.5	33.2	29.2	26.5
14	53.8	38.3	31.5	27.3	25.0
15	51.5	36.4	29.9	25.7	23.7
16	49.5	34.7	28.4	24.4	22.0
17	47.8	33.2	27.1	23.3	21.2
18	46.0	31.8	25.9	22.4	20.4
19	44.3	30.5	24.8	21.5	19.5
20	42.8	29.3	23.8	20.7	18.7
21	41.5	28.2	22.9	19.9	18.0
22	40.3	27.2	22.0	19.2	17.3
23	39.1	26.3	21.2	18.5	16.6
24	37.9	25.5	20.5	17.8	16.0
25	36.7	24.8	19.9	17.2	15.5
26	35.5	24.1	19.3	16.6	15.0
27	34.5	23.4	18.7	16.1	14.5
28	33.7	22.7	18.1	15.7	14.1
29	33.1	22.1	17.5	15.3	13.7
30	32.5	21.6	16.9	14.9	13.3
35	29.3	19.5	15.3	12.9	11.6
40	26.0	17.0	13.5	11.6	10.2
50	21.6	14.3	11.4	9.7	8.6

A11 *Critical values for Grubbs' test (2.5% two-tail, 1.25% one-tail)*
(Reprinted from the *Journal of AOAC INTERNATIONAL* 1995, Vol. 78, pp 143A. Copyright, 1995, by AOAC INTERNATIONAL.)

L labs	*One high or one low*	*Two high or two low*	*One high and one low*
4	86.1	98.9	99.1
5	73.5	90.3	92.7
6	64.0	81.3	84.0
7	57.0	73.1	76.2
8	51.4	66.5	69.6
9	46.8	61.0	64.1
10	42.8	56.4	59.5
11	39.3	52.5	55.5
12	36.1	48.5	51.6
13	33.8	46.1	49.1
14	31.7	43.5	46.5
15	29.9	41.2	44.1
16	28.3	39.2	42.0
17	26.9	37.4	40.1
18	25.7	35.9	38.4
19	24.6	34.5	36.9
20	23.6	33.2	35.4
21	22.7	31.9	34.0
22	21.9	30.7	32.8
23	21.2	29.7	31.8
24	20.5	28.8	30.8
25	19.8	28.0	29.8
26	19.1	27.1	28.9
27	18.4	26.2	28.1
28	17.8	25.4	27.3
29	17.4	24.7	26.6
30	17.1	24.1	26.0
40	13.3	19.1	20.5
50	11.1	16.2	17.3

Subject Index

Acceptance sampling, 93–95
Aliasing, 221–227, 231–239
Analysis of variance – *see* ANOVA
Analytical Methods Committee, 1, 42, 76, 276, 285–286, 304, 386–388
Analytical run, 20
Analytical system, 1
ANOVA
  blocking, 338–343
  control charts, 344–351
  crossed factors, 324–343
  degrees of freedom, 311–314, 327, 331, 340, 346–347, 353–358
  expected mean squares, 312, 315–316, 347, 353–355
  fixed effects, 360–361, 389
  F-test, 315–317, 322, 327–328, 331, 335–336, 340, 355
  hierarchical, 343–361
  honestly significant difference, 320–323, 328–329, 341
  interaction, 327–329, 335–336
  interaction plot, 329, 336
  least significant difference 318–320, 322–323, 328–329, 331
  main effect, 326
  main effect plot, 332
  mean squares, 315–316, 327, 331, 340–341, 346–347, 353–358, 367
  model validation, 317–318, 332–333, 348–351
  multi-factor ANOVA, 324–343
  multiple comparisons, 318–324, 328–329, 331, 341
  nested ANOVA, 343–361, 366–369
  one-way ANOVA, 310–324, 367
  paired *t*-tests, 342–343
  p-values, 317, 322, 327, 331, 334, 340, 342, 355
  random effects, 343–361, 389
  regression, 260–264
  relationship to effects analysis 328, 335–336
  residual analysis – *see* model validation
  Satterthwaite's approximation, 357–358
  statistical model, 314–315, 345, 353
  sums of squares, 311–314, 325–327, 339–340, 346–347, 353–355
  two-way ANOVA, 324–333
  variance components, 343–361, 366–368
  see also: experimental design, factorial designs
AOAC, 366, 368, 372–374
Assigned value, 39, 42, 78–79
Average run length, 61–64
Averaging, 26–29, 66–71, *see also* replicates

Bias, 6–7, 43, 88–93, 103–110, 113–114, 172–179, 375–376, 388–393
Blanks, 65–66, 277

Blocking, 228–229, 338–343, *see also* pairing

Calibration, 264–275
Cause and effect diagram, 8–10
Certified reference materials, 42, 375–376, 388, 390–391
Coefficient of determination, 262–264
Coefficient of variation, 15–17, 282
Collaborative trials, 366–377
Confidence intervals
  confidence level, 106–107
  difference between two means, 140–141
  honestly significance, difference, 320–323, 328–329, 341
  least significant difference, 318–320, 322–323, 328–329, 331
  paired studies, 151–152, 158–159
  regression, 254–260, 267–270, 274–275, 290–292
  repeatability limit, 19–20, 22–24, 72–73, 120
  sample size, 107–108
  significance tests – relationship to, 110, 141
  single mean, 103–114, 196
  standard deviation, 116–120, 356–358
  variance, 116–120, 356–358
  variance components, 356–358
Confounding, 219–228, 231–239
Consumer's risk, 97–98
Contrasts, 221–223, 225–227, 232, 238
Control charts
  action limits, 40
  ANOVA, 344–351
  appropriate measures of precision, 43–45, 60–61, 66–71
  average run length, 61–64
  between-run standard deviation, 43–45, 60–61, 66–71
  blanks, 65–66
  centre line, 42–43
  chart performance, 61–64
  control limits, 39–40, 42–47
  control material, 35, 42
  cusum charts, 76
  data scrutiny, 45–47, 55–59
  EWMA charts, 76
  examples, 36–38, 50–51, 55–61, 155–156
  harmonized guidelines, 41–42, 65–66, 277
  individuals chart, 37–40, 45–47
  moving range, 43–47
  multivariate charts, 76
  nested designs, 343–351
  out of control rules, 37, 40–41, 61–64
  outliers, 45–47, 55–59
  paired comparisons, 155–156
  partial replicates, 51–52
  randomisation, 65
  range charts, 50–52, 55–61
  repeatability standard deviation 72–73
  replicates – nature of, 44–45, 51–52, 66–71
  residual analysis – *see* model validation
  sample size, 47, 64–66, 68–71

Shewhart, 35
standard deviation charts, 52–53
standard error of mean, 40, 66–71
statistical model, 344–345
theory, 39–41
using charts, 74–75
warning limits, 41
within-run standard deviation, 44–45, 50–52, 66–71
X-bar chart, 40, 42–45, 55–61
Control material, 35, 42
Correlation coefficient, 262–264
Critical value, 92, 94–95
see also *t*- and *F*-tests
Curvature, 282–285, 295–304

Data scrutiny, 45–47, 55–59
Defining relation, 222–224, 228
Degrees of freedom, 14–15, 89, 92, 94, 109, 111–112, 117–118, 120, 137–138, 144, 146, 151, 193–196, 208, 252–253, 262, 271–274, 297–298, 300, 311–314, 327, 331, 340, 346–347, 353–358
Detection limit, 275–278
Distributions
chi-square, 117–119
F-distribution, 144–145
rectangular, 380
standard Normal, 11–13
Student's *t*, 89–91
Studentized range, 321
triangular, 381
uniform, 380
Dotplots, 136, 311, 322, 373–374

Effects, 186–187, 190–196, 203–209, 328, 335–336
Error
in significance tests, 96–98, 160–161
measurement error, 1–10
random, 4–8
systematic, 4–8
see also: measurement model
Eurachem, 366, 378
Expectation, 113, 312
Expected mean squares, 312, 315–316, 347, 353–355
Experimental design
appropriate measures of precision, 171–179
blocking, 228–229, 338–343, *see also* pairing
central composite design, 301–304
collaborative trials, 366–376
comparing several independent groups, 310–343 – *see also* factorial designs
comparing two independent groups, 135–147
crossed designs, 324–343
hierarchical designs, 343–361
nested designs, 343–361
one factor at a time designs 186–189
pairing, 148–159, 171–179, 342–343
randomization, 169–171
randomized block designs, 341
representativeness, 179
ruggedness/robustness testing 230–241
run order, 169–171
sequential studies, 233
see also: ANOVA, factorial designs, fractional factorial designs, sample size

External quality control, 76–80

Factorial designs
- 2^2 design, 186–199, 324–329
- 2^3 design, 200–210, 334–336
- 2^5 design, 215–219
- blocking, 338–343
- design matrix, 198–199, 202, 215–216
- effects, 186–187, 190–196, 203–209, 328, 335–336
- factors, 186
- interaction, 187–189, 191–192, 194–196, 203–207, 327–329, 335–336
- interaction plot, 196, 329, 336
- levels, 186
- main effect, 191, 203
- main effect plot, 332
- model validation, 197–198, 210–211
- normal plot of effects, 215–218
- residual analysis – *see* model validation
- sample size, 212–215
- *see also*: ANOVA, experimental design, fractional factorial designs, significance tests

Fishbone diagram, 8–10

Fractional factorial designs
- 2^{3-1} design, 219–223
- 2^{5-1} design, 224–229
- 2^{7-4} design, 231–235
- 2^{12-8}design, 235–239
- aliasing, 221–227, 231–239
- blocking, 228–229
- combining half–fractions, 227–229, 233
- confounding, 221–227, 231–239
- contrasts, 221–223, 225–227, 232, 238
- defining relation, 222–224, 228
- design generator, 222–224, 226, 231–231, 237–238
- design matrix, 220, 222, 224, 227, 231–232, 237
- folding, 233
- Normal plots of effects, 225–227, 239
- ruggedness/robustness testing 230–241
- sequential studies, 233
- *see also*: experimental design, factorial designs, significance tests

Guidelines
- ANSI/ASQC (sampling), 103
- AOAC (collaborative trials), 366, 368, 372–374
- ASTM (outliers), 45
- Eurachem (measurement uncertainty), 366, 378
- harmonized (control charts), 41–42, 65–66, 277
- harmonized (proficiency testing), 77
- ICH (method validation), 19
- ICH (shelf lives), 248
- ISO (collaborative trials), 19–22, 375–377
- ISO (measurement uncertainty), 366, 377–378, 388
- NCCLS (non–linearity), 285
- NCCLS (precision), 72, 351, 356, 358

Heavy tails, 127

Histogram, 3, 122, 129

Honestly significant difference, 320–323, 328–329, 341
Horwitz function, 79

IMEP, 4–6
Interaction, 187–189, 191–192, 194–196, 203–207, 327–329, 335–336
Interaction plot, 196, 329, 336
Interlaboratory study, 4–8, 366–377
Intermediate precision, 22, 73–74, 345
Internal quality control – *see* control charts
Inverse prediction – *see* regression
Ishikawa diagram, 8–10

Laboratory bias, 6–7, 88–93, 103–110, 113–114, 172–179, 388–393
Lack of fit test, 299–301
Least significant difference, 318–320, 322–323, 328–329, 331

Measurement model, 3–4, 10, 113–114, 172–179, 388–393
Method bias, 375–376, 388, 390–392
Method comparison, 392
Method validation, 230–241, 375–376, 390–392
Mixtures, 127
Model validation
  Anderson Darling test, 125–126 – *see also*, Normal probability plots
  ANOVA, 317–318, 332–333, 348–351
  Cochran's test, 371–373
  collaborative trials, 369–374
  comparative studies, 143–145, 147–148
  control charts, 45–47, 55–59
  factorial designs, 197–198, 210–211
  Grubbs' test, 373–374
  Normal probability plots, 121–127, 147–148, 157–158, 197–198, 210–211, 280, 292–293, 317–318, 332–333, 348–350, 370–371
  paired comparisons, 154–159
  regression, 278–285, 292–293, 299–301
  transformations, 128–130
Moving range, 43–47
Multiple comparisons, 318–324

Non-linearity, 282–285, 295–304
Normal probability plot, 121–127, 147–148, 157–158, 197–198, 210–211, 280, 292–293, 317–318, 332–333, 348–350, 370–371

Operating characteristics curve, 103
Outliers – *see* modal validation
Outlier tests, 125–127, 372–374

Power, 101–103, 162–163, 166
Power curve, 101–103, 162–163, 165–166
Precision, 7–29 – *see also* standard deviation
Prediction interval – *see* regression
Producer's risk, 97
Proficiency testing, 76–80
p-value, 125–126, 142–143 – *see also* significance tests, regression

r^2, 262–264
Random numbers, 114–115, 171, 174–176
Randomization, 65, 169–171
Range, 43–47, 50–52, 55–61, 67–68
Regression
ANOVA, 260–264, 299–301
assumptions, 250–251, 304
calibration, 264–275
central composite design, 301–304
changing variance, 281–294
coefficient of determination, 262–264
confidence intervals, 254–260, 267–270, 274–275, 290–292
correlation coefficient, 262–264
curvature, 282–285, 295–304
degrees of freedom, 262, 271–274, 297–298, 300
detection limit, 275–278
fitted values, 251
F-test, 260–262, 299–300
intercept, 249–252, 254–255, 270–272
inverse prediction, 250, 257–258, 266–270, 274–275, 290–292
lack of fit test, 299–301
least squares, 248–249, 286–292
linear functional relationship, 304
mean squares, 260–262, 290, 300
model validation, 278–285, 292–293, 299–301
multiple regression, 295–304
non-linear regression, 295–304
non-linearity, 282–285, 295–304
ordinary least squares, 248–249, 286–292
polynomial regression, 295–304
prediction, 250–278, 290–292, 301–302
prediction interval, 255–260, 267–270, 274–275, 290–292
p-values, 254–255, 271–273, 297–298, 300
r^2, 262–264
regression model, 250–251
residual analysis – *see* model validation
response surfaces, 301–304
sample size, 252
shelf life, 248–264
significance tests, 253–255, 271–272, 298–300
simultaneous inference, 257
slope, 249–255
standard deviation, 252, 262
sums of squares, 261–264, 290, 300
weighted least squares, 286–292
Working Hotelling bounds, 257
zero intercept line, 270–275
see also: model validation
Rejection region, 92, 94 – *see also* *t*- and *F*-tests
Repeatability
conditions, 19
critical difference, 28
error – *see* measurement model
limit, 19–20, 24, 72–73, 120

standard deviation, 19–20, 22–24, 72–73, 120, 345, 353–358, 367
see also: confidence intervals, measurement model, collaborative trials
Replicates
nature of, 44–45, 51–52, 66–71, 100–101, 112–113, 171–179
see also: averaging, confidence intervals, measurement model, sample size, standard error
Reproducibility
collaborative trials, 366–377, 386–388
conditions, 21
limit, 21
standard deviation, 21, 366–369
see also: repeatability
Residuals – *see* model validation
Response surfaces, 301–304
Robustness tests, 230–241
Ruggedness tests, 230–241
Run bias – *see* measurement model
Run chart, 24, 25, 155
Run order, 65, 169–171

Sample size
collaborative trials, 368–369
comparing two means, 159–163, 171–179
comparing two standard deviations, 166–168
confidence intervals, 107–108
consumer's risk, 97–98
control charts, 47, 64–66, 68–71
error – type, 1 96–97, 160–161
error – type, 2 97–98, 160–161
factorial designs, 212–215
OC-curve, 103
paired studies, 164–166
power curve, 101–103, 162–163, 165–166
producer's risk, 97
regression, 252
single means, 96–103, 107–108
tables, 98–100, 161–162, 165, 167
Sample statistic, 14
Sampling, 114–115, 392–393
Sampling distribution, 26–28 – *see also* distributions
Satterthwaite's approximation 357–358
Sequential studies, 233
Shelf life, 248–264
Significance level, 92, 94–95, 97–98 – *see also* *t*- and *F*-tests
Significance tests
acceptance sampling, 93–95
alternative hypothesis, 91, 94–95 – *see also* *t*- and *F*-tests
Anderson Darling test, 125–126 – *see also* model validation
Cochran's test, 371–373
comparing independent means, 137–139, 142–143, 145–146
comparing standard deviations, 143–145
confidence intervals – relationship to, 110, 141
consumer's risk, 97–98
critical value, 92, 94–95 *see also* *t*- and *F*-tests

degrees of freedom, 89, 137–138 – *see also t*- and *F*-tests
effects, 194 – *see also* factorial designs
error – type 1, 96–97, 160–161
error – type 2, 97–98, 160–161
F-test – comparing two SDs, 143–145
– regression, 260–262, 299–300
– ANOVA, 315–317, 322, 327–328, 331, 335–336, 340, 355
Grubbs' test, 373–374
honestly significant difference, 320–323, 328–329, 341
laboratory bias, 88–93, 172–179, 388–393
lack of fit test, 299–301
least significant difference 318–320, 322–323, 328–329, 331
method bias, 388–393
multiple comparisons, 318–324
multiple regression, 295–304
non-linear regression, 295–304
null hypothesis, 91, 94–95 – *see also t*- and *F*-tests
one tail tests, 94–95
operating characteristics curve, 103
outlier tests, 372–375
paired comparisons, 148–159, 164–166, 342–343
power, 101–103, 162–163, 165–166
producer's risk, 97
p-value, 125–126, 142–143 – *see also t*- and *F*-tests, and regression
rejection region, 92, 94 – *see also t*- and *F*-tests
residual analysis – *see* model validation
sample size – *see* sample size
significance level, 92, 94–95, 97–98 – *see also t*- and *F*-tests
test for bias, 88–93, 137–140, 142–143, 145–146, 148–159, 172–179, 388–393
test for curvature, 298
test statistic, 91–95 – *see also t*- and *F*-tests
t-test – comparing two means, 137–139
comparing several means, 318–323
effects, 194
paired comparisons, 148–159, 164–166, 342–343
regression, 253–255, 259–260, 265, 271–273, 298
single sample, 89–95
unequal standard deviations, 145–146
see also: confidence intervals, sample size
two tail tests, 87–93 – *see also t*-tests
Simulation, 105–106, 173–178
Simultaneous inference, 257, 318–324, 328–329, 331, 341
Skewness, 127, 129–130
Standard deviation
between-run, 43–45, 60–61, 66–71, 112–113, 345, 353–360
coefficient of variation, 15–17

combining SDs, 22–24, 137, 193, 208, 319, 339, 347, 354–358, 367, 385
comparing SDs, 143–145
confidence intervals, 116–120, 356–358
control chart, 52–53, 60–61
intermediate precision, 22, 73–74, 345
Normal curve, 10–15
precision performance assessment, 356–358
relationship to sample range, 43, 67–68
relative standard deviation, 15–17, 282
repeatability, 19–20, 22–24, 72–73, 120, 345, 353–358, 367
reproducibility, 21, 366–369
Satterthwaite's approximation, 357–358
variance – relationship to, 24
within-run, 43–45, 50–52, 55–61, 66–73, 112–113, 345, 347, 353–360
see also: ANOVA, confidence intervals, sample size, standard error

Standard error
difference between two means, 137, 151, 177–178
effects, 193–194, 208, 216
mean, 26–28, 40, 43–45, 66–71, 89, 103–104
regression coefficients, 251–252
see also: confidence intervals, significance tests

Statistical model, 3–4, 10, 26–28, 39, 113–114, 121, 137, 143, 150, 172–173, 193, 197, 250–251, 276–278, 314–315, 342–343, 345, 353, 370, 388–393 – *see also*: model validation, regression, transformations, uncertainty of measurement, sums of squares, 261–264, 290, 300, 311–314, 325–327, 339–340, 346–347, 353–355

Transformations, 128–130
True value, 1, 6–7 – *see also* measurement model
Trueness, 375–376

Uncertainty of measurement, 1–2, 365–366, 377–388

Variance, 24 – *see also* ANOVA, standard deviation

Working–Hotelling bounds, 257

Zero intercept calibration 270–275
z score, 77, 122–125
z value, 11–13, 122–125